BEN JONSON

Poems

Oxford University Press, Ely House, London W.1

GLASGOW NEW YORK TORONTO MELBOURNE WELLINGTON
CAPE TOWN IBADAN NAIROBI DAR ES SALAAM LUSAKA ADDIS ABABA
DELHI BOMBAY CALCUTTA MADRAS KARACHI LAHORE DACCA
KUALA LUMPUR SINGAPORE HONG KONG TOKYO

BEN JONSON

Poems

EDITED BY

Ian Donaldson

LONDON
OXFORD UNIVERSITY PRESS
NEW YORK TORONTO
1975

BEN JONSON

Born in Westminster (?), 11 June 1572 (?)
Died 6 August 1637

ISBN 0 19 254166 8
© *Oxford University Press 1975*

This Oxford Standard Authors edition
was first published in 1975

Printed in Great Britain

O.S.A.

CONTENTS

ACKNOWLEDGEMENTS   vi
LIST OF ABBREVIATIONS   vii
TABLE OF DATES   xi
INTRODUCTION   xiii
SELECT BIBLIOGRAPHY   xxi

Epigrams   1
The Forest   85
The Underwood   119
Ungathered Verse   281
Songs and Poems from the Plays and Masques   343
Leges Convivales   369
Dubia   373

SOURCE MATERIAL   385
INDEX OF TITLES AND FIRST LINES   395

LIST OF ILLUSTRATIONS

Title-page of *Epigrams*   3
Title-page of *The Underwood*   121
Frontispiece to Sir Walter Raleigh's *History of the World*   169
Title-page of Coryate's *Crudities*   294

ACKNOWLEDGEMENTS

In preparing this edition I have been greatly helped by the work of all previous editors of Jonson's poems, and by W. D. Briggs's studies of Jonson's source-materials. My deepest debt—too pervasive to acknowledge adequately at every point in this edition—has been to the great Oxford edition of C. H. Herford and Percy and Evelyn Simpson.

I have also been helped by many acts of personal kindness. Lord De L'Isle, Miss K. M. Lea, Dr. Valerie Pearl, Professor Hans Kuhn, and Mrs. Jill Gibbs gave prompt and helpful answers to particular queries. Professor Stephen Orgel made a number of valuable suggestions concerning Jonson's poems on Inigo Jones, and, with Mr. John Harris, drew my attention to the drawing by Jones mentioned in the notes to *Ungathered Verse*, 35. For help of various kinds I am grateful to Miss Ann Duffy, Mr. J. C. Maxwell, Professor Judith K. Gardiner, Mr. Jonathan Price, Mrs. Carol Buckroyd, Mr. Richard Sayce, Mrs. Tamsin Donaldson, and my colleagues Professor Wesley Milgate and Professor J. P. Hardy. Mr. John Buxton and Dr. J. C. Eade kindly read my entire typescript, made many constructive suggestions, and saved me from many errors: those which remain are entirely my own. Finally I must thank the Australian Research Grants Committee for enabling me to spend a short period in Oxford working on this edition in 1971, and the Australian National University for allowing me to return to Oxford for a longer period in 1973–4, to bring it to completion.

C.I.E.D.

The Australian National University,
Canberra.
April 1974.

LIST OF ABBREVIATIONS

(i) *General**

DNB	*Dictionary of National Biography.*
Donne, *Elegies*	John Donne, *The Elegies and the Songs and Sonnets,* edited by Helen Gardner (Oxford, 1965).
Donne, *Satires*	John Donne, *The Satires, Epigrams and Verse Letters,* edited by W. Milgate (Oxford, 1967).
Fleay	F. G. Fleay, *A Biographical Chronicle of the English Drama, 1559–1642,* 2 vols. (London, 1891).
Gifford	*The Works of Ben Jonson,* edited by W. Gifford, 9 vols. (London, 1816).
Gifford/Cunningham	*The Works of Ben Jonson,* edited by W. Gifford, with introduction and appendices by Lieut.-Col. F. Cunningham, 9 vols. (London, 1875).
H & S	*Ben Jonson,* edited by C. H. Herford and Percy and Evelyn Simpson, 11 vols. (Oxford, 1925–52).
Herrick	*The Poetical Works of Robert Herrick,* edited by L. C. Martin (Oxford, 1956).
Hunter	*The Complete Poetry of Ben Jonson,* edited by William B. Hunter, jun. (New York, 1963).
Newdigate	*The Poems of Ben Jonson,* edited by Bernard H. Newdigate (Oxford, 1936).
OED	*Oxford English Dictionary.*
Tilley	M. P. Tilley, *A Dictionary of the Proverbs in England in the Sixteenth and Seventeenth Centuries* (Ann Arbor, 1950).
Trimpi	Wesley Trimpi, *Ben Jonson's Poems. A Study of the Plain Style* (Stanford, 1962).
Whalley	*The Works of Ben Jonson,* edited by Peter Whalley, 7 vols. (London, 1756).

(ii) *Journals*

BJRL	*Bulletin of the John Rylands Library*
CP	*Classical Philology*
Crit. Q.	*Critical Quarterly*
E & S	*Essays and Studies by Members of the English Association*
E in C	*Essays in Criticism*

* Editions of Jonson's poems not listed here are cited more fully when referred to throughout this edition.

ELN	English Language Notes
ELH	ELH, A Journal of English Literary History
ELR	English Literary Renaissance
Eng. Misc.	English Miscellany
Expl.	Explicator
HSNPL	Harvard Studies and Notes in Philology and Literature
JEGP	Journal of English and Germanic Philology
JWCI	Journal of the Warburg and Courtauld Institutes
MLN	Modern Language Notes
MLR	Modern Language Review
MP	Modern Philology
N & Q	Notes and Queries
PMLA	PMLA, Publications of the Modern Language Association of America
PQ	Philological Quarterly
RES	Review of English Studies
RMS	Renaissance and Modern Studies
RQ	Renaissance Quarterly
SEL	Studies in English Literature 1500–1900
SLI	Studies in the Literary Imagination
SP	Studies in Philology
TLS	Times Literary Supplement

(iii) Works by Jonson

Alch.	The Alchemist (1610)
Augurs	The Masque of Augurs (1622)
B.F.	Bartholomew Fair (1614)
Beauty	The Masque of Beauty (1608)
Blackness	The Masque of Blackness (1605)
C. is A.	The Case is Altered (written c. 1598)
C.R.	Cynthia's Revels (1600)
Cat.	Catiline (1611)
Chall. Tilt	A Challenge at Tilt (1613/14)
Chlor.	Chloridia (1631)
Christmas	Christmas his Masque (1616)
Conv. Dr.	Conversations with William Drummond of Hawthornden (1618/19)
D. is A.	The Devil is an Ass (1616)
Disc.	Timber, or Discoveries (1640 Folio)
E.M.I.	Every Man in his Humour (1598 version in 1601 Quarto; revision in 1616 Folio)*
E.M.O.	Every Man out of his Humour (1599)
Eng. Gram.	The English Grammar (1640 Folio)
Ent. K. & Q. Theob.	The Entertainment of the King and Queen at Theobalds (1607)

* All references to the 1616 Folio text, unless otherwise specified.

Ent. Welb.	The King's Entertainment at Welbeck (1633)
Epig.	Epigrams (1616 Folio)
For.	The Forest (1616 Folio)
Fort. Is.	The Fortunate Isles, and their Union (1625)
G.A. Rest.	The Golden Age Restored (1615)
Gyp. Met.	The Gypsies Metamorphosed (1621)
Haddington	The Haddington Masque (1608)
Hym.	Hymenaei (1606)
King's Ent.	The King's Entertainment in Passing to his Coronation (1604)
Leg. Conv.	Leges Convivales
L.R.	Love Restored (1612)
Love's Tr.	Love's Triumph Through Callipolis (1631)
Love's Welc. Bols.	Love's Welcome at Bolsover (1634)
M.L.	The Magnetic Lady (1632)
Merc. Vind.	Mercury Vindicated from the Alchemists at Court (1616)
N.I.	The New Inn (1629)
Nept. Tr.	Neptune's Triumph for the Return of Albion (1623/4)
Oberon	Oberon, the Fairy Prince (1611)
P.H. Barr.	The Speeches at Prince Henry's Barriers (1609)
Panegyre	A Panegyre on the King's Opening of Parliament (1604)
Pleas. Rec.	Pleasure Reconciled to Virtue (1618)
Poet.	Poetaster (1601)
Queens	The Masque of Queens (1609)
Sad. Shep.	The Sad Shepherd (unfinished)
S. of N.	The Staple of News (1626)
S.W.	Epicoene, or The Silent Woman (1609)
Sej.	Sejanus (1603)
Songs	Songs and Poems from the Plays and Masques*
Time Vind.	Time Vindicated to Himself and to his Honours (1623)
T. of T.	A Tale of a Tub (1633)
U.V.	Ungathered Verse
Und.	The Underwood (1640 Folio)
Volp.	Volpone (1605 or 1606)

* This edition.

TABLE OF DATES*

?1572   Jonson born (?11 June), one month after death of his father, 'a grave minister of the gospel'; his mother re-marries, to brick-layer, not long after; family lives near Charing Cross; Jonson attends private school in St. Martin's Church, and later Westminster School: taught by William Camden.

?1588   Leaves Westminster School.

early 1590s   Works as bricklayer; military service in Flanders: 'In his service in the Low Countries he had, in the face of both the camps, killed an enemy and taken *opima spolia* from him' (*Conv. Dr.* 244–6).

1594   Marries Anne Lewis (14 Nov.): 'a shrew yet honest' (*Conv. Dr.* 254). Children of this marriage include: Benjamin (?1596–1603: commemorated in *Epig.* 45); Mary (probably b. soon after 1598, d. aged six months: commemorated in *Epig.* 22); Joseph (b. Dec. 1599); Benjamin (Feb. 1608–18 Nov. 1611). Jonson probably had other children, legitimate and illegitimate.

1597   Mentioned in Henslowe's diary as an actor; likely to have begun this career before this date.
Imprisoned for his share in lost play, *The Isle of Dogs* (Aug.–Oct.).

1598   Indicted (22 Sept.) for killing a fellow-actor, Gabriel Spencer, in a duel; pleads self-defence, given benefit of clergy, branded on thumb, goods confiscated. Converted to Catholicism while in prison: 'thereafter he was 12 years a papist' (*Conv. Dr.* 251). Listed by Francis Meres in *Palladis Tamia* amongst English playwrights 'best for tragedy'.

1600–1   'War of the Theatres', with Dekker and Marston.

1601, 1602   Paid for 'additions' to *The Spanish Tragedy* (25 Sept.; 22 June).

1602   'Ben Jonson the poet now lives upon one Townshend, and scorns the world' (i.e. Sir Robert Townshend; Manningham's diary, Feb.). Jonson later lives five years with Esmé Stewart, Lord Aubigny; separated during this period from his wife.

1603   Death of Queen Elizabeth (24 March); accession of King James I.

1605   Jonson and co-authors imprisoned for libellous references in *Eastward Ho!*
Gunpowder plot (5 Nov.): Jonson assists Privy Council with inquiries.

* For dates of individual works by Jonson, see pp. viii–ix.

1606   Jonson and wife before Consistory Court on charges of
              recusancy.

1612–13   In France as tutor to Sir Walter Raleigh's son, Walter.

1616   Publication of Folio of Jonson's *Works*, containing (*inter alia*)
              Epigrams and *The Forest*.
              Granted royal pension of 100 marks p.a. (Feb.).

1618–19   Journey on foot to Scotland; guest of William Drummond of
              Hawthornden and others.

1619   Honorary degree from Oxford (17 July).

1619–   ?Deputy Professorship of Rhetoric, Gresham College, London.

1623   Fire destroys Jonson's library and MSS. (Nov.).

1625   Death of King James (27 March); accession of King Charles I.

1628   Jonson suffers paralytic stroke.
              Made City Chronologer after death of Thomas Middleton
              (payment withheld 1631–4, because of Jonson's inactivity in
              office).

1630   Pension increased from 100 marks to 100 pounds p.a., plus
              annual tierce of royal sack.

1631   Culmination of feud with Inigo Jones.

1637   Dies (6 Aug.); buried in Westminster Abbey (9 Aug.).

1638   Publication of *Jonsonus Virbius* (memorial tributes).

1640   Publication in Quarto by John Benson, from surreptitiously
              acquired MSS., of *An Execration Upon Vulcan* and 'divers
              epigrams'; re-issued in same year in Duodecimo, with *Ars
              Poetica* and *The Gypsies Metamorphosed*.

1640–1   Publication of Jonson's 2nd Folio, in two volumes, containing
              (*inter alia*) *The Underwood*: collected by Sir Kenelm Digby
              from Jonson's papers.

INTRODUCTION

JONSON is best known and best liked today as a writer of stage comedies. His non-dramatic poetry, which in many ways was of central importance to him, still remains relatively neglected. Jonson's own sense of priorities may be seen in his placing of his *Epigrams* at the head of his 1616 Folio, and in his dedication of them to the Earl of Pembroke as 'the ripest of my studies'; in 1616 Jonson's 'studies' included all the major dramatic works upon which his present-day reputation chiefly rests. It was not as a *playwright* that Jonson wished primarily to be remembered: that very term, indeed, he appears to have invented, and he used it only with contempt.[1] (His own occupation he described as that of 'poet', his plays as 'poems'.) Writing for the public stage was a challenge which Jonson viewed with mixed excitement, hostility, confidence, and disdain. More than once he declared loftily that such work would occupy him no longer:

> Make not thyself a page
> To that strumpet, the stage;
>   But sing high and aloof,
> Safe from the wolf's black jaw, and the dull ass's hoof.
> (*Und.* 23. 33–6)

Jonson's non-dramatic poetry, on the other hand, occupied him more or less continuously, even during those (sometimes lengthy) periods of theatrical abstention, and throughout the final period of his life, when age, sickness, rivalry, and lack of continued popular acclaim had forced him to leave not only 'the loathed stage', but also the court—equally loathed, equally attractive—for which he had been writing masques and entertainments since 1603.

Jonson's non-dramatic poetry was of central importance to him not merely in terms of his own theory and estimate, but also in what it enabled him to accomplish. Above all, it allowed him—as no other form of writing did—to write in an easy and intimate manner about himself. When Jonson approaches anything like an authorial pronouncement or a semi-autobiographical portrait in his plays, his writing tends to become hortatory and stiff. *Poetaster*, for example, opens

---

[1] See *Epig.* 49 n.

forbiddingly with the entry of a 'prologue' who is clad in complete armour, and concludes alarmingly with the entry of the stern and irate figure of 'the author' himself, who proceeds to tell the audience at length and with considerable expenditure of breath that he cares so little for their opinions that he will waste no breath upon them.[2] In Jonson's non-dramatic poetry, on the other hand, not only are the poet's own concerns (predictably) more central: the glimpses of himself which he allows us are also more touching, more humble, more various, more wryly humorous and self-aware than anything we are granted in the plays.

For all this, Jonson's poetry is essentially public and social poetry in a way that (say) Donne's is not. He wished to publish his poems, rather than merely allow them to circulate in manuscript amongst friends. His poems address, assess, commend, and vilify an astonishing range of people. To read through even a single group of poems, such as the *Epigrams*, is to be made vividly aware of the existence of an entire society, headed by the king himself and peopled both by identifiable individuals and semi-fictionalized characters who nevertheless have clear roots in the society of Jonson's time. The high degree of social specificity in Jonson's poetry brings not only its pleasures but also its problems, for editors and readers alike. In his edition of Jonson's *Works* in 1756, Peter Whalley—evidently sensing a possible unease on the part of his readers at the minuteness of annotation into which he had been led by one of Jonson's poems, added a word of apology: 'It is necessary to know such trivial circumstances, as, in these smaller poems, their chief merit often consists in the turns of thought which allude to them.'[3] Despite its slighting air, the remark conveys a truth which bears repetition in an edition such as this one; it need only be added that a knowledge of such 'trivial circumstances' may be important to an understanding not only of Jonson's smaller poems, but of his longer ones as well. Whalley's generation, however, was one which placed special value upon the need for poetry to affect its readers through its power of generality; and at the time at which he wrote these words there was a growing suspicion that Jonson's poetry was too localized, too densely allusive to particular events and people for it to endure beyond the age in which he lived. In the same decade, precisely this criticism was being levelled at Jonson's plays: Thomas Davies had regretfully declared that *Epicoene* was too dependent upon topical allusion for it ever to

---

[2] See *Poetaster*, Prologue 5–10 (where the armour is said to be the author's own, against 'base detractors and illiterate apes'); and Apologetical Dialogue, esp. 202–21.

[3] *The Works of Ben Jonson*, ed. Peter Whalley (London, 1756), vi. 250.

be revived with any likelihood of success, and David Garrick had spoken feelingly of similar doubts and difficulties which assailed him in preparing his adaptation of *Every Man in his Humour* for the stage.[4] 'He was not of an age, but for all time!', Jonson had written in his great tribute to Shakespeare. By the early nineteenth century, Nathan Drake felt that, in respect of Jonson himself, the judgement might well be reversed: he was not for all time, but of an age. '. . . had he paid less attention to the *minutiae* of his own age, and dedicated himself more to universal habits and feelings', wrote Drake, 'his popularity would have nearly equalled that of the poet whom he loved and praised.'[5]

One solution to this problem was simply to forget that Jonson's work had a social context at all. Throughout the nineteenth century and the early part of the twentieth, it became common to submit Jonson's poetry to a more drastic version of the kind of surgery which Garrick had already practised upon some of his plays. Snippets of Jonson's verse (trimmed of offensive local reference), delicate gleanings of songs and catches from the plays and masques (their literary contexts ignored) were brought together with short verses by Robert Herrick, William Browne of Tavistock, and others in slim deckle-edged anthologies. Jonson was seen primarily as a lyric poet, of somewhat indifferent accomplishment; in the verdict of Swinburne, 'one of the singers who could not sing'. From this reputation, Jonson's poetry is still in many ways struggling to escape.[6]

But Jonson's poetry was not merely written—as Nathan Drake thought—for and about the age in which he lived. Dedicating his *Epigrams* to the Earl of Pembroke and thanking him for his protection, Jonson wrote, 'I return you the honour of leading forth so many good and great names as my verses mention on the better part, to their remembrance with posterity . . .'. Occasionally when Jonson dispatches his work to the admiration of posterity, his confidence may seem to be partly the product of chagrin that his work has been slighted: such may appear to be the case (for example) with his similar dedication of the published version of *Catiline*, which had been a disastrous stage failure. The dedication to the *Epigrams*, however, reveals another kind of confidence, in the stability and

---

[4] Thomas Davies, *Dramatic Miscellanies* (London, 1784), ii. 101–2; *The Letters of David Garrick*, ed. David M. Little and George M. Kahrl (London, 1963), i. 303–4.

[5] Nathan Drake, *Shakespeare and his Times* (London, 1817), ii. 580.

[6] See my article, 'Jonson's Ode on Sir Lucius Cary and Sir H. Morison', *SLI*, vi (1973), 139–52. For Swinburne's judgement, see his *A Study of Ben Jonson* (London, 1889), p. 5.

continuity of the moral values which his poems reflect and enact. The qualities shown by such men as Pembroke are qualities which Jonson believes his verse will help to define and to perpetuate; they are qualities which (like Shakespeare's) are not merely of an age, but for all time. Neither will the qualities of a Sir Voluptuous Beast disappear from human society, nor the need to regard such figures with contempt. Behind the highly specific social surface of Jonson's verse is a highly generalized view of the way in which societies—past, present, and future—tend to behave; of 'this strife / Of vice and virtue, wherein all great life, / Almost, is exercised . . .' (*Epig.* 102. 5–7).

It is this confidence about moral continuities (rather than any poverty of creative thought) which allows Jonson to draw so freely upon the writers of the past. In lamenting the degeneracy of his age in the cry, 'O times!' (*Und.* 15. 161; *Und.* 64. 17), Jonson recalls the cry of Cicero: *O tempora!* (*In Catilinam*, I. i. 2). In so doing he reminds us not merely that moral degeneracy—which may seem to present to each generation a unique and unprecedented threat—is always with us, but also that those who oppose themselves to such degeneracy likewise recur; and that society, which may seem to be perpetually on the point of falling about us in ruins, does, after all, survive. Hence Jonson's allusions set up reverberations which are moral as well as purely literary. At times their point lies not in a simple act of recall, but in a significant shift of emphasis:

> I feel my griefs too, and there scarce is ground
> Upon my flesh to inflict another wound.
>
> (*For.* 15. 21–2)

These lines from 'To Heaven' echo the words of Ovid in exile (*Epist. Ex Ponto*, II. vii. 41–2). Ovid's complaint is that he is being crushed by the malevolence of the gods; Jonson significantly refrains from echoing this accusation, making (as it were) an eloquent point by omission.

An attempt has been made in this edition to note most of Jonson's classical allusions, and to supply translations—either at the foot of the page or at the end of the volume—of the more important ones. Lest their very number create an impression that Jonson's verse is heavily bookish, two further points should be made. The first concerns the familiarity of many of the sentiments for which classical sources have been found; often (as the notes also attempt to indicate) their sources are equally to be found in everyday English proverbial lore. The second concerns the extent to which they often take on new life and meaning through their local and contemporary application. One of Jonson's best-known couplets (from the Cary/Morison Ode) has its source in Seneca's ninety-third Epistle:

> In small proportions we just beauty see,
> And in short measures life may perfect be.

'Just as one of small stature can be a perfect man,' wrote Seneca, 'so a life of small compass can be a perfect life.' The point of Jonson's allusion does not lie solely in the fact that Sir Henry Morison had died on the threshold of attaining his majority, but also in the fact that—as the Earl of Clarendon's account of him reveals—Sir Lucius Cary, for all his many gifts, suffered under 'one great disadvantage'—'his stature was low, and smaller than most men' (and yet, adds Clarendon, 'that little person and small stature was quickly found to contain a great heart').[7] Seneca's general remark about short men and short lives is vitalized by the particular context in which it is recalled. So too with Jonson's praise of Robert Sidney's hospitality in 'To Penshurst':

> Where the same beer and bread and self-same wine
> That is his lordship's shall be also mine;
> And I not fain to sit, as some this day
> At great men's tables, and yet dine away.
> (*For.* 2. 63–6)

Of the several classical passages which these lines recall, the most obvious is Martial's rebuke: 'Why is not the same dinner served to me as to you? . . . Why do I dine without you, although, Ponticus, I am dining with you?' (*Epig.* III. lx). But the lines are also charged with the memory of an experience which is not merely literary:

> Being at the end of my Lord Salisbury's table with Inigo Jones, and demanded by my lord why he was not glad: My lord, said he, you promised I should dine with you, but I do not. For he had none of his meat; he esteemed only that his meat which was of his own dish.
> (*Conv. Dr.* 317–21)

That Jonson's remark at table may itself have derived from Martial merely reinforces the general point, that for Jonson the relationship between the worlds of personal experience and classical literature was an intimate one, the movement between those worlds effortless and

---

[7] See note to *Und.* 70. 73–4.

natural. What happened in Martial's day happens again 'this day'. The past imbues the present, which in turn imbues the future. The onus is upon 'posterity' to recognize the vitality of the traditions it inherits.

The text

The text of this edition is a modernized version of that established in Herford and Simpson's major Oxford edition. Occasional departures from Herford and Simpson's text are normally indicated and explained in the notes. In one or two places, words or letters conjecturally inserted by Herford and Simpson have been silently omitted when the sense and metre of the original text upon which those editors draw seem acceptable as they stand.

In the process of modernizing, obsolete spellings have generally been replaced by their modern equivalents: 'tyran' by 'tyrant', 'ghyrlond' by 'garland', and so forth. Occasionally, however—at the expense of rigorous consistency—an obsolete spelling is retained in order to preserve a rhyme ('wull', not 'will', in *Epig*. 90. 17) or a syllable ('babion', not 'baboon', *Epig*. 129. 12); a significant reminder of etymology ('holy-day', not 'holiday', *Und*. 84. ix. 63), or a play upon words ('lanthorn', not 'lantern', *Und*. 43. 10) or upon letters ('Marie', not 'Mary', *U.V*. 41). The flexibility of contemporary spelling sometimes allowed Jonson to play upon homonyms in a way which modern spelling does not permit. Here, for example, is the Folio text of *Epig*. 58, 'To Groom Idiot':

> Ideot, last night, I pray'd thee but forbeare
> To reade my verses; now I must to hear:
> For offring, with thy smiles, my wit to grace,
> Thy ignorance still laughs in the wrong place.
> And so my sharpnesse thou no lesse dis-ioynts,
> Then thou did'st late my sense, loosing my points.
> So haue I seene at CHRIST-masse sports one lost,
> And, hood-wink'd, for a man, embrace a post.

'Loosing my points', or 'losing my points'? Here, as occasionally elsewhere—'waste'/'waist', *Und*. 9. 16, 'modes'/'moods', *U.V*. 2. 3, etc.—it has been necessary to settle for one spelling in the text when two senses are almost certainly intended, and to alert the reader to the ambiguity in a note (which might, after all, equally be required in an old-spelling edition).

Jonson's punctuation—especially in the poems from the 1616 Folio, which he saw carefully through the press—is deliberate and systematic, and as the terse rebuke to Groom Idiot shows: to miss Jonson's points of punctuation is often to miss his points of meaning. Yet to modern eyes and ears, Jonson's punctuation is also at times intolerably heavy. The punctuation of the present text is designed to allow Jonson's lines to move more freely, but also to preserve something of their original weight, measure, and delicacy of transition (as in, for example, 'Inviting a Friend to Supper', which quickens into intimate liveliness after its sober and subtly pausing opening).

Jonson's elaborate system of marking elisions, partial elisions, and syllabification has not been exactly reproduced; a line such as *U.V.* 41. 28: 'As if they'ador'd the Head, whereon th'are fixt' is rendered (for example) 'As if they adored the head whereon they're fixed'.

The notes

Quotations in the notes from the works of other authors are also given in modernized form, with the exception of those from Chaucer. References to modern editions are included in the case of less familiar authors, such as Davenant and Corbett, and of works whose titles, numbering, or scope might give difficulty, e.g. the poems of Herrick. The spelling of these editions is not always reproduced. Quotations from Shakespeare are taken from Peter Alexander's one-volume edition of *The Complete Works* (London and Glasgow, 1951). Classical translations are taken wherever possible from Loeb editions, sometimes with minor emendations; an asterisk in the notes indicates that a translation of a particular passage is to be found under 'Source Material' at the end of the volume.

Words whose meanings may be found in *The Concise Oxford Dictionary* are not glossed in the notes except in cases of possible ambiguity or confusion: e.g. 'lantern' in *For.* 2. 4, 'sack' in *Und.* 70. 12, etc.

SELECT BIBLIOGRAPHY

Modern Editions of Jonson's Poems
Ben Jonson, edited by C. H. Herford and P. and E. Simpson, 11 vols.
(Oxford, 1925–52). Introduction to the poems in vol. ii; text in vol. viii;
commentary in vol. xi.
Ben Jonson, selected, with an introduction and notes, by John Hollander
(The Laurel Poetry Series: New York, 1961).
The Complete Poetry of Ben Jonson, edited with an introduction, notes, and
variants, by William B. Hunter, jun. (New York, 1963).
Poems of Ben Jonson, edited with an introduction by G. B. Johnston (The
Muses' Library: London, 1954).
The Poems of Ben Jonson, edited by Bernard H. Newdigate (Oxford, 1936).

*Criticism and Scholarship**
J. B. Bamborough, *Ben Jonson* (London, 1970).
L. A. Beaurline, 'The Selective Principle in Jonson's Shorter Poems',
Criticism, viii (1966), 64–74.
W. D. Briggs, 'Source-Material for Jonson's *Epigrams* and *Forest*', *CP*, xi
(1916), 169–90.
W. D. Briggs, 'Source-Material for Jonson's *Underwoods* and Miscellaneous
Poems', *MP*, xv (1917), 277–312.
Douglas Bush, *English Literature in the Earlier Seventeenth Century*
(*O.H.E.L.*), 2nd edition revised (Oxford, 1962).
T. S. Eliot, 'Ben Jonson', *Selected Essays* (London, 1951).
W. McC. Evans, *Ben Jonson and Elizabethan Music* (Lancaster, Pa. 1925;
repr. New York, 1965).
H. H. Hudson, *The Epigram in the English Renaissance* (Princeton, 1947).
G. B. Johnston, *Ben Jonson, Poet* (New York, 1945).
F. R. Leavis, *Revaluation*, 3rd impression (London, 1953).
Hugh Maclean, 'Ben Jonson's Poems: Notes on the Ordered Society', in
Essays in English Literature from the Renaissance to the Victorian Age, ed.
Millar MacLure and F. W. Watt (Toronto, 1964), pp. 43–68.
A. F. Marotti, 'All About Jonson's Poetry', *ELH*, xxxix (1972), 208–37.
Earl Miner, *The Cavalier Mode from Jonson to Cotton* (Princeton, 1971).
J. G. Nichols, *The Poetry of Ben Jonson* (London, 1969).
G. A. E. Parfitt, 'The Poetry of Ben Jonson', *E in C*, xviii (1968), 18–31.
Edward Partridge, 'Jonson's *Epigrammes*: the Named and the Nameless',
SLI, vi (1973), 153–98.

* Studies of particular poems are referred to in the notes.

Rufus D. Putney, ' "This So Subtile Sport": Some Aspects of Jonson's Epigrams', *University of Colorado Studies, Series in Language and Literature*, x (1966), 37–56.

Arnold Stein, 'Plain Style, Plain Criticism, Plain Dealing, and Ben Jonson' [review of Trimpi], *ELH*, xxx (1963), 306–16.

W. V. Spanos, 'The Real Toad in the Jonsonian Garden', *JEGP*, lxviii (1969), 1–23.

Wesley Trimpi, *Ben Jonson's Poems. A Study of the Plain Style* (Stanford, 1962).

Geoffrey Walton, 'The Tone of Ben Jonson's Poetry', ch. 2 of *Metaphysical to Augustan* (London, 1955).

David Wykes, 'Ben Jonson's "Chast Booke"—the *Epigrammes*', *RMS*, xiii (1969), 76–87.

EPIGRAMS

[*Title-page*] *Epigrammes. 1. Booke*.: Jonson evidently planned to collect a subsequent book of epigrams; this never materialized.

EPIGRAMMES.

I.

BOOKE.

The Author B. I.

LONDON,

M. DC. XVI.

TO THE
GREAT EXAMPLE OF
HONOUR AND VIRTUE,
THE MOST NOBLE WILLIAM,
EARL OF PEMBROKE,
LORD CHAMBERLAIN,
ETC.

MY LORD: While you cannot change your merit, I dare not
change your title; it was that made it, and not I. Under which
name, I here offer to your Lordship the ripest of my studies, my
Epigrams, which, though they carry danger in the sound, do not
therefore seek your shelter; for when I made them I had nothing   5
in my conscience to expressing of which I did need a cipher. But
if I be fallen into those times wherein, for the likeness of vice
and facts, everyone thinks another's ill deeds objected to him,
and that in their ignorant and guilty mouths the common voice
is, for their security, 'Beware the poet'—confessing therein so   10
much love to their diseases as they would rather make a party for
them than be either rid or told of them—I must expect at your
Lordship's hand the protection of truth and liberty while you
are constant to your own goodness. In thanks whereof I return
you the honour of leading forth so many good and great   15
names as my verses mention on the better part, to their remem-

---

[*Dedication*] *William, Earl of Pembroke*: 1580–1630, son of Sir Philip Sidney's sister,
Mary Herbert, Countess of Pembroke; he sent Jonson £20 every New Year (*Conv. Dr.*
312–13), and is addressed again in *Epig.* 102. Jonson also dedicated to Pembroke
Catiline, which he reckoned the best of his plays. Clarendon gives a more severe
account of Pembroke's moral character (*History of the Rebellion*, i. § 123), but Jonson's
testimony is defended by Dick Taylor, jun., *Studies in the English Renaissance Drama*,
ed. J. W. Bennett et al. (London, 1961), pp. 322–44. 4 *danger in the sound*: see *Epig.*
2. 5–6 A similar insistence that his work is not susceptible to deciphering is to be
found in *Volp.* Epist. Ded., esp. 62–70, and *B.F.* Ind. 135–45. Jonson is probably
following at a distance Martial's Preface to *Epig.* I. Cf. also Jonson's letter of 1605
to Salisbury, H & S, i. 194–6. 8 *facts*: crimes. 8 *objected to*: charged against.
11–12 *make a party for them*: defend them. For the general thought, cf. Seneca,
Epist. lxxxix. 19, and Jonson, *Alch.* Prol. 13–14, *Cat.* IV. 893–4. 15–16 *good
and great names*: this collocation of adjectives recurs throughout Jonson's writings: cf.
e.g. *For.* 14. 34–5, *Und.* 13. 2, 15. 187, 24. 5–6, 70. 105, *Alch.* 'To the Reader', 23, etc.
The significance which Jonson attaches to names throughout the *Epigrams* is discussed
by David Wykes, *RMS*, xii (1969), 76–87, and Edward Partridge, *SLI*, vi (1973), 153–
98. Jonson's insistence on the generality of his satire—perhaps deriving from the
precedent of Martial (*Epig.* X. xxxiii. 10) and Juvenal (*Sat.* i. 147 ff.)—is again

brance with posterity; amongst whom if I have praised, unfor-
tunately, anyone that doth not deserve, or if all answer not in all
numbers the pictures I have made of them, I hope it will be
forgiven me, that they are no ill pieces, though they be not like   20
the persons. But I foresee a nearer fate to my book than this:
that the vices therein will be owned before the virtues (though
there I have avoided all particulars, as I have done names) and
that some will be so ready to discredit me as they will have the
impudence to belie themselves; for if I meant them not, it is so.   25
Nor can I hope otherwise; for why should they remit anything of
their riot, their pride, their self-love, and other inherent graces,
to consider truth or virtue? but with the trade of the world lend
their long ears against men they love not, and hold their dear
mountebank or jester in far better condition than all the study or   30
studiers of humanity. For such, I would rather know them by
their vizards still, than they should publish their faces at their
peril in my theatre, where Cato, if he lived, might enter without
scandal.

Your Lordship's most faithful honourer,   35

Ben Jonson.

---

reminiscent of *Volp*. Epist. Ded.: 'Where have I been particular? Where personal?'
(56–7).   17–18 *praised, unfortunately*: see *Und*. 14. 19–22.   19 *numbers*: parts (Latin
numeri).   28 *trade*: way.   28–9 *lend their long ears against*: listen with their long
(asses') ears for scandal about.   31 *humanity*: the Latin sense is strong here ('learn-
ing or literature concerned with human culture'), but Jonson is also contrasting the
human and the sub-human ('long ears').   33 *Cato*: Marcus Porcius Cato (95–46
B.C.), model of Roman virtue, visited the licentious games of Flora in 55 B.C.; because
of his presence, the usual dance of naked girls was not held. Jonson significantly varies
Martial's words: 'let no Cato enter my theatre, or if he enters, let him look on', *Epig*. I,
Pref. [17–18].

EPIGRAMS

1

To the Reader

Pray thee take care, that tak'st my book in hand,
To read it well; that is, to understand.

2

To My Book

It will be looked for, book, when some but see
    Thy title, *Epigrams*, and named of me,
Thou shouldst be bold, licentious, full of gall,
    Wormwood and sulphur, sharp and toothed withal;
Become a petulant thing, hurl ink and wit   5
    As madmen stones, not caring whom they hit.
Deceive their malice who could wish it so.
    And by thy wiser temper let men know
Thou are not covetous of least self-fame
    Made from the hazard of another's shame;   10
Much less with lewd, profane and beastly phrase,
    To catch the world's loose laughter or vain gaze.
He that departs with his own honesty
    For vulgar praise, doth it too dearly buy.

---

1. Jonson's first four epigrams are loosely modelled on the first four of Martial: see K. A. McEuen, *Classical Influence Upon the Tribe of Ben* (Cedar Rapids, 1939), p. 17. 2 *to understand*: a characteristic stress: cf. *Epig.* 110. 16, *U.V.* 17. 2, *Love's Tr.* 1, *Alch.* 'To the Reader', 1, and (ironically) *B.F.* Ind. 49.

2. 5 *petulant*: insolent. 10 *another's shame*: cf. Martial, *Epig.* VII. xii. 4: 'fame won from another's blush is not dear to me'. 12 *loose laughter*: Horace, *Sat.* I. iv. 82–3. On Jonson's dislike of indiscriminate laughter, see H. W. Baum, *The Satiric and the Didactic in Ben Jonson's Comedy* (Chapel Hill, 1947), pp. 52–4. 13 *departs*: parts. 13–14 Cf. Martial, *Epig.* I, Pref. [6–7]: 'May my fame be bought at a lesser cost, and the last thing to be approved in me be cleverness'.

3

To My Bookseller

Thou that mak'st gain thy end, and wisely well
   Call'st a book good or bad, as it doth sell,
Use mine so, too; I give thee leave; but crave
   · For the luck's sake it thus much favour have:
To lie upon thy stall till it be sought;   5
   Not offered, as it made suit to be bought;
Nor have my title-leaf on posts or walls
   Or in cleft-sticks, advanced to make calls
For termers or some clerk-like serving-man
   Who scarce can spell the hard names; whose knight   10
                              less can.
If, without these vile arts, it will not sell,
   Send it to Bucklersbury: there 'twill, well.

4

To King James

How, best of kings, dost thou a sceptre bear!
   How, best of poets, dost thou laurel wear!
But two things rare the fates had in their store,
   And gave thee both, to show they could no more.
For such a poet, while thy days were green   5
   Thou wert, as chief of them are said to have been.

---

3. Written to the bookseller John Stepneth in 1612. 2 Cf. Martial, *Epig.* XIV, cxciv: 'some there are that say I am no poet; but the bookseller that sells me thinks I am'. 7 *posts*: the origin of the modern 'poster'. Cf. Horace, *Sat.* I. iv. 71–2. 8 *cleft-sticks*: evidently used to hold the pages of a book open to display its title-page. 9 *termers*: those who visit London, for litigation or pleasure, during the legal term. 12 *Bucklersbury*: the street ran south-east from the corner of Cheapside and the Poultry to Walbrook, and 'on both the sides throughout is possessed of grocers and apothecaries' (J. Stow, *A Survey of London*, ed. C. L. Kingsford (Oxford, 1908), i. 260). The book would be used for wrapping-paper: a traditional fantasy from classical times (Catullus, *Carm.* xcv, Horace, *Epist.* II. i. 269–70, Martial, *Epig.* III. ii. 3–5, IV. lxxxvi, etc.) to the present. Cf. Jonson, *Und.* 43. 51 ff., *Poet.* Apol. Dial., 171–2, *Disc.* 587–92.

4. 'No instance could be found of more complete parallelism and balance' (T. K. Whipple, on this epigram: *Martial and the English Epigram* (California, 1925), p. 401). 3 From Florus, *De Qualitate Vitae*, ix*. Cf. Jonson, *E.M.I.* V. v. 38–9, *Panegyre*, 163, *Epig.* 79. 1, *N.I.* Epil. 23–4, *Disc.* 2433. 5 James was 18 when his

And such a prince thou art, we daily see,
  As chief of those still promise they will be.
Whom should my muse then fly to, but the best
  Of kings for grace, of poets for my test?   10

5

On the Union

When was there contract better driven by fate?
  Or celebrated with more truth of state?
The world the temple was, the priest a king,
  The spoused pair two realms, the sea the ring.

6

To Alchemists

If all you boast of your great art be true,
Sure, willing poverty lives most in you.

Essays of a Prentice in the Divine Art of Poesy appeared in 1584, 25 at the appearance of his second volume, *His Majesty's Poetical Exercises at Vacant Hours*, in 1591. In a preface to the second volume he declared apologetically that he 'composed these things in his very young and tender years'. Jonson gives a less flattering opinion of James's poetry in *Conv. Dr.* 561–3.   8 Cf. Pliny's comment on Trajan, *Panegyricus*, 24: 'such a prince as others can only promise to be'.

5. James himself had compared the Union of England and Scotland to a marriage; Jonson echoes the analogy again in *Hym.*, esp. 424–30: see D. J. Gordon, *JWCI*, viii (1945), 107–45. H & S date this poem March 1604. A spate of poems on this subject appeared in that year: see D. H. Willson, *King James VI & I* (London, 1956), pp. 253–4.

7

On the New Hot-House

Where lately harboured many a famous whore,
  A purging bill now fixed upon the door
Tells you it is a hot-house; so it ma',
  And still be a whore-house: they're synonima.

8

On a Robbery

Ridway robbed Duncote of three hundred pound;
  Ridway was ta'en, arraigned, condemned to die;
But for this money was a courtier found,
  Begged Ridway's pardon. Duncote now doth cry,
Robbed both of money and the law's relief: 5
  The courtier is become the greater thief.

9

To All to Whom I Write

May none whose scattered names honour my book
  For strict degrees of rank or title look;
'Tis 'gainst the manners of an epigram;
  And I a poet here, no herald am.

---

7. 3 *hot-house*: a bathing-house with hot or vapour baths; also a brothel (cf. 'bagnio').
Cf. *Measure for Measure*, II. 1. 63–4: 'now she professes a hot-house, which, I think, is
a very ill house too'; and see *Und.* 34. 10.   4 *synonima*: synonyms.

8. 4 *Begged Ridway's pardon*: begged a pardon for Ridway.

9. 4 *no herald*: cf. Sidney, *Arcadia*, Bk. I, ch. ii (see *Complete Works*, ed. A. Feuillerat
(Cambridge, 1939), i. 15): '. . . I am no herald to enquire of men's pedigrees; it
sufficeth me if I know their virtues', Webster, *The Duchess of Malfi*, III. ii. 259–60:
'Will you make yourself a mercenary herald, / Rather to examine men's pedigrees than
virtues?'

10

To My Lord Ignorant

Thou call'st me poet, as a term of shame;
But I have my revenge made in thy name.

11

On Something that Walks Somewhere

At court I met it, in clothes brave enough
    To be a courtier, and looks grave enough
To seem a statesman. As I near it came,
    It made me a great face; I asked the name;
A lord, it cried, buried in flesh and blood,   5
    And such from whom let no man hope least good,
For I will do none; and as little ill,
    For I will dare none. Good Lord, walk dead still.

12

On Lieutenant Shift

Shift, here in town not meanest among squires
    That haunt Pickt-hatch, Marsh-Lambeth, and Whitefriars,
Keeps himself, with half a man, and defrays
    The charge of that state with this charm: God pays.
By that one spell he lives, eats, drinks, arrays   5
    Himself; his whole revenue is, God pays.

---

10. 2 *my revenge*: perhaps pointing forward to *Epig.* 11. 4–5, in the manner of *Und.* 2.
iii. 24 ff.

11. 5 *A lord*: Jonson 'never esteemed a man for the name of a lord' (*Conv. Dr.* 337).

12. 1 *Shift*: to shift is to live by frauds or temporary expediencies; cf. the character of this
name in *E.M.O.*, and Herrick's poem 'Upon Shift', Herrick, p. 144   1 *squires*: apple-
squires, i.e. pimps.   2 *Pickt-hatch, Marsh-Lambeth, and Whitefriars*: notorious areas
of London at this time. Whitefriars, with its privilege of sanctuary, attracted many
criminal types.   6 *revenue*: the word was commonly stressed on the second syllable.

The quarter-day is come; the hostess says
  She must have money; he returns, God pays.
The tailor brings a suit home; he it 'ssays,
  Looks o'er the bill, likes it, and says, God pays.   10
He steals to ordinaries; there he plays
  At dice his borrowed money; which God pays.
Then takes up fresh commodity for days,
  Signs to new bond, forfeits, and cries, God pays.
That lost, he keeps his chamber, reads essays,   15
  Takes physic, tears the papers; still God pays.
Or else by water goes, and so to plays,
  Calls for his stool, adorns the stage; God pays.
To every cause he meets, this voice he brays:
  His only answer is to all, God pays.   20
Not his poor cockatrice but he betrays
  Thus; and for his lechery scores: God pays.
But see! the old bawd hath served him in his trim,
  Lent him a pocky whore. She hath paid him.

13

To Doctor Empiric

When men a dangerous disease did 'scape
  Of old, they gave a cock to Aesculape.
Let me give two, that doubly am got free
  From my disease's danger, and from thee.

9 *'ssays*: essays, tries on. 11 *ordinaries*: taverns, eating houses. 13 *commodity*:
'A parcel of goods sold on credit by a usurer to a needy person, who immediately
raised some cash by re-selling them at a lower price, generally to the usurer himself'
(*OED*). 15 *essays*: Newdigate suggests a possible reference to Florio's translation of
Montaigne's essays, 1603, or to the essays of Sir William Cornwallis, 1600. 16 *papers*:
in which drugs were wrapped. 18 To sit on a stool on the stage itself cost an extra 6*d*.
at Blackfriars at this time: see *C.R.* Ind. 140–6. 21 *cockatrice*: whore. 23 *his trim*:
his own fashion.

13. 2 *Aesculape*: Aesculapius, classical god of medicine. The cock, symbol of vigilance,
was sacred to him. 4 Cf. *Volp.* I. iv. 20–2.

14

To William Camden

Camden, most reverend head, to whom I owe
　All that I am in arts, all that I know,
(How nothing's that?) to whom my country owes
　The great renown and name wherewith she goes;
Than thee the age sees not that thing more grave,　　5
　More high, more holy, that she more would crave.
What name, what skill, what faith hast thou in things!
　What sight in searching the most antique springs!
What weight, and what authority in thy speech!
　Man scarce can make that doubt, but thou canst teach.　10
Pardon free truth, and let thy modesty,
　Which conquers all, be once overcome by thee.
Many of thine this better could than I;
　But for their powers accept my piety.

15

On Court-Worm

All men are worms: but this no man. In silk
　'Twas brought to court first wrapped, and white as milk;
Where afterwards it grew a butterfly,
　Which was a caterpillar. So 'twill die.

14. *William Camden*: the celebrated antiquary (1551–1623), and Jonson's old school-master: *Conv. Dr.* 240. 4 *renown and name*: in 1604 King James had styled himself, controversially, King of Britain (rather than King of England) in his attempt to further the Union of Scotland and England; Jonson suggests that the titles of Camden's works, *Britannia* (1586), and *Remains of a Greater Work Concerning Britain* (1605), had helped to restore the popularity of the name 'Britain'. In 1603, with a gracious aside about the work of his old schoolmaster, Jonson had tactfully greeted James as *Monarchia Britannica* (*King's Ent.* 25 ff.). 5–6 Cf. Pliny, *Epist.* IV. xvii. 4. 7–9 Cf. Pliny on Titius Aristo, *Epist.* I. xxii*. 8 *antique*: then often stressed on the first syllable. 11–12 Cf. Claudian, *De Consulatu Stilichonis*, ii. 329: 'world-conqueror, conquer now thine own diffidence'. 13 *thine*: i.e. thy pupils. 14 *for*: in place of.

15. 1 *All men are worms*: Job 25: 5–6: 'Behold, even the moon hath no brightness, And the stars are not pure in his sight: How much less man, that is a worm! And the son of man, which is a worm!' Cf. Ps. 22 : 6; and *King Lear*, IV. i. 33. 4 *caterpillar*: the word was commonly used of an extortioner or rapacious person: cf. *Richard II*, II. iii. 166, 'caterpillars of the commonwealth'.

16

To Brain-Hardy

Hardy, thy brain is valiant, 'tis confessed;
  Thou more, that with it every day dar'st jest
Thyself into fresh brawls; when called upon,
  Scarce thy week's swearing brings thee off of one.
So in short time thou art in arrearage grown   5
  Some hundred quarrels, yet dost thou fight none;
Nor need'st thou; for those few, by oath released,
  Make good what thou dar'st do in all the rest.
Keep thyself there, and think thy valour right;
  He that dares damn himself dares more than fight.   10

17

To the Learned Critic

May others fear, fly, and traduce thy name
  As guilty men do magistrates; glad I,
That wish my poems a legitimate fame,
  Charge them, for crown, to thy sole censure high;
And but a sprig of bays, given by thee,   5
  Shall outlive garlands stolen from the chaste tree.

16. 8 *Make good*: prove. 10 Cf. Plutarch, *Lysander*, viii. 4*; Sir John Davies, *Epig.* xxviii, *In Sillam*.

17. Jonson's dedication of *Catiline* to William, Earl of Pembroke, places a similar stress on the legitimacy of his work, and draws a similar comparison between the roles of critic and magistrate. 3 *legitimate*: Horace, *Epist.* II. ii. 109 ff. 4 entrust them to your sole, high judgement, to determine whether they deserve the laurel wreath ('crown'); to 'crown' was also to hold a coroner's inquest upon (*OED*, † v.²). 6 *the chaste tree*: the laurel, sacred to Apollo, god of poetry; 'chaste', because of the transformation of Daphne (see *Epig.* 126. 3 n.).

18

To My Mere English Censurer

To thee my way in epigrams seems new,
   When both it is the old way and the true.
Thou sayst that cannot be: for thou hast seen
   Davies and Weever, and the best have been,
And mine come nothing like. I hope so. Yet,   5
   As theirs did with thee, mine might credit get,
If thou'dst but use thy faith as thou didst then
   When thou wert wont to admire, not censure men.
Prithee believe still, and not judge so fast;
   Thy faith is all the knowledge that thou hast.   10

19

On Sir Cod the Perfumed

That Cod can get no widow, yet a knight,
I scent the cause: he woos with an ill sprite.

20

To the Same Sir Cod

The expense in odours is a most vain sin,
Except thou couldst, Sir Cod, wear them within.

---

18. *Mere English Censurer*: i.e., a censurer (critic) who knows only epigrams written in English (cf. the phrase 'merely English', *Eng. Gr.* ch. 4, 237). This poem contains memories of Martial, *Epig.* II. lxxvii and VI. lxv. 4 *Davies and Weever*: the *Epigrams* of Sir John Davies (1569–1626) appeared *c.* 1590; *Epigrams in the Oldest Cut, and Newest Fashion*, by John Weever (1576–1632), were published in 1599. Jonson probably alludes to the latter title in ll. 1–2.

19. 1 *Cod*: civet- or musk-bag; testicles. 1 *yet a knight*: a typical Jonsonian gibe at the ease with which knighthoods were obtained after the accession of James: cf. *Epig.* 3, 46. 2 n., etc. 2 *an ill sprite*: a pun: an evil genius, bad breath. Cf. *E.M.I.* I. iii. 112.

20. Cf. Martial, *Epig.* II. xii.

21

On Reformed Gamester

Lord, how is Gamester changed! His hair close cut,
  His neck fenced round with ruff, his eyes half shut!
His clothes two fashions off, and poor; his sword
  Forbid his side; and nothing but the word
Quick in his lips! Who hath this wonder wrought?   5
  The late-ta'en bastinado. So I thought,
What several ways men to their calling have!
  The body's stripes, I see, the soul may save.

22

On My First Daughter

Here lies, to each her parents' ruth,
Mary, the daughter of their youth;
Yet, all heaven's gifts being heaven's due,
It makes the father less to rue.
At six months' end she parted hence   5
With safety of her innocence;
Whose soul heaven's Queen (whose name she bears),
In comfort of her mother's tears,
Hath placed amongst her virgin train;
Where, while that severed doth remain,   10
This grave partakes the fleshly birth;
Which cover lightly, gentle earth.

21. 1 *close cut*: in Puritan style. Cf. *B.F.* III. vi. 28–30.   4 *the word*: the common Puritan term for the Bible.

22. The date of the poem is unknown, though it probably belongs to Jonson's Catholic period, i.e. between 1598 and 1610. Dorothy Wordsworth found the poem's second line 'affecting': *Journals*, ed. E. de Selincourt (London, 1941), i. 110–11.   10–11 while the soul remains in heaven, severed from the body, the grave retains the body for a while (literally, takes a share in it; it will ultimately be reunited with the soul at the Resurrection).   12 Cf. Martial on the death of Erotion, *Epig.* V. xxxiv. 9–10*; and IX. xxix. 11, XI. xiv; and see R. Lattimore, *Themes in Greek and Latin Epitaphs* (Urbana, 1962), pp. 65 ff.

23

To John Donne

Donne, the delight of Phoebus and each muse,
    Who, to thy one, all other brains refuse;
Whose every work of thy most early wit
    Came forth example, and remains so yet;
Longer a-knowing than most wits do live;   5
    And which no affection praise enough can give!
To it, thy language, letters, arts, best life,
    Which might with half mankind maintain a strife;
All which I meant to praise, and yet I would,
    But leave, because I cannot as I should.

24

To the Parliament

There's reason good that you good laws should make:
Men's manners ne'er were viler, for your sake.

25

On Sir Voluptuous Beast

While Beast instructs his fair and innocent wife
    In the past pleasures of his sensual life,
Telling the motions of each petticoat,
    And how his Ganymede moved, and how his goat,

---

23. Jonson praised Donne as 'the first poet in the world in some things' (*Conv. Dr.* 117–18). The friendship of the two poets was probably of long standing; see R. C. Bald, *John Donne: A Life* (Oxford, 1970), pp. 194–7. *Epig.* 96 is also addressed to Donne.   2 *who*, in favour of your brain, refuse to inspire all other brains.   3 *early*: a use of the adjective not recorded by *OED*; but cf. *Und.* 70. 125, *U.V.* 18. 7.   5 *Longer a-knowing*: longer a skilled practitioner of your art.   7 *To it*: probably, in addition to your wit.   8 *maintain a strife*: vie, contend.

24. 2 *for your sake*: if Fleay's conjectured date of 1604 is correct, the sense is probably 'on account of enmity to you' (see *OED*, 'sake', 5. b).

25. 4 *Ganymede*: catamite.

And now her, hourly, her own cuckquean makes   5
    In varied shapes, which for his lust she takes;
What doth he else but say: leave to be chaste,
    Just wife, and, to change me, make woman's haste?

26

On the Same Beast

Than his chaste wife though Beast now know no more,
He adulters still: his thoughts lie with a whore.

27

On Sir John Roe

In place of scutcheons that should deck thy hearse,
Take better ornaments, my tears and verse.
    If any sword could save from fates, Roe's could;
    If any muse outlive their spite, his can;
    If any friend's tears could restore, his would;   5
    If any pious life e'er lifted man
To heaven, his hath. O happy state! wherein
We, sad for him, may glory and not sin.

5–6 To satisfy her husband sexually, the wife is forced to assume various disguises or roles, allowing him to imagine he is with other partners; thus she becomes hourly her own cuckquean (a female cuckold). Cf. Volpone's invitation to Celia, *Volp.* III. vii. 221 ff., and *D. is A.* V. iii. 18.   8 *to change me*: a pun: to provide variety for me; to exchange me for another. By such tricks, the husband is unwittingly encouraging his wife to commit real adultery.

26. Cf. Seneca, *De Constantia*, vii. 4, Matt. 5 : 27–8, Marston, *The Malcontent*, I. i. 161–7, etc.

27. *Sir John Roe*: 1581–?1606, soldier, poet, and a close friend of Jonson, in whose arms he died of the plague (see *Conv. Dr.* 155–9, 184–7). Also addressed in *Epig.* 32, 33. See Alvaro Ribeiro, *RES*, xxiv (1973), 153–64.   1 *scutcheons*: tablets showing the armorial bearings of a deceased person.   2 *ornaments*: the word was commonly used in relation to poetry, as in the third book of Puttenham's *The Art of English Poesie* (1589), 'Of Ornament'.   3 *If any sword*: cf. *Epig.* 32. 1–2.

28

On Don Surly

Don Surly, to aspire the glorious name
   Of a great man, and to be thought the same,
Makes serious use of all great trade he knows.
   He speaks to men with a rhinocerot's nose,
Which he thinks great; and so reads verses, too;   5
   And that is done as he saw great men do.
He has tympanies of business in his face,
   And can forget men's names with a great grace.
He will both argue and discourse in oaths,
   Both which are great; and laugh at ill-made clothes—   10
That's greater yet—to cry his own up neat.
   He doth, at meals, alone, his pheasant eat,
Which is main greatness. And at his still board
   He drinks to no man; that's, too, like a lord.
He keeps another's wife, which is a spice   15
   Of solemn greatness. And he dares at dice
Blaspheme God, greatly; or some poor hind beat
   That breathes in his dog's way; and this is great.
Nay more, for greatness' sake, he will be one
   May hear my epigrams, but like of none.   20
Surly, use other arts; these only can
   Style thee a most great fool, but no great man.

---

28. The name Surly occurs again in *Epig.* 82, and in *Alch.* 1 *aspire*: attain. 3 *trade*: manners. 4 *rhinocerot's*: rhinoceros's: recalling Martial's description (*Epig.* I. iii. 5–6) of the superciliousness of Rome, where men and boys tilt their noses like rhinoceroses (*nasum rhinocerotis habent*). 7 *tympanies*: swellings. 15 *spice*: species, kind.

29

To Sir Annual Tilter

Tilter, the most may admire thee, though not I;
　And thou, right guiltless, mayst plead to it: why?
For thy late sharp device. I say 'tis fit
　All brains, at times of triumph, should run wit,
For then our water-conduits do run wine;　　　　　　5
　But that's put in, thou'lt say. Why, so is thine.

30

To Person Guilty

Guilty, be wise; and though thou know'st the crimes
　Be thine I tax, yet do not own my rhymes;
'Twere madness in thee to betray thy fame
And person to the world, ere I thy name.

31

On Bank the Usurer

Bank feels no lameness of his knotty gout;
　His moneys travel for him, in and out;
And though the soundest legs go every day,
　He toils to be at hell as soon as they.

29. Fleay (i. 317) considers that 'Annual' implies a King's Day, i.e. the anniversary of James's accession, and suggests a possible date of 24 March 1610, the 'device' of l. 3 possibly being Sir Richard Preston's Elephant (see John Nichols, *The Progresses . . . of King James* (London, 1828), ii. 287).　5 On great occasions the London water-conduits were made to flow with claret.

30. 4 *thy name*: see *Epig*. Ded. 15–16 n., *Epig.* 77.

31. 1 As C. T. Wright has shown (*SP*, xxi (1934), 176–97), usurers in this period are commonly depicted as suffering from gout. Jonson's Sir Moth Interest (*M.L.* III. iv. 37–41) is afflicted by 'A kind of cramp, or hand-gout'. Cf. Horace, *Epist.* I. i. 31, Martial, *Epig.* I. xcviii*.　2 *travel*: a pun: travel/travail.　4 *hell*: possibly with a secondary reference to the debtors' prison known as 'hell', under Westminster Hall. Cf. Middleton, *A Trick to Catch the Old One*, IV. v. 2–3.

32

On Sir John Roe

What two brave perils of the private sword
   Could not effect, not all the furies do
That self-divided Belgia did afford;
   What not the envy of the seas reached to,
The cold of Moscow, and fat Irish air,   5
   His often change of clime (though not of mind)
What could not work; at home in his repair
   Was his blest fate, but our hard lot, to find.
Which shows, wherever death doth please to appear,
   Seas, serenes, swords, shot, sickness, all are there.   10

33

To the Same

I'll not offend thee with a vain tear more,
   Glad-mentioned Roe; thou art but gone before
Whither the world must follow. And I now
   Breathe to expect my when, and make my how;
Which if most gracious heaven grant like thine,   5
   Who wets my grave can be no friend of mine.

---

32. *Sir John Roe*: see *Epig.* 27, n.   3 *self-divided Belgia*: after the Pacification of Ghent in 1576, hostility continued to exist between Protestants in the northern provinces of the Netherlands and Catholics in the south.   5–7 what the cold of Moscow and the moist Irish air ... could not bring about (i.e. his death).   6 Cf. Horace, *Epist.* I. xi. 27: 'they change their clime, not their mind, who rush across the sea'.   7 *repair*: usual place of dwelling.   10 *serenes*: 'A light fall of moisture or fine rain after sunset in hot countries, formerly regarded as a noxious dew or mist' (*OED*). Cf. *Volp*. III. vii. 184.

33. This poem was admired by Elizabeth Barrett Browning (see Bennett Weaver, *PMLA*, lxv (1950), 403–4).   2–3 Cf. Seneca, *Ad Marciam de Consolatione*, xix. 1*; *Ad Polybium de Consolatione*, ix. 9*; *Epist.* lxiii. 16.

34

Of Death

He that fears death, or mourns it in the just,
Shows of the Resurrection little trust.

35

To King James

Who would not be thy subject, James, to obey
  A prince that rules by example more than sway?
Whose manners draw, more than thy powers constrain;
  And in this short time of thy happiest reign
Hast purged thy realms, as we have now no cause   5
  Left us of fear, but first our crimes, then laws.
Like aids 'gainst treasons who hath found before?
  And than in them, how could we know God more?
First thou preserved wert, our king to be,
  And since, the whole land was preserved for thee.   10

36

To the Ghost of Martial

Martial, thou gav'st far nobler epigrams
  To thy Domitian, than I can my James;
But in my royal subject I pass thee:
  Thou flattered'st thine, mine cannot flattered be.

35. Fleay (i. 317) dates the poem April 1604, noticing allusions in ll. 5–8 to the new laws of March 1604, in l. 9 to the treasons of Gowrie (1600) and Ralegh (1603), and in l. 10 to the plague of 1603.   2 Possibly echoing James himself on the qualities of the ideal ruler: see *Basilikon Doron*, ed. James Craigie (Edinburgh, 1944), i. 53. The compliment is applied to James again in *Panegyre*, 125–7, *Haddington*, 216 ff., and *Oberon*, 346–7; see also *C.R.* V. xi. 169–73, *Dubia*, 3. 16–18, and cf. Claudian, *Panegyricus De Quarto Consulatu Honorii Augusti*, 299–301.

36. 2 *Domitian*: despotic and unpopular Roman emperor (A.D. 51–96, emperor from 81). There is perhaps an implied contrast between James's preservation from plotting, mentioned in the previous epigram, and Domitian's murder by conspirators.   4 James had severely castigated flattery ('that filthy vice . . . the pest of all princes, and wrack of republics'), *Basilikon Doron*, ed. Craigie, i. 115. Jonson himself avowed that 'he would not flatter though he saw death' (*Conv. Dr.* 332).

37

On Cheverel the Lawyer

No cause nor client fat will Cheverel leese,
 But as they come, on both sides he takes fees,
And pleaseth both: for while he melts his grease
 For this, that wins, for whom he holds his peace.

38

To Person Guilty

Guilty, because I bade you late be wise,
 And to conceal your ulcers did advise,
You laugh when you are touched, and long before
 Any man else, you clap your hands, and roar,
And cry, Good, good! This quite perverts my sense,   5
 And lies so far from wit, 'tis impudence.
Believe it, Guilty, if you lose your shame,
 I'll lose my modesty, and tell your name.

39

On Old Colt

For all night-sins with others' wives, unknown,
Colt now doth daily penance in his own.

---

37. 1 *Cheverel*: cheverel-, or kid-leather is pliant; so is the lawyer. He reappears in *Epig.* 54. See Tilley, C608.   1 *leese*: lose.   3 *melts his grease*: exerts himself strenuously.

38. 1 *bade you late*: in *Epig.* 30.   8 *tell your name*: see *Epig.* Ded. 15–16 n.

39. *Colt*: the name suggests lasciviousness; cf. *Cymbeline*, II. iv. 133: 'She hath been colted by him'.

40

On Margaret Radcliffe

M arble, weep, for thou dost cover
A dead beauty underneath thee,
R ich as nature could bequeath thee;
G rant, then, no rude hand remove her.
A ll the gazers on the skies   5
R ead not in fair heaven's story
E xpresser truth or truer glory
T han they might in her bright eyes.
R are as wonder was her wit,
A nd like nectar ever flowing;   10
T ill time, strong by her bestowing,
C onquered hath both life and it.
L ife, whose grief was out of fashion
I n these times: few so have rued
F ate, in a brother. To conclude,   15
F or wit, feature, and true passion,
E arth, thou hast not such another.

41

On Gypsy

Gypsy, new bawd, is turned physician,
    And gets more gold than all the College can.
Such her quaint practice is, so it allures,
    For what she gave, a whore, a bawd, she cures.

40. *Margaret Radcliffe*: or Ratcliffe, favourite maid of honour of Queen Elizabeth, and sister of Sir John Radcliffe of *Epig.* 93. Four of her other brothers died within a period of months: Sir William and Sir Alexander were slain in Ireland in 1598 and 1599 respectively, and Edmund and Thomas died of fever in French Flanders in 1599. Margaret herself died of grief on 10 November 1599. The poem is an acrostic; for a scornful view of this form, see *Und.* 43. 39.   16 *feature*: comeliness; figure.

41. 2 *College*: i.e. of Physicians.   3 *quaint*: probably with an obscene sense.

42

On Giles and Joan

Who says that Giles and Joan at discord be?
  The observing neighbours no such mood can see.
Indeed, poor Giles repents he married ever:
  But that his Joan doth too. And Giles would never,
By his free will, be in Joan's company;   5
  No more would Joan he should. Giles riseth early,
And having got him out of doors is glad—
  The like is Joan—but turning home, is sad;
And so is Joan. Oft-times, when Giles doth find
  Harsh sights at home, Giles wisheth he were blind: 10
All this doth Joan. Or that his long-yarned life
  Were quite out-spun; the like wish hath his wife.
The children that he keeps, Giles swears are none
  Of his begetting; and so swears his Joan:
In all affections she concurreth still.   15
  If now, with man and wife, to will and nill
The self-same things a note of concord be,
  I know no couple better can agree!

43

To Robert, Earl of Salisbury

What need hast thou of me, or of my muse,
  Whose actions so themselves do celebrate?
Which should thy country's love to speak refuse,
  Her foes enough would fame thee, in their hate.

---

42. Cf. Martial, *Epig.* VIII. xxxv* and XII. lviii. Whipple (*Martial and the English Epigram*, p. 398 n. 23) cites as the 'immediate original' of this poem an epigram by Parkhurst. Cf. Herrick's 'Jack and Jill', Herrick, p. 163.   16 *to will and nill*: to want and not to want. Cf. Sallust, *Catiline*, xx. 4*.

43. *Robert, Earl of Salisbury*: Robert Cecil (1563–1612), Secretary of State under Elizabeth; created Earl of Salisbury 4 May 1605, when H & S think the poem was written. He is also addressed in *Epig.* 63, 64.   3–4 which, if not spoken of by thy country, out of love, would be made famous enough by her foes, out of hatred.

'Tofore, great men were glad of poets; now 5
   I, not the worst, am covetous of thee;
Yet dare not to my thought least hope allow
   Of adding to thy fame: thine may to me,
When in my book men read but Cecil's name;
   And what I write thereof find far and free 10
From servile flattery (common poets' shame)
   As thou stand'st clear of the necessity.

44

On Chuff, Bank's the Usurer's Kinsman

Chuff, lately rich in name, in chattels, goods,
   And rich in issue to inherit all,
   Ere blacks were bought for his own funeral,
Saw all his race approach the blacker floods.
   He meant they thither should make swift repair, 5
   When he made him executor, might be heir.

45

On My First Son

Farewell, thou child of my right hand, and joy;
   My sin was too much hope of thee, loved boy.
Seven years thou wert lent to me, and I thee pay,
   Exacted by thy fate, on the just day.

---

5 *'Tofore*: heretofore. 10–12 Pliny, *Panegyricus*, 1*.

44. *Chuff*: churl, miser; also a bird in the crow family (chough). Cf. the miser Sordido in *E.M.O.* ('a wretched hobnailed chuff'), and Corvino in *Volp.* (I. ii. 96): a 'gaping crow', awaiting a legacy. Bank appears also in *Epig.* 31. 4 *approach the blacker floods*: the Styx; i.e. die. 5–6 Perhaps: though Chuff had made Bank his executor, Chuff intended rather that he himself should be Bank's heir.

45. *First Son*: also called Benjamin Jonson; he died of the plague, 1603. Jonson had a premonition of the boy's death: see *Conv. Dr.* 261–72. For criticism, see W. D. Kay, *SEL*, xi (1971), 125–36. 1 *right hand, and joy*: in Hebrew, Benjamin means son of the right hand; dextrous, fortunate. In Gen. 35 : 18 the name is implicitly contrasted with the name Ben-oni, 'the son of my sorrow'. 3 *lent*: a common classical notion: see Lattimore, *Themes in Greek and Latin Epitaphs*, pp. 170–1; and cf. *Epig.* 22. 3, 109. 11, *For.* 3. 106, *Und.* 83, 80.

Oh, could I lose all father now! For why   5
  Will man lament the state he should envy?
To have so soon 'scaped world's and flesh's rage,
  And, if no other misery, yet age?
Rest in soft peace, and, asked, say here doth lie
  Ben Jonson his best piece of poetry;   10
For whose sake, henceforth, all his vows be such,
  As what he loves may never like too much.

46

To Sir Luckless Woo-All

Is this the sir, who, some waste wife to win,
  A knighthood bought, to go a-wooing in?
'Tis Luckless he, that took up one on band
  To pay at his day of marriage. By my hand,
The knight-wright's cheated, then: he'll never pay.   5
  Yes, now he wears his knighthood every day.

47

To the Same

Sir Luckless, troth, for luck's sake pass by one:
He that woos every widow will get none.

5 *lose all father*: shake off all paternal feelings.   10 *best piece of poetry*: glancing at the idea of poet as maker.   12 Martial, *Epig.* VI. xxix. 8: 'whatever you love, pray that it may not please you too much'. It was a common classical notion that excessive good fortune aroused the jealousy of the gods; Jonson subtly varies the emphasis. Cf. also Martial, *Epig.* XII. xxxiv. 8–11*.

46. 1 *waste*: *OED*, citing this usage only (*a.* † 8) conjectures 'worthless'; a more likely sense is 'old, cast-off' (cf. 'waste paper'): Sir Luckless is a widow-hunter like Winwife in *B.F.*   2 *A knighthood bought*: see Lawrence Stone's discussion of the abuses arising out of the sale of knighthoods in James's reign: *The Crisis of the Aristocracy* (Oxford, 1965), pp. 74–82.   3 *band*: bond.   5 *knight-wright*: not in *OED*, but Briggs (*Anglia*, xxxix (1916), 315) points to another use of the word in Donne's *Paradoxes* (1652), p. 66.

48

On Mongrel Esquire

His bought arms Mong not liked; for his first day
Of bearing them in field, he threw 'em away;
And hath no honour lost, our duellists say.

49

To Playwright

Playwright me reads, and still my verses damns:
He says I want the tongue of epigrams;
I have no salt: no bawdry, he doth mean;
For witty, in his language, is obscene.
Playwright, I loathe to have thy manners known　　　5
In my chaste book: profess them in thine own.

50

To Sir Cod

Leave, Cod, tobacco-like, burnt gums to take,
Or fumy clysters, thy moist lungs to bake:
Arsenic would thee fit for society make.

---

48. 'The arms were usually portrayed upon the shield; so that on his entering into battle, he flung away his shield, that he might not be encumbered in his flight. This marks him for cowardice' (Whalley). 1 *bought arms*: cf. Sogliardo in *E.M.O.* III. iv. 51 ff., who buys his arms for £30.

49. Jonson's use of the word 'playwright' here and in *Epig.* 68 & 100 predates *OED*'s first recorded example of 1687: it may be an ironical coinage (cf. 'stage-wrights', *Songs*, 14. 35). Jonson habitually referred to himself as a 'poet', and to his plays as 'poems'. 2 *the tongue of epigrams*: echoing Martial's *epigrammaton linguam* ('the language of the epigram'), *Epig.* I, Pref. [10–11]; Martial is justifying the freedom of his language: 3 *salt*: a pun: wit, and salaciousness. 6 Perhaps remembering, and significantly varying, Catullus, *Carm.* xvi: 'for the sacred poet ought to be chaste himself, his verses need not be so'.

50. *Sir Cod*: see *Epig.* 19, 20. 1 *tobacco-like, burnt gums*: the reference is to gums as medicinal drugs. Tobacco, smoked or simply placed in the mouth, was likewise thought to cure various ailments, including venereal disease: see William Barclay, *Nepenthes, or the Virtues of Tobacco* (Edinburgh, 1614), and Sir John Davies, *Epig.* xxxvi; and cf. Bobadill on the subject, *E.M.I.* III. v. 76–95. 2 *clysters*: normally, the pipes used for enemas; here, contemptuously, tobacco pipes. Cf. *Songs*, 25. 32. 3 *Arsenic*: a common cure for venereal disease; but also a lethal poison. Jonson's insult is enhanced by the doubt as to which application he is prescribing.

51

To King James
Upon the Happy False Rumour of His Death,
the Two-and-Twentieth Day of March, 1607

That we thy loss might know, and thou our love,
   Great heaven did well to give ill fame free wing;
Which, though it did but panic terror prove,
   And far beneath least pause of such a king,
Yet give thy jealous subjects leave to doubt,   5
   Who this thy 'scape from rumour gratulate
No less than if from peril; and, devout,
   Do beg thy care unto thy after-state.
For we, that have our eyes still in our ears,
   Look not upon thy dangers, but our fears.   10

52

To Censorious Courtling

Courtling, I rather thou shouldst utterly
   Dispraise my work than praise it frostily:
When I am read thou feign'st a weak applause,
   As if thou wert my friend, but lack'dst a cause.
This but thy judgement fools; the other way   5
   Would both thy folly and thy spite betray.

51. Actually 1606. The rumour that James had been stabbed while hunting came not long after the Gunpowder Plot, and caused widespread alarm. 5 *doubt*: fear. 6 *gratulate*: rejoice over. 8 *after-state*: future (probably a nonce-formation).

52. 1–2 A proverbial sentiment: see Tilley, P538. Courtling is addressed again in *Epig.* 72.

53

To Old-End Gatherer

Long-gathering Old-End, I did fear thee wise
    When, having pilled a book which no man buys,
Thou wert content the author's name to lose;
    But when, in place, thou didst the patron's choose,
It was as if thou printed hadst an oath,   5
    To give the world assurance thou wert both;
And that, as Puritans at baptism do,
    Thou art the father and the witness too.
For, but thyself, where, out of motley, is he
    Could save that line to dedicate to thee?   10

54

On Cheverel

Cheverel cries out my verses libels are,
    And threatens the Star Chamber and the Bar.
What are thy petulant pleadings, Cheverel, then,
    That quitt'st the cause so oft, and rail'st at men?

55

To Francis Beaumont

How I do love thee, Beaumont, and thy muse,
    That unto me dost such religion use!
How I do fear myself, that am not worth
    The least indulgent thought thy pen drops forth!

---

53. *Old-End Gatherer*: i.e. plagiarist; lit., one who collects old ends of material or candles. H & S cf. *Volp*. Prol. 23–4, a possible glance at Marston, who dedicated *The Scourge of Villainy* to himself in 1599.   3 *the author's name*: Old-End Gatherer withholds his own name as well as that of the author from whom he has plagiarized.   4 *patron's*: person to whom the book is dedicated.   6 *both*: author and dedicatee.   8 *witness*: baptismal sponsor or godfather. A Puritan term: see *B.F.* I. iii. 127–8.   9–10 i.e. For who but a fool (apart from thyself) could spare a line to dedicate to thee?

54. *Cheverel*: see *Epig*. 37 and n.   2 Cf. Horace, *Sat*. II. i. 47.   3 *petulant*: insolent.

55. *Francis Beaumont*: ?1584–1616, the dramatist. This is a reply to Beaumont's well-known poem to Jonson from the country, written some time between 1610–13, and printed in H & S, xi, 374–6. A cooler opinion of Beaumont is recorded in *Conv. Dr.* 154.   2 *religion*: pious affection.

At once thou mak'st me happy, and unmak'st;   5
   And giving largely to me, more thou tak'st.
What fate is mine that so itself bereaves?
   What art is thine that so thy friend deceives?
When even there where most thou praisest me
   For writing better, I must envy thee.   10

56

On Poet-Ape

Poor Poet-Ape, that would be thought our chief,
   Whose works are e'en the frippery of wit,
From brokage is become so bold a thief
   As we, the robbed, leave rage, and pity it.
At first he made low shifts, would pick and glean,   5
   Buy the reversion of old plays; now grown
To a little wealth and credit in the scene,
   He takes up all, makes each man's wit his own;
And, told of this, he slights it: Tut, such crimes
   The sluggish gaping auditor devours;   10
He marks not whose 'twas first; and after-times
   May judge it to be his as well as ours.
Fool, as if half-eyes will not know a fleece
   From locks of wool, or shreds from the whole piece!

57

On Bawds and Usurers

If, as their ends, their fruits were so the same,
Bawdry and usury were one kind of game.

---

56. *Poet-Ape*: not identified.   2 *frippery*: old clothes shop.   3 *brokage*: brokerage:
dealing in second-hand goods.   14 *the whole piece*: cf. *U.V.* 42. 17.

57. Cf. *S. of N.* II. v. 100: 'A money-bawd is lightly a flesh-bawd, too'. The associa-
tion was traditional: see Tilley, U31, Herrick's 'Upon Snare, an Usurer', Herrick, p.
220, and H. W. Baum, *The Satiric and the Didactic*, p. 66.

58

To Groom Idiot

Idiot, last night I prayed thee but forbear
     To read my verses; now I must to hear:
For offering with thy smiles my wit to grace,
     Thy ignorance still laughs in the wrong place.
And so my sharpness thou no less disjoints   5
     Than thou didst late my sense, losing my points.
So have I seen at Christmas sports one lost,
     And, hoodwinked, for a man, embrace a post.

59

On Spies

Spies, you are lights in state, but of base stuff,
Who, when you've burnt yourselves down to the snuff,
Stink, and are thrown away. End fair enough.

60

To William, Lord Monteagle

Lo, what my country should have done—have raised
     An obelisk or column to thy name,
Or, if she would but modestly have praised
     Thy fact, in brass or marble writ the same—

---

58. 6 *losing*: Fol. 'loosing': the two words were for a time thought of as interchangeable, and both senses are probably intended here.   6 *points*: punctuation.

59. Jonson may be thinking of the spies who visited him in prison in 1597: see *Conv. Dr.* 256–60. Cf. also *Und.* 43. 187–8; *E.M.I.* I. i. 76–81.

60. *Lord Monteagle*: William Parker, 4th Baron Monteagle, a former Roman Catholic, received in 1605 the anonymous warning letter concerning the Gunpowder Plot which led to its discovery; he was rewarded, and regarded as the saviour of parliament. It has been suggested that he and Jonson may in fact have had earlier knowledge of the Plot: see B. N. De Luna, *Jonson's Romish Plot* (Oxford, 1967), esp. ch. iv.   4 *fact*: action, conduct.

I, that am glad of thy great chance, here do!   5
   And, proud my work shall outlast common deeds,
Durst think it great, and worthy wonder, too:
   But thine, for which I do it, so much exceeds!
My country's parents I have many known,
   But saver of my country thee alone.   10

61

To Fool or Knave

Thy praise or dispraise is to me alike;
One doth not stroke me, nor the other strike.

62

To Fine Lady Would-Be

Fine Madam Would-Be, wherefore should you fear,
   That love to make so well, a child to bear?
The world reputes you barren; but I know
   Your 'pothecary, and his drug says no.
Is it the pain affrights? That's soon forgot.   5
   Or your complexion's loss? You have a pot
That can restore that. Will it hurt your feature?
   To make amends, you're thought a wholesome creature.
What should the cause be? Oh, you live at court:
   And there's both loss of time and loss of sport   10
In a great belly. Write, then, on thy womb:
   Of the not born, yet buried, here's the tomb.

---

61. 1 *praise or dispraise*: cf. *Cat*. 'To the Reader in Ordinary', 4–6; *Queens* 679 ff. (Q, Ff). 2 *stroke*: flatter. The 'stroke' / 'strike' contrast was proverbial: see *OED* 'stroke' *v*.¹, 1. c & e; and cf. *Sej*. I. 413.

62. *Lady Would-Be*: cf. the character of this name in *Volp*. The Collegiate ladies in *S.W.* (IV. iii. 57–61) hold similar views concerning the disagreeableness of child-bearing; see also *Und*. 15. 95–6.   1–2 why should you fear to bear a child when you so enjoy making one (i.e. the sexual act)?

63

To Robert, Earl of Salisbury

Who can consider thy right courses run,
  With what thy virtue on the times hath won,
And not thy fortune; who can clearly see
  The judgement of the king so shine in thee;
And that thou seek'st reward of thy each act   5
  Not from the public voice, but private fact;
Who can behold all envy so declined
  By constant suffering of thy equal mind;
And can to these be silent, Salisbury,
  Without his, thine, and all time's injury?   10
Cursed be his muse that could lie dumb or hid
  To so true worth, though thou thyself forbid.

64

To the Same

Upon the accession of the Treasurership to him

Not glad, like those that have new hopes or suits
  With thy new place, bring I these early fruits
Of love, and what the golden age did hold
  A treasure, art: contemned in the age of gold;
Nor glad as those that old dependents be   5
  To see thy father's rites new laid on thee;
Nor glad for fashion; nor to show a fit
  Of flattery to thy titles; nor of wit.

---

63. *Robert, Earl of Salisbury*: see *Epig.* 43, n.   2–3 Cf. Valerius Maximus on the elder Cato, VIII. xv. 2*. It is characteristic of Jonson to contrast 'virtue' and 'fortune' in this way: cf. *Und.* 23. 16–18, *P. H. Barr.* 363–6, *Sej.* III. 88–9, 321–5, IV. 68–9, etc.   5–6 Cf. Pliny, *Epist.* I. xxii*.   7 *declined*: turned aside.

64. *the Treasurership*: on 6 May 1608.   3–4 *golden age . . . the age of gold*: a traditional paradox. On Jonson's fondness for this theme, see W. Todd Furniss, *Three Studies in the Renaissance* (New Haven, 1958), pp. 89–179.   6 *thy father's rites*: Lord Burghley, Cecil's father, had been made Lord Treasurer in 1592: cf. *Und.* 30.

But I am glad to see that time survive
   Where merit is not sepulchred alive;   10
Where good men's virtues them to honours bring,
   And not to dangers; when so wise a king
Contends to have worth enjoy, from his regard,
   As her own conscience, still the same reward.
These, noblest Cecil, laboured in my thought,   15
   Wherein what wonder, see, thy name hath wrought:
That whilst I meant but thine to gratulate,
   I've sung the greater fortunes of our state.

65

To My Muse

Away, and leave me, thou thing most abhorred,
   That hast betrayed me to a worthless lord,
Made me commit most fierce idolatry
   To a great image through thy luxury.
Be thy next master's more unlucky muse,   5
   And, as thou hast mine, his hours and youth abuse.
Get him the time's long grudge, the court's ill-will;
   And, reconciled, keep him suspected still.
Make him lose all his friends; and, which is worse,
   Almost all ways to any better course.   10
With me thou leav'st an happier muse than thee,
   And which thou brought'st me, welcome poverty;
She shall instruct my after-thoughts to write
   Things manly, and not smelling parasite.
But I repent me: stay. Whoe'er is raised   15
   For worth he has not, he is taxed, not praised.

11–12 Cf. Pliny, *Epist*. V. xiv*.

65. 2 *a worthless lord*: Jonson's placing of this epigram raises the suspicion that Salisbury may be the lord in question. 'Salisbury never cared for any man longer nor he could make use of him' (*Conv. Dr.* 353–4). See De Luna, *Jonson's Romish Plot*, p. 71, and *For.* 2n., 2. 61 ff., n.   3 *most fierce idolatry*: the phrase is used again in *Alch*. IV. i. 39, of Mammon's address to Dol.   4 *luxury*: licentiousness.

66

To Sir Henry Cary

That neither fame nor love might wanting be
    To greatness, Cary, I sing that, and thee;
Whose house, if it no other honour had,
    In only thee might be both great and glad;
Who, to upbraid the sloth of this our time,   5
    Durst valour make almost, but not a crime.
Which deed I know not whether were more high
    Or thou more happy, it to justify
Against thy fortune: when no foe that day
    Could conquer thee but chance, who did betray.   10
Love thy great loss, which a renown hath won
    To live when Broick not stands, nor Ruhr doth run.
Love honours, which of best example be
    When they cost dearest and are done most free;
Though every fortitude deserves applause,   15
    It may be much or little in the cause.
He's valiant'st that dares fight, and not for pay;
    That virtuous is when the reward's away.

67

To Thomas, Earl of Suffolk

Since men have left to do praiseworthy things,
    Most think all praises flatteries. But truth brings
That sound and that authority with her name,
    As to be raised by her is only fame.

---

66. *Sir Henry Cary*: ?1575–1633, created Viscount Falkland 1620; father of Sir Lucius
Cary of *Und.* 70. 6 ff. In October 1605, 1,200 Dutch and English troops under the
command of Maurice of Nassau fled before 400 Italians near the junction of the Ruhr
and the Rhine; Cary was one of four men who stood their ground, and was sub-
sequently captured. Jonson compares, and distinguishes, military capture and civil
arrest. 8 *justify*: in the legal sense: 'to show or maintain sufficient reason in court for
doing that which one is called upon to answer for' (*OED*, 7. a). 12 'The castle and
river near where he was taken' (Jonson's note). 18 Cf. *Cat.* III. 480, and Juvenal,
Sat. x. 141–2*.

67. *Thomas, Earl of Suffolk*: Thomas Howard (1561–1626), Lord Chamberlain 1603–
14, Lord Treasurer 1614–19. In 1605 he rescued Jonson and Chapman from their

Stand high, then, Howard, high in eyes of men,   5
    High in thy blood, thy place, but highest then
When, in men's wishes, so thy virtues wrought
    As all thy honours were by them first sought,
And thou designed to be the same thou art,
    Before thou wert it, in each good man's heart: 10
Which, by no less confirmed than thy king's choice,
    Proves that is God's, which was the people's, voice.

68

On Playwright

Playwright, convict of public wrongs to men,
    Takes private beatings, and begins again.
Two kinds of valour he doth show at once:
    Active in's brain, and passive in his bones.

69

To Pertinax Cob

Cob, thou nor soldier, thief, nor fencer art,
    Yet by thy weapon liv'st! Thou hast one good part.

imprisonment over *Eastward Ho!*: H & S conjecture that this poem may have been
prompted by gratitude for that action. Whalley's suggestion that the poem was written
on Suffolk's appointment as Lord Treasurer perhaps makes more sense of the poem's
concluding lines. (Cf. *Epig.* 133. 193–4 n.)   1 Cf. Pliny, *Epist.* III. xxi. 3. 5–6 Hunter
points out that Jonson is playing on the etymology of 'Howard': William Camden does
likewise in *Remains Concerning Britain* (London, 1870), pp. 148–9, declaring that the
name means 'High Warden or Guardian'.   7–10 Cf. Claudian, *De Consulatu
Stilichonis*, i. 49–50*.   12 Proverbial: *vox populi vox Dei* (Tilley, V95); cf. *Cat.* III.
60–1.

68. *Playwright*: possibly John Marston: see *Conv. Dr.* 160; and cf. *Epig.* 49, n.

69. *Pertinax Cob*: *pertinax*: obstinate, stiff (Lat.); *cob*: a large and lumpish person or
thing.   2 *weapon . . . part*: with a play on the sexual meanings. Cf. Martial, *Epig.* IX.
lxiii. 2.

70

To William Roe

When Nature bids us leave to live, 'tis late
  Then to begin, my Roe; he makes a state
In life that can employ it, and takes hold
  On the true causes ere they grow too old.
Delay is bad, doubt worse, depending worst;   5
  Each best day of our life escapes us first.
Then, since we (more than many) these truths know,
  Though life be short, let us not make it so.

71

On Court-Parrot

To pluck down mine, Poll sets up new wits still;
Still, 'tis his luck to praise me 'gainst his will.

72

To Courtling

I grieve not, Courtling, thou are started up
  A chamber-critic, and dost dine and sup
At madam's table, where thou mak'st all wit
  Go high or low as thou wilt value it.
'Tis not thy judgement breeds the prejudice: 5
  Thy person only, Courtling, is the vice.

70. *William Roe*: b. 1585, brother of Sir John Roe (see *Epig.* 27, 32, 33), and cousin of Sir Thomas Roe (*Epig.* 98, 99). Jonson appeared on his behalf in a law-suit (H & S, i. 223–30). *Epig.* 128 is also addressed to him.   1–2 Cf. Seneca, *De Brevitate Vitae*, iii. 5*.   2 *makes a state*: i.e., achieves an admirable state.   5 Seneca, op. cit. ix. 1*.   6 Cf. Virgil, *Georgics*, iii. 66–7*, and Seneca's comment on those lines, *Epist.* cviii. 24 ff. For the general thought, cf. Horace, *Epist.* I. ii. 41 ff., Martial, *Epig.* I. xv, V. xx, etc.

71. *Court-Parrot*: Mark Eccles suggests that this may be Henry Parrot, a contemporary writer of epigrams, and detects a possible allusion to Parrot's *The Mousetrap*, *Epig.* xcvii, 1606 (*RES*, xiii (1937), 388).

72. *Courtling*: also addressed in *Epig.* 52.   2 *chamber-critic*: a critic who gives his opinions in private; the phrase also connotes 'effeminacy or wantonness' (*OED*).

73

To Fine Grand

What is't, fine Grand, makes thee my friendship fly,
Or take an epigram so fearfully,
As 'twere a challenge, or a borrower's letter?
The world must know your greatness is my debtor.
In primis, Grand, you owe me for a jest   5
I lent you, on mere acquaintance, at a feast;
Item, a tale or two some fortnight after,
That yet maintains you and your house in laughter;
Item, the Babylonian song you sing;
Item, a fair Greek posy for a ring,   10
With which a learned madam you belie.
Item, a charm surrounding fearfully
Your *partie-per-pale* picture, one half drawn
In solemn cypress, the other cobweb-lawn;
Item, a gulling impres' for you, at tilt;   15
Item, your mistress' anagram, i' your hilt;
Item, your own, sewed in your mistress' smock;
Item, an epitaph on my lord's cock,
In most vile verses, and cost me more pain
Than had I made 'em good, to fit your vain.   20
Forty things more, dear Grand, which you know true:
For which or pay me quickly or I'll pay you.

74

To Thomas, Lord Chancellor Egerton

Whilst thy weighed judgements, Egerton, I hear,
And know thee then a judge not of one year;
Whilst I behold thee live with purest hands;
That no affection in thy voice commands;

73. 5 In primis: first. 9 *Babylonian*: incomprehensible. 13 partie-per-pale: 'A term in heraldry denoting that the field, on which the figures making up a coat of arms are represented, is divided into two equal parts by a perpendicular line' (H & S). 14 *cypress*: black crape. 14 *cobweb-lawn*: white linen. 15 *impres'*: impresa, device with motto, here probably on a shield. 20 *vain*: vanity; punning on 'vein'.

74. *Thomas, Lord Chancellor Egerton*: ?1540–1617; created Baron Ellesmere and Lord Chancellor in 1603; addressed again in *Und.* 31, 32. 2 *not of one year*: imitating Horace's *consul non unius anni*, *Odes*, IV. ix. 39. 4 *affection*: animosity.

That still thou'rt present to the better cause,   5
    And no less wise than skilful in the laws;
Whilst thou art certain to thy words, once gone,
    As is thy conscience, which is always one:
The Virgin, long since fled from earth, I see,
    To our times returned, hath made her heaven in thee.   10

75

On Lippe, the Teacher

I cannot think there's that antipathy
    'Twixt Puritans and players as some cry;
Though Lippe, at Paul's, ran from his text away
    To inveigh 'gainst plays, what did he then but play?

76

On Lucy, Countess of Bedford

This morning, timely rapt with holy fire,
    I thought to form unto my zealous muse
What kind of creature I could most desire
    To honour, serve and love, as poets use.
I meant to make her fair, and free, and wise,   5
    Of greatest blood, and yet more good than great;

---

5 *present to*: favourably attentive to; cf. *Cat.* II. 368.   7 while you remain true to your words, once spoken.   9 *The Virgin*: Astraea, goddess of justice, who dwelt on earth during the golden age, and later fled to heaven: see Virgil, *Ecl.* iv. 6; and contrast *Und.* 15. 39.

75. *Lippe*: suggesting both English 'lip' and Latin *lippus*, blear-eyed. Cf. 'Upon Trap', Herrick, p. 325. 2 *Puritans and players*: the Puritans, traditionally hostile to the theatre, were popularly known as 'hypocrites', a word deriving from the Greek for 'actors'. Cf. *B.F.* V. v. 50–1: 'I know no fitter match than a puppet to commit with an hypocrite'.

76. *Lucy, Countess of Bedford*: ?1581–1627, celebrated patroness of poets: she befriended Jonson, Donne, Drayton, Daniel, and others. She danced in several of Jonson's masques, and is addressed again in *Epig.* 84, 94. George Eliot admired the ideal of womanhood expressed in this poem: 'that grander feminine type—at once sweet, strong, large-thoughted' (*Letters*, ed. G. Haight (London, 1956), vi. 360). 5 *free*: noble. 6 *more good than great*: see *Epig.* Ded. 15–16 n.

I meant the day-star should not brighter rise,
   Nor lend like influence from his lucent seat.
I meant she should be courteous, facile, sweet,
   Hating that solemn vice of greatness, pride;   10
I meant each softest virtue there should meet,
   Fit in that softer bosom to reside.
Only a learned and a manly soul
   I purposed her, that should, with even powers,
The rock, the spindle and the shears control   15
   Of destiny, and spin her own free hours.
Such when I meant to feign and wished to see,
   My muse bade, *Bedford* write, and that was she.

77

To One that Desired Me Not to Name Him

Be safe, nor fear thyself so good a fame
   That any way my book should speak thy name:
For if thou shame, ranked with my friends, to go,
   I'm more ashamed to have thee thought my foe.

78

To Hornet

Hornet, thou hast thy wife dressed for the stall
To draw thee custom: but herself gets all.

---

7 *day-star*: morning-star: see *Epig.* 94. 2 n.   8 *lucent*: playing on the derivation of Lucy's name, from *lux, lucis*, light; cf. *Epig.* 94. 1, *For.* 12. 66, and the play on Lucius Cary's name, *Und.* 70. 97.   10 Cf. Claudian, *De Consulatu Stilichonis*, ii. 160–2*.   15 *The rock, the spindle and the shears*: emblems of the three Parcae or Fates: the distaff ('rock') of Clotho, who determined the moment of a person's birth; the spindle of Lachesis, who spun the events of his life; and the shears of Atropos, who finally cut the thread of his life. For a similar insistence on the superiority of virtue to fate (and fortune), see *For.* 11. 1–2, and *Epig.* 63. 2–3 n.

77. Cf. *Epig.* Ded. 23, *Epig.* 30.

78. Imitated by Herrick, 'Upon Slouch', p. 253. 1 *stall*: market booth.

79

To Elizabeth, Countess of Rutland

That poets are far rarer births than kings
    Your noblest father proved: like whom, before
Or then or since, about our muses' springs,
    Came not that soul exhausted so their store.
Hence was it that the destinies decreed   5
    (Save that most masculine issue of his brain)
No male unto him, who could so exceed
    Nature, they thought, in all that he would feign;
At which she, happily displeased, made you:
    On whom, if he were living now, to look,   10
He should those rare and absolute numbers view
    As he would burn, or better far, his book.

80

Of Life and Death

The ports of death are sins; of life, good deeds,
    Through which our merit leads us to our meeds.
How wilful blind is he, then, that would stray,
    And hath it in his powers to make his way!
This world death's region is, the other life's;   5
    And here it should be one of our first strifes
So to front death, as men might judge us past it:
    For good men but see death, the wicked taste it.

---

79. *Elizabeth, Countess of Rutland*: ?1584–1612, daughter of Sir Philip Sidney; Jonson reckoned her 'nothing inferior to her father' in poetry (*Conv. Dr.* 213), and addresses her again in *For.* 12 and *Und.* 50.   1 See *Epig.* 4. 3 n.   3–4 Jonson works up his hyperbole from Martial's more restrained figure, *Epig.* VIII. lxx. 3–4*. 6 *The Arcadia.* 11 *absolute numbers*: perfect verses.

80. 1 *ports*: gates. 8 See Matt. 16 : 28.

81

To Prowl the Plagiary

Forbear to tempt me, Prowl, I will not show
  A line unto thee till the world it know,
Or that I've by two good sufficient men
  To be the wealthy witness of my pen:
For all thou hear'st, thou swear'st thyself didst do;   5
  Thy wit lives by it, Prowl, and belly too.
Which if thou leave not soon (though I am loath)
  I must a libel make, and cozen both.

82

On Cashiered Captain Surly

Surly's old whore in her new silks doth swim:
He cast, yet keeps her well! No, she keeps him.

83

To a Friend

To put out the word 'whore' thou dost me woo,
Throughout my book. Troth, put out 'woman' too.

84

To Lucy, Countess of Bedford

Madam, I told you late how I repented,
  I asked a lord a buck, and he denied me;
And, ere I could ask you, I was prevented:
  For your most noble offer had supplied me.

---

81. *Prowl*: to 'prowl' was to pilfer. Cf. Martial's poem on a plagiarist, *Epig*. I. lxiii*.
4 *wealthy witness*: 'This is a pure Latinism: *testis locuples* is the Roman phrase for a
full and sufficient evidence' (Whalley).   8 *a libel*: probably, a defamatory epigram.

82. 2 *cast*: i.e. discarded her; playing on the fact that Surly himself has been 'cast', i.e.
cashiered.

83. 1 *put out*: delete.

84. *Lucy, Countess of Bedford*: see *Epig*. 76, n. The present epigram was evidently one
of Jonson's favourites: see *Conv. Dr*. 93.   3 *prevented*: anticipated.

Straight went I home; and there, most like a poet, 5
 I fancied to myself what wine, what wit
I would have spent; how every muse should know it,
 And Phoebus' self should be at eating it.
O madam, if your grant did thus transfer me,
 Make it your gift: see whither that will bear me. 10

85

To Sir Henry Goodyere

Goodyere, I'm glad and grateful to report
 Myself a witness of thy few days' sport,
Where I both learned why wise men hawking follow,
 And why that bird was sacred to Apollo:
She doth instruct men by her gallant flight 5
 That they to knowledge so should tower upright,
And never stoop but to strike ignorance;
 Which, if they miss, they yet should re-advance
To former height, and there in circle tarry
 Till they be sure to make the fool their quarry. 10
Now, in whose pleasures I have this discerned,
 What would his serious actions me have learned?

86

To the Same

When I would know thee, Goodyere, my thought looks
 Upon thy well-made choice of friends and books;
Then do I love thee, and behold thy ends
 In making thy friends books, and thy books friends.

---

9 *transfer*: transport; with a pun on the legal sense. 9, 10 *grant ... gift*: Jonson distinguishes between the promise and its execution.

85. *Sir Henry Goodyere*: d. 1628; close friend of Donne, who also speaks of his fondness for hawking: see Donne, *Satires*, pp. 78–9. He took part in Jonson's *Hymenaei* in 1606. 4 *sacred to Apollo*: Apollo in his role as augur: the hawk was reckoned a bird of good omen (see Jonson's note to *Augurs*, 350). The bird's action was also likened to that of the intellect: see Ripa, *Iconologia* (Hertel edn.), ed. E. A. Maser (New York, 1971), no. 184. 6, 7 *tower ... stoop*: hawking terms; respectively, soar, swoop. 11–12 what would I have learnt from the serious actions of one whose pleasures have taught me this?

Now I must give thy life and deed the voice   5
  Attending such a study, such a choice:
Where, though it be love that to thy praise doth move,
  It was a knowledge that begat that love.

87

On Captain Hazard the Cheater

Touched with the sin of false play in his punk,
  Hazard a month forsware his, and grew drunk
Each night, to drown his cares. But when the gain
  Of what she had wrought came in, and waked his brain,
Upon the account, hers grew the quicker trade;   5
  Since when, he's sober again, and all play's made.

88

On English Monsieur

Would you believe, when you this monsieur see,
  That his whole body should speak French, not he?
That so much scarf of France, and hat, and feather,
  And shoe, and tie, and garter should come hither
And land on one whose face durst never be   5
  Toward the sea, farther than half-way tree?
That he, untravelled, should be French so much,
  As Frenchmen in his company should seem Dutch?
Or had his father, when he did him get,
  The French disease, with which he labours yet?   10
Or hung some monsieur's picture on the wall,
  By which his dam conceived him, clothes and all?
Or is it some French statue? No: it doth move,
  And stoop, and cringe. Oh, then it needs must prove
The new French tailor's motion, monthly made,   15
  Daily to turn in Paul's and help the trade.

---

86. 7–8 Cf. *Und.* 70. 107–10.

88. 6 *half-way tree*: probably a landmark on the Dover road.   10 *The French disease*:
the pox.   15 *motion*: puppet.   16 *Paul's*: the middle aisle of St. Paul's Cathedral was a
fashionable area in which to promenade. In *E.M.O.* III. v, Fungoso, one who 'follows
the fashion afar off, like a spy', visits St. Paul's with his tailor to observe fashions.

89

To Edward Alleyn

If Rome so great, and in her wisest age,
    Feared not to boast the glories of her stage,
As skilful Roscius and grave Aesop, men
    Yet crowned with honours as with riches then,
Who had no less a trumpet of their name   5
    Than Cicero, whose every breath was fame:
How can so great example die in me,
    That, Alleyn, I should pause to publish thee?
Who both their graces in thyself hast more
    Out-stripped, than they did all that went before;   10
And present worth in all dost so contract,
    As other speak, but only thou dost act.
Wear this renown. 'Tis just that who did give
    So many poets life, by one should live.

90

On Mill, My Lady's Woman

When Mill first came to court, the unprofiting fool,
    Unworthy such a mistress, such a school,
Was dull and long ere she would go to man.
    At last ease, appetite and example wan
The nicer thing to taste her lady's page;   5
    And, finding good security in his age,
Went on; and proving him still day by day,
    Discerned no difference of his years or play,
Not though that hair grew brown which once was amber,
    And he, grown youth, was called to his lady's chamber.   10

89. *Edward Alleyn*: 1566–1626, the famous actor; he had taken leading roles in Marlowe's plays, and in 1608 acted in an entertainment by Jonson at Salisbury House (see Scott McMillan in *Renaissance Drama*, N.S. i (Evanston, 1968), 160). 3 *Roscius . . . Aesop*: distinguished Roman actors, in comedy and tragedy respectively. Jonson is echoing Horace, *Epist.* II. i. 82, 'what stately Aesopus and learned Roscius once acted'. Several of Jonson's contemporaries compared Alleyn, and his great rival, Burbage, to one or both of these actors. 6 *Cicero*: see *De Oratore*, I. xxviii. 129–30; I. lxi. 258; *Pro Sestio*, lvii. 121–2, lviii. 123. 12 Playing on the proverb 'not words but deeds' (Tilley, W820).

90. 4 *wan*: won.

Still Mill continued; nay, his face growing worse,
    And he removed to gent'man of the horse,
Mill was the same. Since, both his body and face
    Blown up, and he (too unwieldy for that place)
Hath got the steward's chair, he will not tarry   15
    Longer a day, but with his Mill will marry.
And it is hoped that she, like Milo, wull,
    First bearing him a calf, bear him a bull.

91

To Sir Horace Vere

Which of thy names I take, not only bears
    A Roman sound, but Roman virtue wears:
Illustrous Vere, or Horace, fit to be
    Sung by a Horace, or a muse as free;
Which thou art to thyself, whose fame was won   5
    In the eye of Europe, where thy deeds were done,
When on thy trumpet she did sound a blast
    Whose relish to eternity shall last.
I leave thy acts, which should I prosecute
    Throughout, might flattery seem; and to be mute   10
To anyone were envy, which would live
    Against my grave, and time could not forgive.
I speak thy other graces, not less shown
    Nor less in practice, but less marked, less known:
Humanity and piety, which are   15
    As noble in great chiefs as they are rare;
And best become the valiant man to wear,
    Who more should seek men's reverence than fear.

---

17 *Milo*: famous athlete of Crotona in Italy, d. *c.* 500 B.C., said to have developed his strength by lifting a calf every day until it became a bull. 17 *wull*: will. 18 A familiar proverb, 'applied unto those that, falling into small offences when they are young, commit great sins when they are of perfect age' (Baret, 1580; see Tilley, B711). Petronius applies the proverb in a sexual context, *Satyricon*, 25.

91. *Sir Horace Vere*: 1565–1635, distinguished soldier of distinguished family; exploits at Mulheim (1605), Mannheim (1625); created Baron Vere of Tilbury, 1625. 3 *illustrous*: illustrious. 3 *Vere*: 'truly' in Latin; possibly Jonson is also thinking of Horace's friends Lucius Rufus Varius, the poet, or Quintilius Varus, the critic, both of whom are mentioned in *Ars Poetica* (Jonson's trs., ll. 79, 623). 8 *relish*: musical embellishment. 9 *prosecute*: treat in detail. 15 *Humanity*: cf. *Epig.* Ded. 31. 15–16 Cf. Seneca, *De Clementia*, I. v. 4. 18 Cf. Pliny, *Panegyricus*, xlvi. 1.

92

The New Cry

Ere Cherries ripe! and Strawberries! be gone,
   Unto the cries of London I'll add one:
Ripe statesmen, ripe! They grow in every street;
   At six-and-twenty, ripe; you shall 'em meet
And have 'em yield no savour but of state.   5
   Ripe are their ruffs, their cuffs, their beards, their gait,
And grave as ripe, like mellow as their faces.
   They know the states of Christendom, not the places;
Yet have they seen the maps, and bought 'em too,
   And understand 'em, as most chapmen do.   10
The counsels, projects, practices they know,
   And what each prince doth for intelligence owe,
And unto whom; they are the almanacs
   For twelve years yet to come, what each state lacks.
They carry in their pockets Tacitus,   15
   And the *Gazetti*, or *Gallo-Belgicus*;
And talk reserved, locked up, and full of fear,
   Nay, ask you how the day goes, in your ear;
Keep a Star-Chamber sentence close, twelve days,
   And whisper what a proclamation says.   20
They meet in sixes, and at every mart
   Are sure to con the catalogue by heart;
Or every day someone at Rimee's looks,
   Or Bill's, and there he buys the names of books.
They all get Porta, for the sundry ways   25
   To write in cipher, and the several keys

---

92. 1 *Cherries ripe*: the common street-cry; cf. Campion's and Herrick's poems, both entitled 'Cherry-ripe'.   3 *statesmen*: observers of affairs of state, as well as statesmen proper; cf. *S.W.* II. ii. 114–15.   4 *At six-and-twenty, ripe*: 'refers, I think, to the peerage bestowed on Carr 1611, Mar. 25' (Fleay, i. 319).   15 *Tacitus*: whom Sir John Daw, another pretender to knowledge, also tries to master, *S.W.* IV. v. 50.   16 Gazetti: 'running reports, daily news, idle intelligences, or flim-flam tales that are daily written from Italy, namely Rome and Venice' (Florio, *Queen Anna's New World of Words*, 1616).   16 Gallo-Belgicus: the famous register of news, *Mercurii Gallo-Belgici*, published at Cologne.   17 *talk reserved, locked up*: cf. Sir Politic Would-Be (*Volp.* IV. i. 13), who also shows an interest in ciphers (*Volp.* II. i. 67–89; and ll. 25 ff. below). H & S (i. 254) wonder if there may be a covert allusion to Sir Thomas Monson, master of armoury at the Tower.   18 *in your ear*: cf. Martial's repeated phrase, *in aurem*, *Epig.* I. lxxxix.   23, 24 *Rimee's . . . Bill's*: James Rimee (or Rymer) and John Bill were London booksellers.   25 *Porta*: Giovanni Battista della Porta's book on ciphers, *De Furtivis Literarum Notis, vulgo de Ziferis Libri IV*, published in Naples in 1563.

To ope the character. They've found the sleight
  With juice of lemons, onions, piss, to write,
To break up seals, and close 'em. And they know,
  If the States make peace, how it will go   30
With England. All forbidden books they get;
  And of the Powder Plot they will talk yet.
At naming the French king, their heads they shake,
  And at the Pope and Spain slight faces make;
Or 'gainst the bishops, for the brethren, rail,   35
  Much like those brethren, thinking to prevail
With ignorance on us, as they have done
  On them; and therefore do not only shun
Others more modest, but contemn us too,
  That know not so much state, wrong, as they do.   40

93

To Sir John Radcliffe

How like a column, Radcliffe, left alone
  For the great mark of virtue—those being gone
Who did, alike with thee, thy house up-bear—
  Stand'st thou, to show the times what you all were!
Two bravely in the battle fell, and died,   5
  Upbraiding rebels' arms and barbarous pride;
And two that would have fallen as great as they,
  The Belgic fever ravished away.
Thou, that art all their valour, all their spirit,
  And thine own goodness to increase thy merit;   10

30 *If the States make peace*: complex peace negotiations between the States General of
the United Provinces of the Low Countries and Spain began in 1604; a truce was
achieved in 1609.   32 *Powder Plot*: Gunpowder Plot, November 1605.   33 *French
king*: Henry IV was assassinated in 1610.   34 *the Pope and Spain*: relations with the
Papacy were particularly hostile after the introduction of the Oath of Allegiance in
1606. Though England and Spain made peace in 1604, the two countries remained on
uneasy terms for some years.   35-8 John Whitgift, Archbishop of Canterbury from
1583 to 1604, had taken a severe line with the Puritans, which was maintained by his
successor, Richard Bancroft. James I made his famous statement against the Puritans
and in favour of the bishops ('No bishop, no king') at the Hampton Court Conference
of 1604.

93. *Sir John Radcliffe*: ?1580–1627, the remaining brother of Margaret Radcliffe,
commemorated in *Epig*. 40 (see note to that poem). He was knighted by Essex in
Ireland 24 Sept. 1599.   5-6 'In Ireland' (Jonson's note).

Than whose I do not know a whiter soul,
    Nor could I, had I seen all Nature's roll:
Thou yet remain'st, unhurt in peace or war,
    Though not unproved: which shows thy fortunes are
Willing to expiate the fault in thee   15
    Wherewith, against thy blood, they offenders be.

94

To Lucy, Countess of Bedford, with Mr. Donne's Satires

Lucy, you brightness of our sphere, who are
    Life of the muses' day, their morning-star!
If works, not the authors, their own grace should look,
    Whose poems would not wish to be your book?
But these, desired by you, the maker's ends   5
    Crown with their own. Rare poems ask rare friends.
Yet satires, since the most of mankind be
    Their unavoided subject, fewest see:
For none e'er took that pleasure in sin's sense
    But, when they heard it taxed, took more offence.   10
They, then, that living where the matter is bred
    Dare for these poems yet both ask and read,
And like them, too, must needfully, though few,
    Be of the best; and 'mongst those, best are you.
Lucy, you brightness of our sphere, who are   15
    The muses' evening- as their morning-star.

11 *a whiter soul*: Horace, *Sat.* I. v. 41–2.   14–16 thy fortunes, which have offended against thy family, are willing to expiate that fault in thee.

94. Donne's *Satires* were first published in 1633, but were written considerably earlier. Jonson's poem was probably not written before 1607: see R. C. Bald, *John Donne: A Life* (Oxford, 1970), pp. 172–4; Donne, *Satires*, p. lix. For Lucy, Countess of Bedford, see *Epig.* 76, n.   1 *brightness of our sphere*: playing on the meaning of the name Lucy, as in *Epig.* 76. 8 and *For.* 12. 66.   2 *morning-star*: so in *Epig.* 76. 7, 'day-star'. The planet Venus is called Lucifer ('light-bearing') when it appears in the morning before the sun, Hesperus when it appears after the setting of the sun (cf. l. 16).   3 *look*: have regard to.   7–8 Cf. Horace, *Sat.* I. iv. 22 ff.   8 *unavoided*: inevitable.

95

To Sir Henry Savile

If, my religion safe, I durst embrace
    That stranger doctrine of Pythagoras,
I should believe the soul of Tacitus
    In thee, most weighty Savile, lived to us:
So hast thou rendered him in all his bounds   5
    And all his numbers, both of sense and sounds.
But when I read that special piece, restored,
    Where Nero falls and Galba is adored,
To thine own proper I ascribe then more,
    And gratulate the breach I grieved before: 10
Which fate, it seems, caused in the history
    Only to boast thy merit in supply.
Oh, wouldst thou add like hand to all the rest!
    Or (better work!) were thy glad country blest
To have her story woven in thy thread,   15
    Minerva's loom was never richer spread.
For who can master those great parts like thee,
    That liv'st from hope, from fear, from faction free;
That hast thy breast so clear of present crimes
    Thou need'st not shrink at voice of after-times;   20
Whose knowledge claimeth at the helm to stand,
    But wisely thrusts not forth a forward hand,
No more than Sallust in the Roman state.
    As, then, his cause, his glory emulate.

---

95. *Sir Henry Savile*: 1549–1622, Fellow and later Warden of Merton College, Oxford; subsequently Provost of Eton. Translated Tacitus's *Histories*, 1591, with an original section, *The End of Nero and Beginning of Galba*; edited Chrysostom in 8 volumes, 1610–13. Jonson commends his learning and gravity in *Disc.* 912. 2 *doctrine of Pythagoras*: transmigration of souls: cf. *Volp.* I. ii, *Conv. Dr.* 130 ff. 6 *numbers*: parts. 9 *proper*: genius, personal gifts. 10 *gratulate*: rejoice over. 14–15 'It was then imagined, that Sir Henry Savile intended to have compiled a general history of England: but he gave over the design, and engaged in that excellent edition of Chrysostom, which he afterwards published' (Whalley). 16 *Minerva's loom*: Minerva was the goddess of weaving as well as of wisdom. But Jonson is probably also comparing the history which Savile might write with Minerva's great mantle (πέπλος), which was embroidered each year with the actions and figures of naval heroes and gods, and carried through the streets of Athens to her temple. See Briggs, *CP*, xi (1916), 178. 23 *Sallust*: After serving in Africa with Julius Caesar, Sallust (86–35 B.C.) retired to his pleasure grounds to write his historical works, *Catiline*, *Jugurtha*, and the *Historiae* of the period 78–67 B.C. Ll. 17–26 of Jonson's poem contain several echoes of

Although to write be lesser than to do,   25
   It is the next deed, and a great one too.
We need a man that knows the several graces
   Of history, and how to apt their places:
Where brevity, where splendour, and where height,
   Where sweetness is required, and where weight;   30
We need a man can speak of the intents,
   The counsels, actions, orders and events
Of state, and censure them; we need his pen
   Can write the things, the causes, and the men.
But most we need his faith (and all have you)   35
   That dares nor write things false, nor hide things true.

96

To John Donne

Who shall doubt, Donne, whe'er I a poet be,
   When I dare send my epigrams to thee?
That so alone canst judge, so alone dost make;
   And in thy censures, evenly dost take
As free simplicity to disavow   5
   As thou hast best authority to allow.
Read all I send; and if I find but one
   Marked by thy hand, and with the better stone,
My title's sealed. Those that for claps do write,
   Let puisnes', porters', players' praise delight,   10
And till they burst, their backs, like asses', load:
   A man should seek great glory, and not broad.

Sallust's *Catiline*, iii and iv.   25–6 Cf. Sallust, *Catiline*, iii. 1–2.   28 *apt*: make
fit.   29–30 Cf. Pliny on Pompeius Saturninus, the pleader: *Epist.* I. xvi*.   31–6
Cicero's essentials of history, *De Oratore*, II. xv. 62–3. As W. J. Courthope observed,
Jonson makes 'the exceptional *moral* character of Sir H. Savile the crown of his
qualifications, rather than his mere technical skill' (*A History of English Poetry* (N.Y.,
1903), iii. 181 n).   33 *censure*: give an opinion of.

96. *John Donne*: also addressed in *Epig.* 23.   1 *whe'er*: whether.   4 *censures*: judge-
ments.   8 *the better stone*: the Roman manner of marking happy days: see Persius,
Sat. ii. 1, Martial, *Epig.* IX. lii. 5, and Jonson's note to *King's Ent.* 289.   10 *puisnes'*:
novices'.

97

On the New Motion

See you yond motion? Not the old fa-ding,
   Nor Captain Pod, nor yet the Eltham thing,
But one more rare, and in the case so new:
   His cloak with orient velvet quite lined through,
His rosy ties and garters so o'er-blown,   5
   By his each glorious parcel to be known!
He wont was to encounter me, aloud,
   Where'er he met me; now he's dumb, or proud.
Know you the cause? He has neither land nor lease,
   Nor bawdy stock that travails for increase,   10
Nor office in the town, nor place in court
   Nor 'bout the bears, nor noise to make lords sport.
He is no favourite's favourite, no dear trust
   Of any madam's, hath need o' squires and must.
Nor did the king of Denmark him salute   15
   When he was here. Nor hath he got a suit,
Since he was gone, more than the one he wears.
   Nor are the Queen's most honoured maids by the ears
About his form. What then so swells each limb?
   Only his clothes have over-leavened him.   20

---

97. 1–2 *motion*: a piece of moving mechanism (*OED*, 14). But Jonson plays with several different senses of the word: a *fading* is an Irish dance (cf. 'motion', *OED*, 12. a); *Captain Pod* was a master of puppets, commonly known as 'motions' (see *B.F.* V. i. 8, for Pod as 'a master of motions', and cf. *E.M.O.* IV. v. 62, *Epig.* 129. 16); the *Eltham thing* was a perpetual motion device: cf. *S.W.* V. iii. 63, and see *England As Seen By Foreigners in the Days of Elizabeth & James I*, ed. W. B. Rye (London, 1865/N.Y., 1967), pp. 232–42. 3 *case*: i.e. clothing. 6 *parcel*: i.e. part of clothing. 10 *bawdy stock*: prostitutes; the 'increase' being merely financial. 12 *'bout the bears*: cf. *Und.* 47. 50. 12 *noise*: band of musicians. 14 A crux. 1616 Folio reads, 'Of any *Madames*, hath neadd squires and must'. H & S, citing dubious precedents, conjecture that 'neadd' may be an abbreviated past participle form, 'needed', and take the sense to be: '[Of any madam] who has required pimps and must have them': a reading which seems at once compressed and tautologous. Whalley's emendation, 'nead o'', accounts for the curious form 'neadd', but 'of' is not otherwise required (see *OED*, 'need', *sb*. 7. c, and cf. *Paradise Lost*, ii. 413–14: 'Here he had need / All circumspection'). A play on words may be intended: *squires* are apple-squires, i.e. pimps; *must* may be a noun: apple-juice undergoing alcoholic fermentation (see *OED*, *sb*.², 2. a). But the line remains obscure. 15–16 During 1606.

98

To Sir Thomas Roe

Thou hast begun well, Roe, which stand well to,
    And I know nothing more thou hast to do.
He that is round within himself, and straight,
    Need seek no other strength, no other height;
Fortune upon him breaks herself, if ill,   5
    And what would hurt his virtue makes it still.
That thou at once, then, nobly mayst defend
    With thine own course the judgement of thy friend,
Be always to thy gathered self the same,
    And study conscience more than thou wouldst fame.   10
Though both be good, the latter yet is worst,
    And ever is ill-got without the first.

99

To the Same

That thou hast kept thy love, increased thy will,
    Bettered thy trust to letters; that, thy skill;
Hast taught thyself worthy thy pen to tread,
    And that to write things worthy to be read:
How much of great example wert thou, Roe,   5
    If time to facts, as unto men, would owe?
But much it now avails what's done, of whom:
    The self-same deeds, as diversely they come
From place or fortune, are made high or low,
    And even the praiser's judgement suffers so.   10
Well, though thy name less than our great ones be,
    Thy fact is more: let truth encourage thee.

---

98. *Sir Thomas Roe*: 1581–1644, knighted 1603, later James I's ambassador to India and Constantinople. He was a cousin of William and Sir John Roe (see *Epig.* 70 n.), a friend of Donne and Jonson, and probably the lover of Cecilia Bulstrode (*Und.* 49, *U.V.* 9). 3 *round within himself*: self-knowing and self-contained. Jonson is imitating Horace, *Sat.* II. vii. 86–8*. On his fondness for this idea, see Thomas M. Greene, *SEL*, x (1970), 325–48. Cf. Herrick, 'But to live round and close and wisely true / To thine own self, and known to few', 'A Country Life', 135–6, Herrick, p. 38. 11–12 These lines also form the conclusion of *Catiline* (1611).

99. 2 *that, thy skill*: i.e. thy trust to letters has in turn bettered thy skill in writing. 3–4 Cf. Pliny, *Epist.* VI. xvi. 3. 6 if time would pay as much respect to noble deeds ('facts') as to persons. 7–9 Cf. Pliny, *Epist.* VI. xxiv. 1.

100

On Playwright

Playwright, by chance, hearing some toys I'd writ,
Cried to my face, they were the elixir of wit;
And I must now believe him: for today
Five of my jests, then stolen, passed him a play.

101

Inviting a Friend to Supper

Tonight, grave sir, both my poor house and I
Do equally desire your company;
Not that we think us worthy such a guest,
But that your worth will dignify our feast
With those that come; whose grace may make that seem   5
Something, which else could hope for no esteem.
It is the fair acceptance, sir, creates
The entertainment perfect, not the cates.
Yet shall you have, to rectify your palate,
An olive, capers, or some better salad   10
Ushering the mutton; with a short-legged hen,
If we can get her, full of eggs, and then
Lemons, and wine for sauce; to these, a coney
Is not to be despaired of, for our money;
And though fowl now be scarce, yet there are clerks,   15
The sky not falling, think we may have larks.
I'll tell you of more, and lie, so you will come:
Of partridge, pheasant, woodcock, of which some
May yet be there; and godwit, if we can;
Knat, rail and ruff, too. Howsoe'er, my man   20

100. 4 *passed him a play*: i.e. enabled him to make a passable play.

101. Three of Martial's poems of invitation lie behind this poem: *Epig.* V. lxxviii, X. xlviii, XI. lii. H & S suggest that the poem may be addressed to such a guest as William Camden (see *Epig.* 14 n.). 7–8 Cf. Martial, *Epig.* V. lxxviii. 16, 'the wine you will make good by drinking it': a common Roman formula; and 'Content, not cates', Herrick, p. 124. 16 Varying the proverb, 'If the sky falls, we shall have larks' (Tilley, S517). 17 Cf. Martial, *Epig.* XI. lii. 13, 'I shall lie, so you will come'. 19, 20 *godwit . . . Knat, rail and ruff*: edible birds: the godwit is a marsh bird, like a curlew; the knat

Shall read a piece of Virgil, Tacitus,
    Livy, or of some better book to us,
Of which we'll speak our minds, amidst our meat;
    And I'll profess no verses to repeat;
To this, if aught appear which I not know of,   25
    That will the pastry, not my paper, show of.
Digestive cheese and fruit there sure will be;
    But that which most doth take my muse and me
Is a pure cup of rich Canary wine,
    Which is the Mermaid's now, but shall be mine;   30
Of which had Horace or Anacreon tasted,
    Their lives, as do their lines, till now had lasted.
Tobacco, nectar, or the Thespian spring
    Are all but Luther's beer to this I sing.
Of this we will sup free, but moderately;   35
    And we will have no Poley or Parrot by;
Nor shall our cups make any guilty men,
    But at our parting we will be as when
We innocently met. No simple word
    That shall be uttered at our mirthful board   40
Shall make us sad next morning, or affright
    The liberty that we'll enjoy tonight.

(or knot) and ruff are in the sandpiper family; the rail is otherwise known as the
corncrake. Sir Epicure Mammon (*Alch.* II. ii. 80–1) reckons knat and godwit suitable
only for his footboy.   20–4 It was a contemporary as well as a classical custom to
listen to recitations at table. See Juvenal, *Sat.* xi. 179–82, and cf. Martial's assurances
to his guests that he will not bore them by reading (*Epig.* V. lxxviii. 25) or by recitation
(*Epig.* XI. lii. 16 ff.); and cf. *Leg. Conv.* 18. Jonson's 'man' may be Richard Brome (see
B.F. Ind. 8).   26 i.e. the cook may produce a pastry surprise (cf. *Nept. Tr.* 89 ff.) but
Jonson will not produce a surprise reading.   30 *Mermaid's*: famous tavern in Bread
Street, Cheapside.   30–2 These lines were admired by Leigh Hunt as an example of
Jonson in his 'entirely happy, familiar, unmisgiving, self-referential and yet not self-
loving' mood. 'Why did he not write more such?' ('On Poems of Joyous Impulse',
1854). Both Horace and Anacreon wrote in praise of wine; see esp. Horace's ironical
quotation of the sentiment, 'no poems can please long, nor live, which are written by
water-drinkers', *Epist.* I. xix. 1–3.   33 *Tobacco*: sometimes said to be 'drunk', rather
than smoked.   33 *Thespian spring*: at the foot of Mount Helicon, sacred to the
muses.   34 *Luther's beer*: continental beer, made with hops, was thought inferior to
the stronger old English ale, made with malt, yeast, and water.   36 Robert Poley (or
Pooly) was a government spy, who was present at the fatal stabbing of Marlowe at
Deptford, 'after supper, as Jonson may have remembered' (Mark Eccles, *RES*, xiii
(1937), 386). Parrot was another informer; it is not known whether he is the Henry
Parrot of *Epig.* 71: Franklin B. Williams, jun., suggests not, *HSNPL*, xx (1938), 19.
Poley and Parrot may have been the spies who harassed Jonson in prison: see *Epig.*
59 n.   37–42 See Martial, *Epig.* X. xlviii. 21–4*; cf. *Leg. Conv.* 17, 24.

102

To William, Earl of Pembroke

I do but name thee, Pembroke, and I find
  It is an epigram on all mankind,
Against the bad, but of and to the good;
  Both which are asked, to have thee understood.
Nor could the age have missed thee, in this strife   5
  Of vice and virtue, wherein all great life,
Almost, is exercised; and scarce one knows
  To which yet of the sides himself he owes.
They follow virtue for reward today,
  Tomorrow vice, if she give better pay;   10
And are so good and bad, just at a price,
  As nothing else discerns the virtue or vice.
But thou, whose noblesse keeps one stature still,
  And one true posture, though besieged with ill
Of what ambition, faction, pride can raise,   15
  Whose life even they that envy it must praise,
That art so reverenced, as thy coming in
  But in the view doth interrupt their sin:
Thou must draw more; and they that hope to see
  The commonwealth still safe must study thee.   20

103

To Mary, Lady Wroth

How well, fair crown of your fair sex, might he
  That but the twilight of your sprite did see,
And noted for what flesh such souls were framed,
  Know you to be a Sidney, though unnamed?

102. *William, Earl of Pembroke*: see *Epig*. Ded. n.   3–4 Cf. Seneca, *De Tranquillitate Animi*, vii. 5*.   9–10 Cf. Seneca, *Epist*. cxv. 10*.   13 Cf. Seneca, *Epist*. lxxi. 8: 'virtue, however, cannot be increased or decreased; its stature is uniform'.   16 Cf. Seneca, *De Beneficiis*, IV. xvii. 2: 'and so attractive is virtue that even the wicked instinctively approve of the better course'.   17–18 Again recalling, and reversing, the sentiment of Martial's preface to his *Epigrams* ('let no Cato enter my theatre', etc.): see *Epig*. Ded. 33 n. and Valerius Maximus, II. x. 8.   20 *study*: as E. B. Partridge points out (*SLI*, vi (1973), 156), the word is significantly repeated throughout the *Epigrams*, with reference both to art and to persons: cf. *Epig*. 86, 122. 9. Cf. Valerius Maximus, III. ii. 17: 'let those who wish the republic to be secure, follow me'.

103. *Mary, Lady Wroth*: born *c*. 1586, wife of Sir Robert Wroth (see *For*. 3 n.); her mother was Barbara, née Gamage; her father was Robert Sidney, first Earl of

And, being named, how little doth that name   5
    Need any muse's praise to give it fame?
Which is itself the imprese of the great,
    And glory of them all, but to repeat!
Forgive me, then, if mine but say you are
    A Sidney; but in that extend as far   10
As loudest praisers, who perhaps would find
    For every part a character assigned.
My praise is plain, and wheresoe'er professed
    Becomes none more than you, who need it least.

104

To Susan, Countess of Montgomery

Were they that named you prophets? Did they see,
    Even in the dew of grace, what you would be?
Or did our times require it, to behold
    A new Susanna equal to that old?
Or, because some scarce think that story true,   5
    To make those faithful did the fates send you?
And to your scene lent no less dignity
    Of birth, of match, of form, of chastity?
Or, more than born for the comparison
    Of former age, or glory of our own,   10

Leicester, brother of Sir Philip Sidney (see *For.* 2). Her romance, *Urania*, in the tradition of her uncle's *Arcadia*, was published 1621, and addressed to the Countess of Montgomery (see *Epig.* 104). She is also addressed in *Epig.* 105, *Und.* 28, and *Alch.* Ded. 5 ff. Hunter notes that Camden in his *Remains* glosses 'Mary' as 'exalted'. But Jonson is clearly saying that 'Sidney' is the only name he intends to invoke, though others may find significance in 'every part' (l. 12) of Lady Mary's name. In the Ded. to *Alch.* she is addressed as 'the lady most deserving her name': 'Wroth' was alternatively spelt 'Worth'. 7 *imprese*: emblems or devices with mottos.

104. *Susan, Countess of Montgomery*: 1587–1629, daughter of Edward de Vere, Earl of Oxford, and granddaughter of Lord Burghley; in 1604 she married the Earl of Pembroke's brother, Philip Herbert, who became the Earl of Montgomery the following year. She danced in several of Jonson's masques. 4 *Susanna*: the story of whose chastity is told in the Apocrypha. 6 *faithful*: believers.

Were you advanced past those times, to be
    The light and mark unto posterity?
Judge they that can; here I have raised to show
    A picture which the world for yours must know,
And like it, too, if they look equally;   15
    If not, 'tis fit for you some should envy.

105

To Mary, Lady Wroth

Madam, had all antiquity been lost,
    All history sealed up and fables crossed;
That we had left us, nor by time nor place,
    Least mention of a nymph, a muse, a grace,
But even their names were to be made anew: 5
    Who could not but create them all from you?
He that but saw you wear the wheaten hat
    Would call you more than Ceres, if not that;
And, dressed in shepherd's 'tire, who would not say
    You were the bright Oenone, Flora, or May?   10
If dancing, all would cry the Idalian queen
    Were leading forth the graces on the green;
And, armed to the chase, so bare her bow
    Diana alone, so hit, and hunted so.
There's none so dull that for your style would ask   15
    That saw you put on Pallas' plumed casque;
Or, keeping your due state, that would not cry
    There Juno sat, and yet no peacock by.
So are you nature's index, and restore
    In yourself all treasure lost of the age before.   20

15 *equally*: impartially.

105. *Mary, Lady Wroth*: see *Epig.* 103 n. Cf. the style of compliment paid in *Epig.* 122, 125. 2 *crossed*: struck out. 8 *Ceres*: Roman goddess of corn. 10 *Oenone*: nymph of Mount Ida, and wife of Paris. 10 *Flora*: Roman goddess of flowers. 10 *May*: or Maia, Roman goddess of growth and increase. A reference, perhaps, to Lady Mary's association with Jonson's lost pastoral, *The May Lord*: see *Conv. Dr.* 393–8. 11 *Idalian queen*: Venus. Lady Mary had danced in Jonson's *Masque of Blackness* in 1605. 16 *Pallas' plumed casque*: Pallas Athene, classical goddess of war, wisdom, and all the liberal arts, was traditionally represented wearing a plumed helmet. 18 *peacock*: sacred to Juno; here absent, for it is also an emblem of vanity. 19 *index*: probably a pun: guiding principle (*OED*, 4), and list of names (*OED*, 5).

106

To Sir Edward Herbert

If men get name for some one virtue, then
    What man art thou, that art so many men,
All-virtuous Herbert! on whose every part
    Truth might spend all her voice, fame all her art.
Whether thy learning they would take, or wit,   5
    Or valour, or thy judgement seasoning it,
Thy standing upright to thyself, thy ends
    Like straight, thy piety to God and friends:
Their latter praise would still the greatest be,
    And yet they, all together, less than thee.   10

107

To Captain Hungry

Do what you come for, Captain, with your news,
    That's sit and eat; do not my ears abuse.
I oft look on false coin, to know't from true:
    Not that I love it more than I will you.
Tell the gross Dutch those grosser tales of yours,   5
    How great you were with their two emperors,
And yet are with their princes; fill them full
    Of your Moravian horse, Venetian bull.

---

106. *Sir Edward Herbert*: 1583–1648, later Lord Herbert of Cherbury; courtier, soldier, diplomat, historian, poet, and philosopher, and brother of the poet George Herbert.

107. *Captain Hungry*: Newdigate suggests this may be Captain Thomas Gainford (or Gainsford), a veteran of the Irish wars, and later a prolific newsmonger; he is the Captain Pamphlet of *Und.* 43. 79–80. The poem is based on Martial, *Epig.* IX. xxxv; Martial's Philomusus is primarily an artful political gossip, rather than a braggadocio. 6 *two emperors*: probably Charles V (1500–58), Holy Roman Emperor, and his son Philip II of Spain, to whom Charles resigned the Netherlands in 1555. 7 *their princes*: the stadholders of the seven provinces of the Dutch republic, the most important of whom was Maurice, Prince of Orange (1567–1625), military leader in the war of independence against Spain. 8 *Moravian horse*: a famous breed. 8 *Venetian bull*: the captain's confusion as to the city from which a bull (*OED*, *sb.*^{2}, 2), or papal edict, emanates, is itself a bull (*OED sb.*^{4}, 2), or ludicrous blunder.

Tell them what parts you've ta'en, whence run away,
    What states you've gulled, and which yet keeps you in pay.   10
Give them your services and embassies
    In Ireland, Holland, Sweden (pompous lies),
In Hungary and Poland, Turkey too;
    What at Leghorn, Rome, Florence you did do;
And, in some year, all these together heaped,   15
    For which there must more sea and land be leaped
—If but to be believed you have the hap—
    Than can a flea at twice skip in the map.
Give your young statesmen (that first make you drunk
    And then lie with you, closer than a punk,   20
For news) your Villeroys and Silleries,
    Janins, your nuncios and your Tuileries,
Your archduke's agents and your Beringhams,
    That are your words of credit. Keep your names
Of Hannow, Shieter-huissen, Popenheim,   25
    Hans-spiegle, Rotteinberg and Boutersheim
For your next meal: this you are sure of. Why
    Will you part with them here, unthriftily?
Nay, now you puff, tusk, and draw up your chin,
    Twirl the poor chain you run a-feasting in.   30
Come, be not angry, you are Hungry: eat;
    Do what you come for, Captain: there's your meat.

18 *at twice*: twice.   21 *Villeroys*: Nicholas IV de Neufville (1543–1617), Seigneur, and later Marquis, de Villeroy, negotiator for Henry III during the Wars of Religion, and Secretary of State under Henry IV.   21 *Silleries*: Nicholas Bruslaut (1544–1624), Marquis de Sillery, Chancellor of France under Henry IV and minister of Louis XIII.   22 *Janins*: Pierre Jeannin (?1541–1623), French statesman and diplomatic negotiator.   22 *Tuileries*: palace of the French sovereign in Paris.   23 *Beringhams*: Hunter suggests Pierre de Beringhen (d. 1619), Valet de Chambre and Commissaire des Guerres for Henry IV.   25–6 *Hannow* and *Rotteinberg* are probably the German cities Hanau and Rotenburg; the other names are probably invented: *Shieter-huissen* = shit-house; *Boutersheim* possibly alludes to the legendary fondness of the Dutch for butter.   29 *tusk*: show the teeth.

108

To True Soldiers

Strength of my country, whilst I bring to view
    Such as are mis-called captains, and wrong you
And your high names, I do desire that thence
    Be nor put on you, nor you take offence.
I swear by your true friend, my muse, I love   5
    Your great profession; which I once did prove,
And did not shame it with my actions then,
    No more than I dare now do with my pen.
He that not trusts me, having vowed thus much,
    But's angry for the Captain still, is such.   10

109

To Sir Henry Neville

Who now calls on thee, Neville, is a muse
    That serves nor fame nor titles, but doth choose
Where virtue makes them both, and that's in thee,
    Where all is fair, beside thy pedigree.
Thou art not one seek'st miseries with hope,   5
    Wrestlest with dignities, or feign'st a scope
Of service to the public when the end
    Is private gain, which hath long guilt to friend.
Thou rather striv'st the matter to possess,
    And elements of honour, than the dress;   10
To make thy lent life good against the fates;
    And first to know thine own state, then the State's;
To be the same in root thou art in height,
    And that thy soul should give thy flesh her weight.

---

108. Quoted in *Poet*. Apol. Dial. 131–40; both this and the preceding epigram were thus probably written in 1601 or shortly before.   6 See Jonson's brief account of his service in the Low Countries, *Conv. Dr*. 244 ff.   10 *such*: i.e. as he.

109. *Sir Henry Neville*: ?1564–1615, M.P. from 1584 to 1614, had been imprisoned in the Tower and heavily fined for his involvement in Essex's plot; though released in 1603, he remained out of royal favour.   11 *lent life*: see *Epig*. 45. 3 n.   13 Cf. Virgil, *Georgics*, ii. 290–2.

Go on, and doubt not what posterity,   15
   Now I have sung thee thus, shall judge of thee.
Thy deeds unto thy name will prove new wombs,
   Whilst others toil for titles to their tombs.

110

To Clement Edmondes, on His
Caesar's Commentaries Observed and Translated

Not Caesar's deeds, nor all his honours won
   In these west parts; nor when that war was done,
The name of Pompey for an enemy,
   Cato's to boot, Rome and her liberty
All yielding to his fortune; nor, the while,   5
   To have engraved these acts with his own style,
And that so strong and deep as't might be thought
   He wrote with the same spirit that he fought;
Nor that his work lived in the hands of foes,
   Unargued then, and yet hath fame from those: 10
Not all these, Edmondes, or what else put to,
   Can so speak Caesar as thy labours do.
For, where his person lived but one just age,
   And that midst envy and parts, then fell by rage:
His deeds too dying, but in books (whose good   15
   How few have read, how fewer understood!)
Thy learned hand and true Promethean art
   (As by a new creation) part by part,
In every counsel, stratagem, design,
   Action or engine worth a note of thine,   20
To all future time not only doth restore
   His life, but makes that he can die no more.

110. *Clement Edmondes*: ?1564–1622, published his *Observations* on Caesar's *De Bello Gallico* between 1600 and 1604; this and the following poem were prefixed to the 1609 edition. 6 *style*: with a pun on stylus. 14 *parts*: factions.

III

To the Same, on the Same

Who, Edmondes, reads thy book and doth not see
    What the antique soldiers were, the modern be?
Wherein thou show'st how much the latter are
    Beholding to this master of the war;
And that in action there is nothing new,   5
    More than to vary what our elders knew:
Which all but ignorant captains will confess,
    Nor to give Caesar this, makes ours the less.
Yet thou, perhaps, shalt meet some tongues will grutch
    That to the world thou shouldst reveal so much,   10
And thence deprave thee and thy work. To those
    Caesar stands up, as from his urn late rose,
By thy great help, and doth proclaim by me,
    They murder him again that envy thee.

112

To a Weak Gamester in Poetry

With thy small stock, why art thou venturing still
    At this so subtle sport, and play'st so ill?
Think'st thou it is mere fortune that can win?
    Or thy rank setting? That thou dar'st put in
Thy all, at all; and whatsoe'er I do,   5
    Art still at that, and think'st to blow me up too?
I cannot for the stage a drama lay,
    Tragic or comic, but thou writ'st the play.
I leave thee there, and, giving way, intend
    An epic poem: thou hast the same end.   10

---

111. Edmondes had compared ancient and modern methods of warfare in his *Observations*.   8 i.e. not to concede this knowledge to Caesar is to diminish the value of our own actions.   9 *grutch*: complain, grudge.   11 *deprave*: disparage.

112. Cf. Martial, *Epig.* XII. xciv. Jonson introduces a new metaphor from the card game primero.   4 *rank setting*: excessive betting.   6 *blow me up*: ruin; cf. *For.* 3. 79, *Alch.* I. ii. 78 ff., *M.L.* III. vi. 135.   7 *lay*: plan.   10 *epic poem*: Jonson mentions this intention to Drummond, *Conv. Dr.* 1.

I modestly quit that, and think to write,
    Next morn, an ode: thou mak'st a song ere night.
I pass to elegies: thou meet'st me there;
    To satires: and thou dost pursue me. Where,
Where shall I 'scape thee? in an epigram?   15
    Oh, (thou criest out) that is thy proper game.
Troth, if it be, I pity thy ill luck,
    That both for wit and sense so oft dost pluck,
And never art encountered, I confess;
    Nor scarce dost colour for it, which is less.   20
Prithee yet save thy rest; give o'er in time:
    There's no vexation that can make thee prime.

113

To Sir Thomas Overbury

So Phoebus makes me worthy of his bays
    As but to speak thee, Overbury, is praise;
So, where thou liv'st, thou mak'st life understood,
    Where what makes others great doth keep thee good!
I think the fate of court thy coming craved,   5
    That the wit there and manners might be saved;
For since, what ignorance, what pride is fled,
    And letters and humanity in the stead!
Repent thee not of thy fair precedent,
    Could make such men and such a place repent;   10
Nor may any fear to lose of their degree
    Who in such ambition can but follow thee.

---

18 *pluck*: draw cards from a pack.   19 ff. *encountered . . . colour . . . rest . . . prime*: all terms from primero; 'colour' also perhaps picks up the sense of 'blush' (the gamester's brazenness contrasting with the author's modesty, l. 11); 'rest' is a stake kept in reserve; 'prime' is a hand consisting of a card from each of the four suits, and the word also means in a more general sense 'first-rate'.

113. *Sir Thomas Overbury*: 1581–1613, secretary and close adviser to James I's favourite, Robert Carr; poisoned at the instigation of Lady Essex, whose marriage with Carr he had opposed. Jonson's own relationship with Overbury underwent a reversal after Overbury had attempted to use him as a go-between in his suit to the Countess of Rutland: see *Conv. Dr.* 170, 213–18. Gifford conjectured that this poem was written *c.* 1610, when Overbury had returned from his travels.

114

To Mistress Philip Sidney

I must believe some miracles still be
  When Sidney's name I hear, or face I see;
For Cupid, who at first took vain delight
  In mere out-forms, until he lost his sight,
Hath changed his soul, and made his object you: 5
  Where finding so much beauty met with virtue,
He hath not only gained himself his eyes,
  But, in your love, made all his servants wise.

115

On the Town's Honest Man

You wonder who this is, and why I name
  Him not aloud that boasts so good a fame,
Naming so many, too! But this is one
  Suffers no name but a description:
Being no vicious person, but the vice   5
  About the town; and known too, at that price.
A subtle thing that doth affections win
  By speaking well o' the company it's in.
Talks loud and bawdy, has a gathered deal
  Of news and noise to strow out a long meal.   10

---

114. *Mistress Philip Sidney*: 1594–1620, daughter of Sir Robert Sidney; she married Sir John Hobart. She was christened Philip, evidently after her famous uncle. For biographical details, see L. C. John, *JEGP*, xlv (1946), 214–17.   4 *out-forms*: external forms.

115. Probably aimed at the architect Inigo Jones (1573–1652), Jonson's collaborator in his court masques.   3 *Naming so many*: see *Epig*. Ded. 15–16 and n.   5 Cf. Martial, *Epig*. XI. xcii: 'He lies who says you are vicious, Zoilus: you are not a vicious man, Zoilus, but vice'. Jonson alludes also to the character of the Vice or Iniquity (see l. 27, and cf. *U.V*. 34. 26–8) in the interludes.   7–8 Cf. *Und*. 47. 19–20.   10 *strow out*: intersperse. H & S's emendation of Folio 'sow': cf. *S. of N*. III. ii. 183, *Und*. 47. 28.

Can come from Tripoli, leap stools, and wink,
   Do all that 'longs to the anarchy of drink,
Except the duel. Can sing songs and catches,
   Give everyone his dose of mirth; and watches
Whose name's unwelcome to the present ear,   15
   And him it lays on—if he be not there.
Tells of him all the tales itself then makes;
   But if it shall be questioned, undertakes
It will deny all, and forswear it, too:
   Not that it fears, but will not have to do   20
With such a one. And therein keeps its word.
   'Twill see its sister naked, ere a sword.
At every meal where it doth dine or sup,
   The cloth's no sooner gone but it gets up
And, shifting of its faces, doth play more   25
   Parts than the Italian could do with his door.
Acts old Iniquity, and in the fit
   Of miming gets the opinion of a wit.
Executes men in picture. By defect,
   From friendship, is its own fame's architect.   30
An engineer in slanders of all fashions,
   That, seeming praises, are yet accusations.
Described, it's thus; defined would you it have?
   Then the town's honest man's her arrant'st knave.

---

11 *come from Tripoli*: 'To vault and tumble with activity. It was, I believe, first applied to the tricks of an ape or monkey, which might be supposed to come from that part of the world' (Robert Nares, *A Glossary*, rev. J. O. Halliwell and T. Wright (2 vols. London, 1859)). Cf. *S.W.* V. i. 44–6.   12 *the anarchy of drink*: cf. *Und.* 47. 10.   16 *lays on*: assails.   25 *shifting of its faces*: cf. *Sej.* I. 7.   26 Probably referring to an Italian Zanni actor performing a quick change *lazzo*, or bobbing in briefly from behind a stage door; see also K. M. Lea, *Italian Popular Comedy* (Oxford, 1934), i. 246, 252. Whalley saw a possible allusion to Scoto of Mantua. The Folio reading is 'dore': hence 'dor', meaning 'fool' or 'mockery' is a plausible alternative sense (cf. *C.R.* V. i. 19, V. ii *passim*).   27 *old Iniquity*: see l. 5 n.   29 *in picture*: by his descriptions. 29 *defect*: probably, disparagement (cf. *OED*, v. 4).   30, 31 *architect . . . engineer*: alluding to Jones's profession; but an 'engineer' is also a plotter or layer of snares: cf. *Sej.* I. 4.

116

To Sir William Jephson

Jephson, thou man of men, to whose loved name
  All gentry yet owe part of their best flame!
So did thy virtue inform, thy wit sustain
  That age when thou stood'st up the master-brain;
Thou wert the first mad'st merit know her strength,   5
  And those that lacked it to suspect at length
'Twas not entailed on title; that some word
  Might be found out as good, and not My Lord;
That Nature no such difference had impressed
  In men, but every bravest was the best;   10
That blood not minds, but minds did blood adorn;
  And to live great was better than great born.
These were thy knowing arts: which who doth now
  Virtuously practise must at least allow
Them in, if not from, thee; or must commit   15
  A desperate solecism in truth and wit.

117

On Groin

Groin, come of age, his 'state sold out of hand
For his whore: Groin doth still occupy his land.

118

On Gut

Gut eats all day, and lechers all the night,
  So all his meat he tasteth over, twice;
And striving so to double his delight,
  He makes himself a thoroughfare of vice.
Thus in his belly can he change a sin: 5
  Lust it comes out, that gluttony went in.

---

116. *Sir William Jephson*: of Froyle in Hampshire; knighted 1603. 8 *My Lord*: see
Epig. 11. 5 n. 9–10 See Sallust, *Jugurtha*, lxxxv. 15*.

117. 2 *occupy*: the word also had a sexual sense: cf. *Disc.* 1546, and *2 Henry IV*, II. iv.
137–40.

118. 2 *meat*: cf. the slang term 'mutton' = prostitute. 2 *tasteth*: playing on the sexual
sense of the word; cf. *Cymbeline*, II. iv. 57. 6 Cf. *Und.* 47. 13.

119

To Sir Ralph Sheldon

Not he that flies the court for want of clothes,
  At hunting rails, having no gift in oaths,
Cries out 'gainst cocking, since he cannot bet,
  Shuns press for two main causes, pox and debt,
With me can merit more than that good man,   5
  Whose dice not doing well, to a pulpit ran.
No, Sheldon, give me thee, canst want all these,
  But dost it out of judgement, not disease;
Dar'st breathe in any air, and with safe skill,
  Till thou canst find the best, choose the least ill;   10
That to the vulgar canst thyself apply,
  Treading a better path, not contrary;
And in their error's maze thine own way know:
  Which is to live to conscience, not to show.
He that but living half his age, dies such,   15
  Makes the whole longer than 'twas given him, much.

120

Epitaph on Salomon Pavy, a Child of Queen Elizabeth's Chapel

    Weep with me all you that read
      This little story,
  And know, for whom a tear you shed,
      Death's self is sorry.
  'Twas a child that so did thrive   5
      In grace and feature,
  As heaven and nature seemed to strive
      Which owned the creature.

119. *Sir Ralph Sheldon*: of Beoley in Worcester; a wealthy recusant, said to have been implicated in the plot to assassinate Queen Elizabeth in 1594; knighted 1607. 4 *press*: crowds. 8 *disease*: annoyance. 11–12 Cf. Seneca, *Epist*. v. 3: 'let us try to maintain a higher standard of life than that of the multitude, but not a contrary standard'; and contrast Lovel's dictum in *N.I.*, IV. iv. 213–15. 14 Cf. Pliny, *Epist*. I. xxii. 5: '[Titius Aristo] places no part of his happiness in ostentation, but refers the whole of it to conscience'. 15–16 From Martial, *Epig*. VIII. lxxvii. 7–8. The context of Martial's sentiment is quite different: see Jonson's translation, *Und*. 89. For Swinburne's scorn of this couplet, see *A Study of Ben Jonson* (New York, 1889), p. 113.

120. *Salomon Pavy*: who had acted in Jonson's *C.R.* in 1600 and *Poet*. in 1601. For an account of his brief career, see G. E. Bentley, *TLS*, 30 May 1942, p. 276.

Years he numbered scarce thirteen
    When fates turned cruel,   10
Yet three filled zodiacs had he been
    The stage's jewel,
And did act (what now we moan)
    Old men so duly
As, sooth, the Parcae thought him one,   15
    He played so truly.
So, by error, to his fate
    They all consented,
But viewing him since (alas, too late)
    They have repented;   20
And have sought, to give new birth,
    In baths to steep him;
But being so much too good for earth,
    Heaven vows to keep him.

121

To Benjamin Rudyerd

Rudyerd, as lesser dames to great ones use,
    My lighter comes to kiss thy learned muse,
Whose better studies while she emulates,
    She learns to know long difference of their states.
Yet is the office not to be despised,   5
    If only love should make the action prized;
Nor he for friendship to be thought unfit
    That strives his manners should precede his wit.

15–18 Cf. Martial's lines on the death of a young and successful charioteer: 'Me, snatched away in my ninth three-year span, jealous Lachesis, counting my victories, deemed old in years'. For Lachesis, one of the Parcae, see *Epig.* 76. 15 n. For the tradition of the *puer senex* glanced at here, see E. R. Curtius, *European Literature and the Latin Middle Ages*, tr. W. R. Trask (London, 1953), pp. 98 ff. 22 *baths*: as Jupiter restored to life the young Pelops, and as Medea rejuvenated the aged Aeson. Cf. 'To his Mistresses', 9–10: 'Aeson had (as poets feign) / Baths that made him young again', Herrick, p. 10.

121. *Benjamin Rudyerd*: 1572–1658, knighted 1618; politician, poet, and friend of the Earl of Pembroke, whose poems were published with Rudyerd's in 1660.

122

To the Same

If I would wish, for truth and not for show,
  The aged Saturn's age and rites to know;
If I would strive to bring back times, and try
  The world's pure gold and wise simplicity;
If I would virtue set as she was young,   5
  And hear her speak with one and her first tongue;
If holiest friendship naked to the touch
  I would restore, and keep it ever such;
I need no other arts but study thee,
  Who prov'st all these were, and again may be.   10

123

To the Same

Writing thyself or judging others' writ,
  I know not which thou'st most, candour or wit;
But both thou'st so, as who affects the state
  Of the best writer and judge should emulate.

124

Epitaph on Elizabeth, L.H.

Wouldst thou hear what man can say
  In a little? Reader, stay.
Underneath this stone doth lie
  As much beauty as could die;

---

122. This poem may be modelled in part on Martial, *Epig*. I. xxxix. 2 *Saturn's age*:
on Jonson's fondness for this notion, see *Epig*. 64. 3–4 n. 'Show' in l. 1 may allude to
the fact that Jonson used the theme in several court masques, e.g. *The Golden Age
Restored*, 1615.

123. 1 *writ*: writing. 2 *candour*: impartiality.

124. *Elizabeth, L.H.*: not firmly identified. Fleay and Newdigate guess that she may
be Elizabeth, Lady Hatton, wife of Sir Edward Coke; but as this lady was still living in
1616, the identification seems unlikely. James McKenzie, *N & Q*, ix (1962), 210,

Which in life did harbour give   5
   To more virtue than doth live.
If at all she had a fault,
   Leave it buried in this vault.
One name was Elizabeth,
   The other let it sleep with death: 10
Fitter where it died to tell,
   Than that it lived at all. Farewell.

125

To Sir William Uvedale

Uvedale, thou piece of the first times, a man
   Made for what nature could, or virtue can;
Both whose dimensions, lost, the world might find
   Restored in thy body and thy mind!
Who sees a soul in such a body set   5
   Might love the treasure for the cabinet.
But I, no child, no fool, respect the kind,
   The full, the flowing graces there enshrined;
Which (would the world not miscall't flattery)
   I could adore, almost to idolatry.   10

argues for Elizabeth, Lady Hunsdon (formerly Lady Elizabeth Spencer of Althorpe), the date of whose death is unknown. S. E. Tabachnick speculates that 'Elizabeth died pregnant with an illegitimate child': this is guesswork (*Explic.* xxix (1971), item 77). See also O. B. Hardison, jun., *The Enduring Monument* (Chapel Hill, 1962), pp. 124–6. 11–12 Howard S. Babb paraphrases: 'It is a more appropriate thing to record this lady's death than her life, for she was much too good for this world' (*JEGP*, lxii (1963), 741). But Jonson seems to be referring more specifically to the death of the lady's name; it is tempting to think that her husband may have suffered some disgrace. For Jonson's attitude to names throughout the *Epigrams*, see Ded. 15–16 n.

125. *Sir William Uvedale*: of Wickham, Hants; knighted 1613; later Treasurer of the Chamber and Treasurer-at-Arms of the army of the north.   1 *piece*: example.   10 Cf. Jonson's praise of Shakespeare, *Disc.* 655.

126

To His Lady, then Mistress Cary

Retired, with purpose your fair worth to praise,
  'Mongst Hampton shades and Phoebus' grove of bays,
I plucked a branch: the jealous god did frown,
  And bade me lay the usurped laurel down;
Said I wronged him, and, which was more, his love.   5
  I answered, Daphne now no pain can prove.
Phoebus replied, Bold head, it is not she:
  Cary my love is, Daphne but my tree.

127

To Esmé, Lord Aubigny

Is there a hope that man would thankful be
  If I should fail in gratitude to thee,
To whom I am so bound, loved Aubigny?
  No; I do therefore call posterity
Into the debt, and reckon on her head   5
  How full of want, how swallowed up, how dead
I and this muse had been if thou hadst not
  Lent timely succours, and new life begot:
So all reward or name that grows to me
  By her attempt, shall still be owing thee.   10
And than this same I know no abler way
  To thank thy benefits: which is, to pay.

---

126. *Mistress Cary*: daughter of Sir Edmund Cary of Devon.   3 *plucked a branch*: Daphne, pursued by Phoebus Apollo, was turned into a laurel; the tree was thereafter sacred to Phoebus and associated with his art, poetry.   6 *prove*: experience.

127. *Esmé, Lord Aubigny*: Esmé Stuart, Seigneur d'Aubigny (1574–1624), with whom Jonson lodged for five years (*Conv. Dr.* 254–5), and to whom he dedicated *Sejanus*.   4 *posterity*: a characteristic invocation: cf. *Epig.* Ded. 17, 109. 15, *Cat.* Ded., etc.

128

To William Roe

Roe (and my joy to name) thou'rt now to go
    Countries and climes, manners and men to know,
To extract and choose the best of all these known,
    And those to turn to blood and make thine own.
May winds as soft as breath of kissing friends   5
    Attend thee hence; and there may all thy ends,
As the beginnings here, prove purely sweet,
    And perfect in a circle always meet.
So when we, blest with thy return, shall see
    Thyself, with thy first thoughts, brought home by thee,   10
We each to other may this voice inspire:
    This is that good Aeneas, passed through fire,
Through seas, storms, tempests; and embarked for hell,
    Came back untouched. This man hath travailed well.

129

To Mime

That not a pair of friends each other see,
    But the first question is, when one saw thee?
That there's no journey set or thought upon
    To Brentford, Hackney, Bow, but thou mak'st one;
That scarce the town designeth any feast   5
    To which thou'rt not a week bespoke a guest;
That still thou'rt made the supper's flag, the drum,
    The very call, to make all others come:

---

128. *William Roe*: see *Epig.* 70 n.   2 *manners and men*: a favourite collocation: cf. *Und.*
14. 33, 50. 22, 77. 18, etc.   8 *circle*: the emblem of perfection; and cf. *Epig.* 98. 3 n.
14 *travailed*: a pun. Cf. *Epig.* 31. 2.

129. *Mime*: identified by H & S as Inigo Jones; they cf. similar touches in *Epig.* 115.
25–9, and in the portrait of Lanthorn Leatherhead, *B.F.* III. iv. 123–9.   4 *Brentford*:
a town in Middlesex on the junction of the Brent and the Thames, 8 miles west of
London.   4 *Hackney*: a village north of London, 2 miles from St. Paul's.   4 *Bow*: or
Stratford-at-Bow, a suburb of London on the Lea, 4½ miles north-east of St.
Paul's.   7 *flag*: flags were flown from theatres on days of performance.

Think'st thou, Mime, this is great? or that they strive
    Whose noise shall keep thy miming most alive   10
Whilst thou dost raise some player from the grave,
    Out-dance the babion, or out-boast the brave?
Or (mounted on a stool) thy face doth hit
    On some new gesture, that's imputed wit?
Oh, run not proud of this. Yet take thy due: 15
    Thou dost out-zany Cokeley, Pod, nay, Gue,
And thine own Coryate too. But, wouldst thou see,
    Men love thee not for this: they laugh at thee.

130

To Alphonso Ferrabosco, on His Book

To urge, my loved Alphonso, that bold fame
    Of building towns, and making wild beasts tame,
Which music had; or speak her known effects:
    That she removeth cares, sadness ejects,
Declineth anger, persuades clemency,   5
    Doth sweeten mirth and heighten piety,
And is to a body, often ill-inclined,
    No less a sovereign cure than to the mind;
To allege that greatest men were not ashamed
    Of old, even by her practice to be famed;   10

---

12 *babion*: baboon.   12 *brave*: bully.   13 (*mounted on a stool*): i.e. like a mountebank. Jones is called 'mountebank' in *U.V.* 34. 16. The activities of Venetian mountebanks had been described in 1611 by Coryate (see l. 17 n); and cf. *Volp.* II. ii.   14 *gesture*: facial expression.   16 *Cokeley*: a jester who improvised at entertainments; cf. *B.F.* III. iv. 126, *D. is A.* I. i. 93.   16 *Pod*: a puppet-master; see *Epig.* 97. 1–2 n.   16 *Gue*: a showman.   17 *Coryate*: Thomas Coryate (?1577–1617) published in 1611 his *Crudities*, an account of his travels in France and Italy in 1608. See *U.V.* 10, 11, 12.

130. *Alphonso Ferrabosco*: ?1575–1628, composer, violist, lutenist, and musical instructor to Prince Henry. This poem was prefixed to the volume of Ferrabosco's *Airs* published in 1609, which included music for songs from *Volpone* and several of Jonson's masques.   2 Amphion, the legendary inventor of music, is said to have built the walls of Thebes by the sound of his lyre; Orpheus's music is reputed to have calmed wild beasts. See Horace, *Ars Poetica*, 391–5.   3 *known effects*: discussed by John C. Meagher, *Method and Meaning in Jonson's Masques* (Notre Dame, 1966), pp. 69–80.   5 *Declineth*: turns aside.   5 *persuades*: encourages the practice of.

To say indeed she were the soul of heaven,
    That the eighth sphere, no less than planets seven,
Moved by her order, and the ninth more high,
    Including all, were thence called harmony:
I yet had uttered nothing on thy part,   15
    When these were but the praises of the art.
But when I have said, The proofs of all these be
    Shed in thy songs, 'tis true: but short of thee.

131

To the Same

When we do give, Alphonso, to the light
    A work of ours, we part with our own right;
For then all mouths will judge, and their own way:
    The learn'd have no more privilege than the lay.
And though we could all men, all censures hear,   5
    We ought not give them taste we had an ear.
For if the humorous world will talk at large,
    They should be fools, for me, at their own charge.
Say this or that man they to thee prefer;
    Even those for whom they do this know they err: 10
And would, being asked the truth, ashamed say
    They were not to be named on the same day.
Then stand unto thyself, not seek without
    For fame, with breath soon kindled, soon blown out.

12–14 In Ptolemaic astronomy, the earth was surrounded by seven spheres carrying the sun, moon, and five known planets, by an eighth sphere carrying the fixed stars, and a ninth, or crystalline, sphere; according to the Pythagoreans, the movements of the spheres created harmonious music.   18 *Shed*: imparted.

131. Printed with Ferrabosco's *Lessons* in 1609. 2 *our own right*: i.e. to be sole judge of the work. Cf. *Queens*, 679 f. (Q, Ff), *B.F.* Ind. 87, *Cat.* 'To the Reader', 3; Horace, *Epist.* I. xx. 6. 5–10 'And even if we were able to hear all men and their criticisms, we ought not to give them any indication that we listened to them. For the capricious world may talk as much as they like, they can be fools to their own cost so far as I am concerned. Say they prefer this or that man to you; even those for whose sake they indulge in such praise know that they err . . .'. 13 Cf. Persius, *Sat.* i. 7, 'look to no one outside yourself'.

132

To Mr. Joshua Sylvester

If to admire were to commend, my praise
   Might then both thee, thy work and merit raise;
But as it is (the child of ignorance,
   And utter stranger to all air of France)
How can I speak of thy great pains, but err,   5
   Since they can only judge that can confer?
Behold! the reverend shade of Bartas stands
   Before my thought, and, in thy right, commands
That to the world I publish for him this:
   Bartas doth wish thy English now were his.   10
So well in that are his inventions wrought
   As his will now be the translation thought,
Thine the original; and France shall boast
   No more those maiden glories she hath lost.

133

On the Famous Voyage

No more let Greece her bolder fables tell
   Of Hercules or Theseus going to hell,
Orpheus, Ulysses; or the Latin muse
   With tales of Troy's just knight our faiths abuse:
We have a Sheldon and a Heydon got,   5
   Had power to act what they to feign had not.

---

132. *Joshua Sylvester*: 1563–1618, groom of Prince Henry's chamber. Jonson's poem was prefixed to Sylvester's translation of Guillaume du Bartas's *Divine Weeks and Works* in 1605. 6 *confer*: compare (languages). Later Jonson confessed to Drummond 'that Sylvester's translation of Du Bartas was not well done, and that he wrote his verses before it ere he understood to confer' (*Conv. Dr.* 29–31).

133. 4 *Troy's just knight*: Aeneas. 5 *Sheldon*: usually identified as Sir Ralph Sheldon of *Epig.* 119, but on grounds of age his grandson William (b. 1589) might seem a more likely candidate. William's father had married a cousin of Sir John Harington: see l. 195 and n. below, and E. S. Donno's edition of *A New Discourse* (London, 1962), pp. 58, 179. 5 *Heydon*: probably Sir Christopher Heydon, knighted 1596, d. 1623, a writer on astrology (see Whalley, and H & S index).

All that they boast of Styx, of Acheron,
   Cocytus, Phlegethon, our have proved in one:
The filth, stench, noise; save only what was there
   Subtly distinguished, was confused here.   10
Their wherry had no sail, too; ours had none;
   And in it two more horrid knaves than Charon.
Arses were heard to croak instead of frogs,
   And for one Cerberus, the whole coast was dogs.
Furies there wanted not; each scold was ten;   15
   And for the cries of ghosts, women and men
Laden with plague-sores and their sins were heard,
   Lashed by their consciences; to die, afeard.
Then let the former age with this content her:
   She brought the poets forth, but ours the adventer.   20

The Voyage Itself

I sing the brave adventure of two wights,
And pity 'tis I cannot call 'em knights:
One was; and he for brawn and brain right able
To have been styled of King Arthur's table.
The other was a squire of fair degree,   25
But in the action greater man than he,
Who gave, to take at his return from hell,
His three for one. Now, lordings, listen well.
   It was the day, what time the powerful moon
Makes the poor Bankside creature wet it' shoon   30
In it' own hall, when these (in worthy scorn
Of those that put out moneys on return

7–8 *Styx . . . Acheron . . . Cocytus, Phlegethon*: four of the five rivers of Hades.   8 *our*:
ours.   8 *in one*: the Fleet Ditch, a stream rising in the Highgate and Hampstead hills,
and flowing into the Thames at Blackfriars. Once easily navigable as far as Holborn
Bridge, it had become a common sewer by the sixteenth century, and was now choked
with refuse. See John Ashton, *The Fleet: Its Rivers, Prison, and Marriages* (London,
1888).   11 *wherry*: light rowing boat.   12 *Charon*: god of hell, who ferried the souls of
the dead across the rivers Styx and Acheron.   13 *frogs*: who form a croaking chorus in
Aristophanes' comedy of that name, in which Charon ferries Dionysus to Hades.   14
Cerberus: watchdog of Hades; he had three (or, according to some, fifty) heads.   20
adventer: adventure.   28 *three for one*: a form of wagering on one's chances of success-
ful completion of travel; a wagerer stood to treble, or forfeit, his stake.   29 *powerful
moon*: i.e. a spring tide.

From Venice, Paris, or some inland passage
Of six times to and fro without embassage,
Or him that backward went to Berwick, or which   35
Did dance the famous morris unto Norwich)
At Bread Street's Mermaid having dined, and merry,
Proposed to go to Holborn in a wherry:
A harder task than either his to Bristo',
Or his to Antwerp. Therefore, once more, list ho!   40
   A dock there is that called is Avernus,
Of some, Bridewell, and may in time concern us
All that are readers; but methinks 'tis odd
That all this while I have forgot some god
Or goddess to invoke, to stuff my verse,   45
And with both bombard-style and phrase rehearse
The many perils of this port, and how
Sans help of sibyl or a golden bough
Or magic sacrifice, they passed along.
Alcides, be thou succouring to my song!   50
Thou hast seen hell, some say, and know'st all nooks there,
Canst tell me best how every fury looks there,
And art a god, if fame thee not abuses,
Always at hand to aid the merry muses.
Great Club-fist, though thy back and bones be sore   55
Still, with thy former labours, yet once more
Act a brave work, call it thy last adventry;
But hold my torch while I describe the entry
To this dire passage. Say thou stop thy nose:
'Tis but light pains: indeed this dock's no rose.   60

---

34 *without embassage*: without being ambassadors.   35 *backward went to Berwick*: in
1589 Sir Robert Carey won £2,000 by walking from London to Berwick in 12 days
(*Memoirs* (Edinburgh, 1808), p. 20). 'Backward' may be simply a gibe, or may refer to
an actual attempt to outdo this feat.   36 *the famous morris*: danced by the actor
William Kemp in 1599.   39-40 Two of several daring small boat enterprises of the
time. The wherry journey to Bristol was undertaken by Richard Ferris and two
companions in 1590.   41 *Avernus*: sulphurous lake in Campania, thought to be the
entrance to hell (see *Aeneid*, vi. 126 ff.).   42 *Bridewell*: Bridewell dock, at the north
end of what is now Blackfriars Bridge, was the outlet of the Fleet Ditch; Bridewell
prison was situated here.   46 *bombard-style*: bombast.   48 *sibyl or a golden bough*:
which assisted Aeneas in the underworld (*Aeneid*, vi).   50 *Alcides*: Hercules, whose
twelfth labour was to bring Cerberus from hell.   55 *Club-fist*: Hercules inadvertently
killed a man in Calydon with a blow of his fist.   57 *adventry*: adventure (a nonce-
formation).   60 *dock's no rose*: punning on the name of the dock plant; a proverbial
saying (*Oxford Dictionary of English Proverbs*: cf. Tilley, D420).

In the first jaws appeared that ugly monster
Ycleped Mud, which when their oars did once stir,
Belched forth an air as hot as at the muster
Of all your night-tubs, when the carts do cluster,
Who shall discharge first his merd-urinous load: 65
Thorough her womb they make their famous road
Between two walls, where on one side, to scar men
Were seen your ugly centaurs ye call car-men,
Gorgonian scolds and harpies; on the other
Hung stench, diseases, and old filth, their mother,   70
With famine, wants and sorrows many a dozen,
The least of which was to the plague a cousin.
But they unfrighted pass, though many a privy
Spake to 'em louder than the ox in Livy,
And many a sink poured out her rage anenst 'em;   75
But still their valour and their virtue fenced 'em,
And on they went, like Castor brave and Pollux,
Ploughing the main. When see (the worst of all lucks)
They met the second prodigy, would fear a
Man that had never heard of a Chimaera.   80
One said it was bold Briareus, or the beadle
(Who hath the hundred hands when he doth meddle);
The other thought it Hydra, or the rock
Made of the trull that cut her father's lock;
But coming near, they found it but a lighter,   85
So huge, it seemed, they could by no means quite her.
Back, cried their brace of Charons; they cried, No,
No going back! On still, you rogues, and row.
How hight the place? A voice was heard: Cocytus.
Row close then, slaves! Alas, they will beshite us.   90

65 *merd-urinous*: of dung and urine.   67 *scar*: scare.   68 *car-men*: carriers, carters.   74
ox in Livy: see Livy, I. vii. 4–7, XXXV. xx. 4–5.   75 *anenst*: beside.   77 *Castor brave
and Pollux*: who sailed with Jason in search of the golden fleece, and later cleared the
Hellespont and neighbouring seas of pirates.   80 *Chimaera*: a monster which was lion
in its foreparts, dragon aft, and goat amidships.   81 *Briareus*: a giant with 100 hands
and 50 heads; encountered by Aeneas, *Aeneid*, vi. 287.   83 *Hydra*: a monster with 100
(or 50, or 9) heads; when one was cut off, two more would grow in its place.   84 *the
trull*: Scylla, daughter of Typhon, was changed by the jealous Circe into a monster
with barking nether parts, 12 feet, and 6 heads; dismayed, she threw herself into the
sea, and was changed again, into rocks. Jonson, like several classical authors, confuses
her with the Scylla who cut off the hair of her father, Nisus, king of Megara.   86
quite: avoid.   89 *Cocytus*: the unsavoury river 'of lamentation' in the classical under-
world.

No matter, stinkards, row! What croaking sound
Is this we hear? of frogs? No, guts wind-bound,
Over your heads; well, row. At this a loud
Crack did report itself, as if a cloud
Had burst with storm, and down fell *ab excelsis*   95
Poor Mercury, crying out on Paracelsus
And all his followers, that had so abused him,
And in so shitten sort so long had used him;
For (where he was the god of eloquence,
And subtlety of metals) they dispense   100
His spirits now in pills and eke in potions,
Suppositories, cataplasms and lotions.
But many moons there shall not wane, quoth he,
(In the meantime, let 'em imprison me)
But I will speak—and know I shall be heard—   105
Touching this cause, where they will be afeard
To answer me. And sure, it was the intent
Of the grave fart, late let in parliament,
Had it been seconded, and not in fume
Vanished away: as you must all presume   110
Their Mercury did now. By this, the stem
Of the hulk touched and, as by Polypheme
The sly Ulysses stole in a sheepskin,
The well-greased wherry now had got between,
And bade her farewell sough unto the lurden.   115
Never did bottom more betray her burden:
The meat-boat of Bears' College, Paris Garden,
Stunk not so ill; nor, when she kissed, Kate Arden.

95 ab excelsis: from on high.   96 *Parcelsus*: Theophrastus Bombastus von Hohen-
heim (?1490–1541), physician and alchemist, laid stress on the necessary balance
of mercury, sulphur, and salt in the human constitution; he and his followers
essayed 'strange cures with mineral physic' (*Alch*. II. iii. 231). Cf. the complaints of
Mercury in *Merc. Vind*.   108 *the grave fart*: 'The peculiar manner in which Henry
Ludlow said "no" to a message brought by the Serjeant from the Lords' in 1607 forms
the subject of a poem written before 1610 and published in *Musarum Deliciae* in 1656.
(Cf. *Alch*. II. ii. 63.) From this evidence H & S date Jonson's poem *c*. 1610; see,
however, ll. 193–4 below, and note.   109 *in fume*: cf. *Alch*. Argument 12, IV. v. 58.   112
Polypheme: *Odyssey*, ix. 431–4.   115 *sough*: deep sigh.   115 *lurden*: sluggard, i.e. the
slow-moving boat.   116 *bottom*: the lower hull of a boat; a pun.   117 Butchers' offal
was taken by boat to Paris Garden on the Bankside, immediately opposite the outlet of
Fleet Ditch; bull- and bear-baiting were held here. For 'Bear's College' cf. *Poet*. Apol.
Dial. 45, *Gyp. Met*. 1358 (*Songs*, 25. 30).   118 *Kate Arden*: whose charms are also
remembered in *Und*. 43. 148–9.

Yet one day in the year for sweet 'tis voiced,
And that is when it is the Lord Mayor's foist.   120
    By this time had they reached the Stygian pool
By which the masters swear when, on the stool
Of worship, they their nodding chins do hit
Against their breasts. Here several ghosts did flit
About the shore, of farts but late departed,   125
White, black, blue, green, and in more forms out-started
Than all those atomi ridiculous
Whereof old Democrite and Hill Nicholas,
One said, the other swore, the world consists.
These be the cause of those thick frequent mists   130
Arising in that place, through which who goes
Must try the unused valour of a nose:
And that ours did. For yet no nare was tainted,
Nor thumb nor finger to the stop acquainted,
But open and unarmed encountered all.   135
Whether it languishing stuck upon the wall
Or were precipitated down the jakes,
And after swom abroad in ample flakes,
Or that it lay heaped like an usurer's mass,
All was to them the same: they were to pass,   140
And so they did, from Styx to Acheron,
The ever-boiling flood; whose banks upon
Your Fleet Lane furies and hot cooks do dwell,
That with still-scalding steams make the place hell.
The sinks ran grease, and hair of measled hogs,   145
The heads, houghs, entrails, and the hides of dogs:
For, to say truth, what scullion is so nasty
To put the skins and offal in a pasty?
Cats there lay divers had been flayed and roasted
And, after mouldy grown, again were toasted;   150
Then selling not, a dish was ta'en to mince 'em,
But still, it seemed, the rankness did convince 'em.

---

120 *foist*: (i) barge; (ii) fart.   128 *Democrite*: Democritus, born *c.* 460 B.C., developed the theory of the atomic nature of the universe.   128 *Hill Nicholas*: Nicholas Hill (?1570–1610), fellow of St. John's, Oxford, published in 1601 his *Philosophia, Epicurea, Democritiana, Theophrastica, proposita simpliciter, non edocta.*   133 *nare*: nostril.   143 *Fleet Lane*: now Farringdon Street; then chiefly occupied by taverns and cookshops.   144 *hell*: the association of cooks with hell is traditional: cf. *Alch.* III. i. 17 ff., *B.F.* II. ii. 44–5.   145 *measled*: spotty, infected.   152 *convince*: convict, give away.

For here they were thrown in wi' the melted pewter,
Yet drowned they not. They had five lives in future.
But 'mongst these Tiberts, who d'you think there was?   155
Old Banks the juggler, our Pythagoras,
Grave tutor to the learned horse: both which
Being, beyond sea, burned for one witch,
Their spirits transmigrated to a cat;
And now, above the pool, a face right fat,   160
With great grey eyes, are lifted up, and mewed;
Thrice did it spit, thrice dived. At last it viewed
Our brave heroes with a milder glare,
And in a piteous tune began: How dare
Your dainty nostrils (in so hot a season,   165
When every clerk eats artichokes and peason,
Laxative lettuce, and such windy meat)
'Tempt such a passage? When each privy's seat
Is filled with buttock, and the walls do sweat
Urine and plasters? When the noise doth beat   170
Upon your ears of discords so unsweet,
And outcries of the damned in the Fleet?
Cannot the plague-bill keep you back? nor bells
Of loud Sepulchre's, with their hourly knells,
But you will visit grisly Pluto's hall?   175
Behold where Cerberus, reared on the wall
Of Holborn (three sergeants' heads) looks o'er,
And stays but till you come unto the door!
Tempt not his fury; Pluto is away,
And Madam Caesar, great Proserpina,   180

---

155 *Tiberts*: cats; from the name of the cat in *Reynard the Fox* (trs. Caxton, 1481).   156 *Banks*: *fl.* 1588–1637, a showman, owner of the famous horse Morocco (or Marocco). The story of their both being burned by the Pope in Rome is probably untrue (see *DNB*). G. B. Johnston (*Ben Jonson: Poet* (New York, 1945), p. 26 n.) compares Banks's situation to that of Alessio Interminei in Dante's *Inferno* (xviii. 115–26).   156 *Pythagoras*: see *Epig.* 95. 2 n.   162 *Thrice … thrice*: an expected number on epic (and mock-epic) occasions: cf. Virgil, *Aeneid*, ii. 792–4, vi. 700–2; Pope, *The Rape of the Lock*, iii. 137–8; *The Dunciad* (A), i. 203–4; etc.   166 *peason*: peas.   172 *the damned in the Fleet*: the prisoners in Fleet prison. New prisoners were taken by boat along Fleet Ditch, and entered the prison through a water-gate.   174 *Sepulchre's*: the bells of St. Sepulchre's tolled for criminals proceeding to execution, and also, with great frequency, for the dead in time of plague.   175 *grisly Pluto's hall*: probably Fleet prison. Pluto is king of the classical underworld.   176–7 The reference is obscure; but the 'sergeants' are sergeants-at-law in the near-by Inns of Court on Holborn Hill.   180 *Madam Caesar*: a brothel-keeper. Cf. *Alch.* V. iv. 142.

Is now from home. You lose your labours quite,
Were you Jove's sons, or had Alcides' might.
They cried out, Puss! He told them he was Banks,
That had so often showed 'em merry pranks.
They laughed at his laugh-worthy fate; and passed   185
The triple head without a sop. At last,
Calling for Rhadamanthus, that dwelt by,
A soap-boiler, and Aeacus him nigh,
Who kept an ale-house, with my little Minos,
An ancient purblind fletcher with a high nose,   190
They took 'em all to witness of their action,
And so went bravely back, without protraction.
    In memory of which most liquid deed,
The city since hath raised a pyramid.
And I could wish for their eternized sakes,   195
My muse had ploughed with his that sung A-jax.

186 *a sop*: Aeneas pacified Cerberus with a drugged honey-cake, *Aeneid*, vi. 417–25. 187–9 *Rhadamanthus . . . Aeacus . . . Minos*: the three judges of the classical underworld. Cf. *Poet*. III. i. 145 ff.   190 *fletcher*: arrow-maker.   190 *high*: long.   193–4 Possibly referring to Sir Hugh Myddelton's New River scheme, begun in 1609 and ceremoniously opened in 1613; the 'pyramid' may be the water-house and tower at New River Head, Clerkenwell (Dr. Valerie Pearl, private communication). See illustration facing p. 58 in J. W. Gough, *Sir Hugh Myddelton, Entrepreneur and Engineer* (Oxford, 1964). The identification may require a slightly later dating for the poem than that proposed by H & S (see l. 108 and n. above), and a modification of the idea that none of the poems in the 1616 Folio was written later than 1612 (see H & S, viii. 16).   196 *A-jax*: Sir John Harington's treatise on the jakes (1596), *A New Discourse of a Stale Subject, Called the Metamorphosis of Ajax*. But the reference is also, perhaps, to Homer, as G. B. Johnston suggests (*Ben Jonson: Poet*, p. 28). B. R. Smith (*SEL*, xiv (1974), 109) sees a further gibe at Harington's longwindedness as an epigrammatist; cf. *Conv. Dr.* 37–40.

THE FOREST

THE FOREST

I

Why I Write Not of Love

Some act of Love's bound to rehearse,
I thought to bind him in my verse;
Which when he felt, Away! quoth he,
Can poets hope to fetter me?
It is enough they once did get   5
Mars and my mother in their net:
I wear not these my wings in vain.
With which he fled me; and again
Into my rhymes could ne'er be got
By any art. Then wonder not   10
That since, my numbers are so cold,
When Love is fled, and I grow old.

2

To Penshurst

Thou art not, Penshurst, built to envious show
  Of touch or marble, nor canst boast a row
Of polished pillars, or a roof of gold;
  Thou hast no lantern whereof tales are told,
Or stair, or courts; but stand'st an ancient pile,   5
  And these grudged at, art reverenced the while.

The Forest: see Jonson's explanation of this title in his note 'To the Reader' prefacing
Underwood; and n. (p. 123).

1. 4–5 Vulcan caught his wife Venus and her lover Mars in a net, and exposed them to
the ridicule of the gods. His informant was Apollo, god of poetry: Cupid therefore
blames and suspects poets in general.   12 *old*: Jonson was not yet forty. Contrast *Und.*
2. i. 1–5.

2. *Penshurst*: near Tonbridge, Kent; home of the Sidney family since 1552. The 'great
lord' of l. 91 is Robert Sidney (1563–1626), second son of Sir Henry Sidney and
younger brother of Sir Philip Sidney; knighted, 1586; Baron Sidney of Penshurst,
1603; Viscount Lisle, 1605; Earl of Leicester, 1618. The poem was written before the
death of Prince Henry in November 1612: see l. 77. J. C. A. Rathmell, *ELR*, i (1971),
250–60, suggests that Jonson had an intimate knowledge both of the estate and of Lord
Lisle's difficult financial circumstances, and finds a further possible clue to dating in

Thou joy'st in better marks, of soil, of air,
Of wood, of water; therein thou art fair.
Thou hast thy walks for health as well as sport:
Thy Mount, to which the dryads do resort,   10
Where Pan and Bacchus their high feasts have made,
Beneath the broad beech and the chestnut shade;
That taller tree, which of a nut was set
At his great birth, where all the muses met.
There, in the writhed bark, are cut the names   15
Of many a sylvan taken with his flames;
And thence the ruddy satyrs oft provoke
The lighter fauns to reach thy lady's oak.
Thy copse, too, named of Gamage, thou hast there,
That never fails to serve thee seasoned deer   20
When thou wouldst feast or exercise thy friends.
The lower land, that to the river bends,
Thy sheep, thy bullocks, kine and calves do feed;
The middle grounds thy mares and horses breed.

l. 45: walls were built at Penshurst in May 1612 with stone from a local quarry. G. R. Hibbard, *JWCI*, xix (1956), 159–74, discusses the poem in relation to other 'country-house' poems. Other important studies by P. M. Cubeta, *PQ*, xlii (1963), 14–24, and A. Fowler, *RES*, xxiv (1973), 266–82.   1–6 Jonson may be contrasting Penshurst with a particular, ostentatious, house such as nearby Knole, or—as John Carey suggests (Fowler, p. 268)—Theobalds, a house with which Jonson was familiar (see ll. 61 ff., n., below), and which boasted the architectural features mentioned here: see Daniel Lysons, *The Environs of London* (London, 1796), pp. 29–39. Salisbury's fondness for building—like that of his father, Lord Burghley—was notorious: see Jonson's *Ent. K. & Q. Theob.* 83, and P. M. Handover, *The Second Cecil* (London, 1959), p. 276. By this date, however, Theobalds was owned by King James, and too overt a reference would, in any case, have been indiscreet. G. E. Wilson, *SEL*, viii (1968), 77–89, sees similarities with the description of Solomon's temple, 1 Kings, 6–7.   2 *touch*: black marble.   4 *lantern*: glassed turret to admit light.   5 *ancient pile*: Penshurst was built about 1340, over 200 years before the Sidney family came to live there; a 'pile' may be a castle or stronghold (*OED*, *sb.*²) or a lofty mass of buildings (*OED*, *sb.*³, 4).   7 *marks*: distinctive features.   10 *Mount*: a piece of high ground in the park, still called by this name.   10 ff. Modelled on Martial's idealized description of a grove of trees planted by Julius Caesar on his estate at Tartessus, *Epig.* IX. lxi. 11–16*.   13–14 An oak which still stands today at Penshurst is said to have grown from an acorn planted on the day of Sir Philip Sidney's birth, 30 November 1554. Cf. Suetonius on the poplar planted at Virgil's birth, *Vita Virgili*, 5.   16 *his flames*: 'either "his passion" or "Sidney's passion: the same passion as Sidney's" ' (Fowler).   18–19 'There is an old tradition that a Lady Leicester (the wife undoubtedly of Sir Robert Sidney) was taken in travail under an oak in Penshurst Park, which was afterwards called *My Lady's Oak*' (Gifford). The same lady, Barbara Gamage, fed deer in a copse in the grounds, which became known as Lady Gamage's Bower.

Each bank doth yield thee conies, and the tops,   25
    Fertile of wood, Ashour and Sidney's copse,
To crown thy open table, doth provide
    The purpled pheasant with the speckled side;
The painted partridge lies in every field,
    And for thy mess is willing to be killed.   30
And if the high-swoll'n Medway fail thy dish,
    Thou hast thy ponds that pay thee tribute fish:
Fat, aged carps, that run into thy net;
    And pikes, now weary their own kind to eat,
As loath the second draught or cast to stay,   35
    Officiously, at first, themselves betray;
Bright eels, that emulate them, and leap on land
    Before the fisher, or into his hand.
Then hath thy orchard fruit, thy garden flowers,
    Fresh as the air and new as are the hours: 40
The early cherry, with the later plum,
    Fig, grape and quince, each in his time doth come;
The blushing apricot and woolly peach
    Hang on thy walls, that every child may reach.
And though thy walls be of the country stone,   45
    They're reared with no man's ruin, no man's groan;
There's none that dwell about them wish them down,
    But all come in, the farmer and the clown,
And no one empty-handed, to salute
    Thy lord and lady, though they have no suit.   50
Some bring a capon, some a rural cake,
    Some nuts, some apples; some that think they make
The better cheeses, bring 'em; or else send
    By their ripe daughters, whom they would commend

---

25 ff. As in many a 'pleasant place', Nature actively offers itself up to man: cf. stanza 5 of Marvell's 'The Garden', Dryden's description of Eden in *The State of Innocence*, III. i. 35–8, and Carew's 'To Saxham', 21–8. With Jonson's fish, cf. those of Martial, *Epig.* X. xxx. 21–4*, and Juvenal, *Sat.* iv. 69: 'The fish himself wanted to be caught' (flatterer's hyperbole). For the fruit, see Virgil, *Georgics*, ii. 501–2*.   26 These woods still exist.   29 *painted partridge*: Martial's *picta perdix*, *Epig.* III. lviii. 15.   31 *Medway*: the near-by river was notoriously 'hard to be fished' (Rathmell, 255).   36 *Officiously*: obligingly.   40 *hours*: the Horae, the three female divinities which presided over the three seasons of the ancient year; hence, the seasons themselves. Cf. *U.V.* 44. 1.   45–7 Cf. 'A Panegyric to Sir Lewis Pemberton', 115 ff., Herrick, pp. 146–9.   48 *all come in*: as in Carew's 'To Saxham', 49–58, and Herrick's 'Panegyric', 12 ff.   49 ff. Cf. Martial, *Epig.* III. lviii. 33–40*.

This way to husbands; and whose baskets bear   55
    An emblem of themselves, in plum or pear.
But what can this (more than express their love)
    Add to thy free provisions, far above
The need of such? whose liberal board doth flow
    With all that hospitality doth know!   60
Where comes no guest but is allowed to eat
    Without his fear, and of thy lord's own meat;
Where the same beer and bread and self-same wine
    That is his lordship's shall be also mine;
And I not fain to sit, as some this day   65
    At great men's tables, and yet dine away.
Here no man tells my cups, nor, standing by,
    A waiter, doth my gluttony envy,
But gives me what I call, and lets me eat;
    He knows below he shall find plenty of meat,   70
Thy tables hoard not up for the next day.
    Nor, when I take my lodging, need I pray
For fire or lights or livery: all is there,
    As if thou then wert mine, or I reigned here;
There's nothing I can wish, for which I stay.   75
    That found King James, when, hunting late this way
With his brave son, the Prince, they saw thy fires
    Shine bright on every hearth as the desires
Of thy Penates had been set on flame
    To entertain them; or the country came   80
With all their zeal to warm their welcome here.
    What (great, I will not say, but) sudden cheer
Didst thou then make 'em! and what praise was heaped
    On thy good lady then! who therein reaped

---

56 *emblem*: of sexual maturity; cf. *Venus and Adonis*, 527–8, for the plum. The pear was sometimes associated with Venus (in place of the more usual apple), and thus with marriage; see Guy de Tervarent, *Attributs et symboles dans l'art profane 1450–1600* (Geneva, 1958–9), ii. 309, J. Brand, *Popular Antiquities of Great Britain*, ed. W. C. Hazlitt (London, 1870), ii. 97–8.   61 ff. Cf. Lucian, 'Laws for Banquets', *Saturnalia**, Juvenal, *Sat.* v. 24 ff., and Martial, *Epig.* III. lx. 1–2, 9*. Jonson reacted in a similar way to Salisbury's hospitality: 'My lord, said he, you promised I should dine with you, but I do not', etc. (*Conv. Dr.* 317–21). G. E. Wilson (p. 85) sees similarities with the Lord's Supper. The ungrudging waiter is from Martial, *Epig.* III. lviii. 43–4*; cf. Carew, 'To Saxham', 47–8 and Herrick, 'Panegyric', 47ff.   73 *livery*: provisions; appointments.   82 *sudden*: prompt.

The just reward of her high housewifery: 85
    To have her linen, plate, and all things nigh
When she was far; and not a room but dressed
    As if it had expected such a guest!
These, Penshurst, are thy praise, and yet not all.
    Thy lady's noble, fruitful, chaste withal;   90
His children thy great lord may call his own,
    A fortune in this age but rarely known.
They are and have been taught religion; thence
    Their gentler spirits have sucked innocence.
Each morn and even they are taught to pray   95
    With the whole household, and may every day
Read in their virtuous parents' noble parts
    The mysteries of manners, arms and arts.
Now, Penshurst, they that will proportion thee
    With other edifices, when they see   100
Those proud, ambitious heaps, and nothing else,
    May say, their lords have built, but thy lord dwells.

3

To Sir Robert Wroth

How blest art thou canst love the country, Wroth,
    Whether by choice, or fate, or both;
And, though so near the city and the court,
    Art ta'en with neither's vice nor sport;
That at great times art no ambitious guest   5
    Of sheriff's dinner or mayor's feast;
Nor com'st to view the better cloth of state,
    The richer hangings, or crown-plate;
Nor throng'st, when masquing is, to have a sight
    Of the short bravery of the night,   10

90 *fruitful*: Lady Lisle bore at least ten children.   95–6 Rathmell (p. 254) finds
provision for these regular devotions in the Sidney papers.   98 *mysteries*: skills; but
the word acquires resonance from its alternative meaning, religious truths.   99 *proportion*: compare or estimate proportionately.   100–2 Cf. Martial, *Epig.* XII. l. 8, on the
builder of a great house: 'How well you are—not housed!' See also *For.* 3. 94.

3. *Sir Robert Wroth*: 1576–1614; knighted, 1601; married Sir Robert Sidney's daughter, Lady Mary (see *Epig.* 103 n., 105, *Und.* 28), 1604. Jonson later judged her to be
'unworthily married on a jealous husband' (*Conv. Dr.* 355–6). Durrants, in the parish
of Enfield, was one of his estates. He was sheriff of Essex in 1613.

To view the jewels, stuffs, the pains, the wit
　There wasted, some not paid for yet!
But canst at home in thy securer rest
　Live with unbought provision blest;
Free from proud porches or their gilded roofs,　15
　'Mongst lowing herds and solid hoofs;
Alongst the curled woods and painted meads,
　Through which a serpent river leads
To some cool, courteous shade, which he calls his,
　And makes sleep softer than it is!　20
Or, if thou list the night in watch to break,
　A-bed canst hear the loud stag speak,
In spring oft roused for thy master's sport,
　Who for it makes thy house his court;
Or with thy friends the heart of all the year　25
　Divid'st upon the lesser deer;
In autumn at the partridge makes a flight,
　And giv'st thy gladder guests the sight;
And in the winter hunt'st the flying hare,
　More for thy exercise than fare;　30
While all that follow, their glad ears apply
　To the full greatness of the cry;
Or hawking at the river, or the bush,
　Or shooting at the greedy thrush,
Thou dost with some delight the day out-wear,　35
　Although the coldest of the year!
The whilst, the several seasons thou hast seen
　Of flowery fields, of copses green,
The mowed meadows, with the fleeced sheep,
　And feasts that either shearers keep;　40

12 *not paid for yet*: cf. *L.R.* 11–12.　13 ff. Cf. Virgil, *Georgics*, ii. 458–71*.　14 *unbought provision*: Horace's *dapes inemptas*, *Epodes*, ii. 48 (see Jonson's translation, *Und.* 85. 48), cf. Martial, *Epig.* I. lv. 12, IV. lxvi. 5; Virgil, *Georgics*, iv. 133.　16 *lowing*: playing, perhaps, on words (contrast 'proud', l. 15).　17 *Alongst*: close by.　19 *courteous shade*: the force of the adjective lies in its derivation: manners appropriate to the court. Cf. the effect of Marvell's 'courteous briars', 'Upon Appleton House', 616.　23 ff. Wroth was a keen sportsman, and King James occasionally visited Durrants for hunting. But the passage also contains memories of Horace, *Epodes*, ii, and Martial, *Epig.* I. xlix.　25 *heart*: middle; i.e. summer. Punning on 'hart'.　26 *lesser deer*: smaller animals.　32 *full greatness of the cry*: cf. Theseus on the harmony of his pack, *A Midsummer Night's Dream*, IV. i. 120–1.　37 ff. Cf. the seasonal tasks and delights of Virgil, *Georgics*, ii. 516 ff.　40 *either shearers*: i.e. those who 'shear' the meadows or the sheep.

The ripened ears, yet humble in their height,
    And furrows laden with their weight;
The apple-harvest, that doth longer last;
    The hogs returned home fat from mast;
The trees cut out in log, and those boughs made   45
    A fire now, that lent a shade!
Thus Pan and Silvane having had their rites,
    Comus puts in for new delights,
And fills thy open hall with mirth and cheer,
    As if in Saturn's reign it were;   50
Apollo's harp and Hermes' lyre resound,
    Nor are the muses strangers found.
The rout of rural folk come thronging in
    (Their rudeness then is thought no sin);
Thy noblest spouse affords them welcome grace,   55
    And the great heroes of her race
Sit mixed with loss of state or reverence:
    Freedom doth with degree dispense.
The jolly wassail walks the often round,
    And in their cups their cares are drowned;   60
They think not then which side the cause shall leese,
    Nor how to get the lawyer fees.
Such and no other was that age of old,
    Which boasts to have had the head of gold.
And such, since thou canst make thine own content,   65
    Strive, Wroth, to live long innocent.
Let others watch in guilty arms, and stand
    The fury of a rash command,
Go enter breaches, meet the cannon's rage,
    That they may sleep with scars in age,   70
And show their feathers shot, and colours torn,
    And brag that they were therefore born.

45–6 Cf. Martial, *Epig*. I. xlix. 27: 'To your very hearth shall come down the neighbouring wood'.   47 *Pan and Silvane*: as in Virgil, *Georgics*, ii. 493–4.   48 *Comus*: 'the god of cheer, or the belly', *Pleas. Rec.* 6.   50 *Saturn's reign*: the golden age: a favourite Jonsonian theme; cf. also Virgil, *Georgics*, ii. 538.   53 ff. Cf. Martial, *Epig*. I. xlix. 29–30.   54 *rudeness*: want of refinement.   56 ff. Cf. Statius, *Silvae*, I. vi. 43–5, 'One table serves every class alike, children, women, people, knights, and senators: freedom has loosed the bonds of awe.'   59 *walks*: circulates.   59 *often*: frequent.   61 *leese*: lose.   67 ff. Perhaps remembering several classical passages: Virgil, *Georgics*, ii. 503 ff., Tibullus, I. x. 29–32, Martial, *Epig*. I. xlix. 37–40, Horace, *Epodes*, ii. 4–8 (see *Und*. 85. 4–8).

Let this man sweat and wrangle at the bar
    For every price in every jar,
And change possessions oftener with his breath   75
    Than either money, war or death;
Let him than hardest sires more disinherit,
    And each-where boast it as his merit
To blow up orphans, widows, and their states,
    And think his power doth equal fate's.   80
Let that go heap a mass of wretched wealth
    Purchased by rapine, worse than stealth,
And brooding o'er it sit, with broadest eyes,
    Not doing good, scarce when he dies.
Let thousands more go flatter vice, and win   85
    By being organs to great sin,
Get place and honour, and be glad to keep
    The secrets that shall break their sleep;
And, so they ride in purple, eat in plate,
    Though poison, think it a great fate.   90
But thou, my Wroth, if I can truth apply,
    Shalt neither that nor this envy:
Thy peace is made; and when man's state is well,
    'Tis better if he there can dwell.
God wisheth none should wrack on a strange shelf;   95
    To him man's dearer than to himself;
And howsoever we may think things sweet,
    He always gives what he knows meet,
Which who can use is happy: such be thou.
    Thy morning's and thy evening's vow   100
Be thanks to him, and earnest prayer, to find
    A body sound, with sounder mind;
To do thy country service, thyself right;
    That neither want do thee affright,
Nor death; but when thy latest sand is spent,   105
    Thou mayst think life a thing but lent.

---

79 *blow up*: ruin; cf. *Epig.* 112. 6, *Alch.* I. ii. 78 ff., *M.L.* III. vi. 135. 84 From Pithou, *Epigrammata* (1590): 'The miser does nothing aright, save when he dies.' 86 Cf. *Sej.* I. 23–7. 94 *dwell*: cf. *For.* 2. 102. 95–106 Cf. Juvenal, *Sat.* x. 349–59*. 106 *lent*: cf. *Epig.* 45. 3 n.

4

To the World
A Farewell for a Gentlewoman, Virtuous and Noble

False world, good night. Since thou hast brought
    That hour upon my morn of age,
Henceforth I quit thee from my thought;
    My part is ended on thy stage.
Do not once hope that thou canst tempt   5
    A spirit so resolved to tread
Upon thy throat and live exempt
    From all the nets that thou canst spread.
I know thy forms are studied arts,
    Thy subtle ways be narrow straits,   10
Thy courtesy but sudden starts,
    And what thou call'st thy gifts are baits.
I know too, though thou strut and paint,
    Yet art thou both shrunk up and old;
That only fools make thee a saint,   15
    And all thy good is to be sold.
I know thou whole art but a shop
    Of toys and trifles, traps and snares,
To take the weak, or make them stop;
    Yet art thou falser than thy wares.   20
And knowing this, should I yet stay,
    Like such as blow away their lives
And never will redeem a day,
    Enamoured of their golden gyves?
Or, having 'scaped, shall I return   25
    And thrust my neck into the noose
From whence so lately I did burn
    With all my powers myself to loose?
What bird or beast is known so dull
    That, fled his cage, or broke his chain,   30
And tasting air and freedom, wull
    Render his head in there again?

4. 14 For the currency of this notion, see M. Macklem, *The Anatomy of the World* (Minneapolis, 1958), V. Harris, *All Coherence Gone* (Chicago, 1949); cf. *Disc.* 124–8, *Songs*, 19. 4. 18 Cf. *Disc.* 1437–44. 25 ff. Cf. Horace, *Sat.* II. vii. 68–71*. 31 *wull*: will.

If these, who have but sense, can shun
    The engines that have them annoyed,
Little for me had reason done,   35
    If I could not thy gins avoid.
Yes, threaten, do. Alas, I fear
    As little as I hope from thee;
I know thou canst nor show nor bear
    More hatred than thou hast to me.   40
My tender, first, and simple years
    Thou didst abuse, and then betray;
Since stirredst up jealousies and fears,
    When all the causes were away.
Then in a soil hast planted me   45
    Where breathe the basest of thy fools,
Where envious arts professed be,
    And pride and ignorance the schools;
Where nothing is examined, weighed,
    But, as 'tis rumoured, so believed;   50
Where every freedom is betrayed,
    And every goodness taxed or grieved.
But what we're born for we must bear:
    Our frail condition it is such
That, what to all may happen here,   55
    If 't chance to me, I must not grutch.
Else I my state should much mistake,
    To harbour a divided thought
From all my kind; that, for my sake,
    There should a miracle be wrought.   60
No, I do know that I was born
    To age, misfortune, sickness, grief;
But I will bear these with that scorn
    As shall not need thy false relief.
Nor for my peace will I go far,   65
    As wanderers do that still do roam,
But make my strengths, such as they are,
    Here in my bosom, and at home.

33 *sense*: physical senses.   45 *soil*: probably the court.   48 i.e. pride and ignorance are the schools in which the 'envious arts' are practised.   56 *grutch*: complain.   68 *at home*: cf. *Und*. 14. 30 and 78. 8; *For*. 3. 13.

5

Song
To Celia

Come, my Celia, let us prove,
While we may, the sports of love;
Time will not be ours for ever;
He at length our good will sever.
Spend not then his gifts in vain.   5
Suns that set may rise again;
But if once we lose this light,
'Tis with us perpetual night.
Why should we defer our joys?
Fame and rumour are but toys.   10
Cannot we delude the eyes
Of a few poor household spies?
Or his easier ears beguile,
So removed by our wile?
'Tis no sin love's fruit to steal,   15
But the sweet theft to reveal:
To be taken, to be seen,
These have crimes accounted been.

5. Sung to Celia by Volpone (with a few small verbal differences), *Volp.* III. vii. 166–
83; in dramatic context, a more sinister invitation than its Catullan model, *Carm.* v. A
musical setting for the song was printed in Alphonso Ferrabosco's *Airs*, 1609 (see *Epig.*
130). See F. W. Sternfeld in *Studies in the English Renaissance Drama*, ed. J. W.
Bennett et al. (London, 1961), pp. 310–21; J. P. Cutts, *N & Q*, cciii (1958), 217–
19.   6–8 Catullus, *Carm.* v. 4–6: 'Suns may set and rise again. For us, when the short
light has once set, remains to be slept the sleep of one unbroken night.'   12 *household
spies*: cf. 'His Parting From Her', attributed to Donne, l. 41; Jon Stallworthy, *Between
the Lines* (Oxford, 1963), p. 143, compares Yeats's 'Parting', l. 3.   13 *his*: those of
Corvino, Celia's husband.

6

To the Same

Kiss me, sweet: the wary lover
Can your favours keep and cover,
When the common courting jay
All your bounties will betray.
Kiss again: no creature comes.   5
Kiss, and score up wealthy sums
On my lips, thus hardly sundered,
While you breathe. First give a hundred,
Then a thousand, then another
Hundred, then unto the tother   10
Add a thousand, and so more,
Till you equal with the store
All the grass that Romney yields,
Or the sands in Chelsea fields,
Or the drops in silver Thames,   15
Or the stars that gild his streams
In the silent summer nights,
When youths ply their stol'n delights:
That the curious may not know
How to tell 'em as they flow,   20
And the envious, when they find
What their number is, be pined.

7

Song
That Women Are but Men's Shadows

Follow a shadow, it still flies you;
    Seem to fly it, it will pursue:
So court a mistress, she denies you;
    Let her alone, she will court you.

6. Lines 19–22 are sung by Volpone later in the same scene, *Volp*. III. vii. 236–9; Catullus is again the model. On the 'arithmetical' poem, see J. B. Leishman, *The Art of Marvell's Poetry* (London, 1966), pp. 73 ff.   6–11 Cf. Catullus, *Carm.* v. 7–13*.   12–22 Cf. Catullus, *Carm.* vii*.   13 *Romney*: on the east coast of Kent, surrounded by marshy grazing land.   14 *Chelsea*: the name is said to derive from O.E. *ceosel* = sand, pebbles.   22 *pined*: pained, vexed.

7. 'Pembroke and his lady discoursing, the earl said that women were men's shadows,

Say, are not women truly then 5
  Styled but the shadows of us men?

At morn and even shades are longest;
  At noon they are or short or none:
So men at weakest, they are strongest,
  But grant us perfect, they're not known. 10
Say, are not women truly then
  Styled but the shadows of us men?

8

To Sickness

Why, disease, dost thou molest
Ladies, and of them the best?
Do not men enough of rites
To thy altars, by their nights
Spent in surfeits, and their days 5
And nights too, in worser ways?
Take heed, sickness, what you do:
I shall fear you'll surfeit too.
Live not we as all thy stalls,
Spitals, pest-house, hospitals, 10
Scarce will take our present store?
And this age will build no more;
Pray thee feed contented, then,
Sickness, only on us men.
Or if needs thy lust will taste 15
Woman-kind, devour the waste
Livers, round about the town.
But, forgive me, with thy crown
They maintain the truest trade,
And have more diseases made. 20

and she maintained them. Both appealing to Jonson, he affirmed it true, for which my
lady gave a penance to prove it in verse: hence his epigram' (*Conv. Dr.* 364–7).
Jonson's poem is based on a Latin original by the sixteenth-century poet
Bartholomaeus Anulus (Barthélemi Aneau): see Oswald Wallace, *N & Q*, 3rd series,
viii (1865), 187. The thought is proverbial; see Tilley, L518: 'Woman like a shadow
flies one following and pursues one fleeing.'

8. Cf. *Und.* 34, and *Conv. Dr.* 348–9, on Lady Sidney's smallpox.   9 *stalls*: lodgings in
almshouses.   10 *Spitals*: lazar-houses.   18 *crown*: a pun: the coin (five shillings) and
the pox (cf. *Measure for Measure*, I. ii. 50); 'the truest trade', because one crown is
given for another.

What should yet thy palate please?
Daintiness, and softer ease,
Sleeked limbs, and finest blood?
If thy leanness love such food,
There are those that for thy sake   25
Do enough, and who would take
Any pains—yea, think it price
To become thy sacrifice;
That distil their husbands' land
In decoctions; and are manned   30
With ten emp'rics in their chamber,
Lying for the spirit of amber;
That for the oil of talc dare spend
More than citizens dare lend
Them, and all their officers;   35
That, to make all pleasure theirs,
Will by coach and water go,
Every stew in town to know;
Dare entail their loves on any,
Bald or blind or ne'er so many;   40
And for thee, at common game
Play away health, wealth and fame.
These, disease, will thee deserve;
And will, long ere thou shouldst starve,
On their beds, most prostitute,   45
Move it as their humblest suit,
In thy justice to molest
None but them, and leave the rest.

27 *price*: admirable. 29–30 Cf. *S.W.* II. ii. 109–11, *For.* 13. 73–4. 31 *emp'rics*: quacks. Sir Epicure Mammon promises these delights to Dol Common, *Alch.* IV. i. 135–6. 32 *spirit of amber*: succinic acid. 33 *oil of talc*: used as a wash for the complexion; recommended by Subtle for faces 'decayed / Beyond all cure of paintings', *Alch.* III. ii. 34–6. 41 *common game*: promiscuous sex. 45 *prostitute*: gathering in a secondary sense of the word (*OED*, 4), prostrate.

9

Song
To Celia

Drink to me only with thine eyes,
　　And I will pledge with mine;
Or leave a kiss but in the cup,
　　And I'll not look for wine.
The thirst that from the soul doth rise　　5
　　Doth ask a drink divine;
But might I of Jove's nectar sup,
　　I would not change for thine.
I sent thee late a rosy wreath,
　　Not so much honouring thee　　10
As giving it a hope that there
　　It could not withered be.
But thou thereon didst only breathe,
　　And sent'st it back to me;
Since when it grows, and smells, I swear,　　15
　　Not of itself, but thee.

10

And must I sing? What subject shall I choose?
Or whose great name in poets' heaven use
For the more countenance to my active muse?

9. As Richard Cumberland first pointed out in *The Observer* (1788), no. cix, p. 316, Jonson's lyric is based on a number of separate passages from the *Epistles* of Philostratus*. A. D. Fitton-Brown examines Jonson's revision of the poem and the debt to Philostratus in *MLR*, liv (1959), 554–7, and compares also *Greek Anthology*, v. 91: 'I send thee sweet perfume, not so much honouring thee as it; for thou canst perfume the perfume.' Cf. also 'Upon a Virgin Kissing a Rose', Herrick, p. 51.   7–8 For the view held by Yvor Winters and others, that the speaker would prefer Jove's nectar to Celia's kisses, see M. van Deusen, *E in C* vii (1957), 95–103. This reading has little to commend it: see Fitton Brown, loc. cit. p. 557, W. Empson, *Seven Types of Ambiguity* (London, 1953, 3rd edition revised), p. 242, J. G. Nichols, *The Poetry of Ben Jonson* (London, 1969), pp. 163–4.

10. This poem, along with *For.* 11, *U.V.* 4 and 5, Shakespeare's *The Phoenix and Turtle*, and poems by Marston, Chapman, and 'Ignoto', was first printed amongst the 'Diverse Poetical Essays' on the subject of *The Phoenix and Turtle* appended to Robert

Hercules? Alas, his bones are yet sore
With his old earthly labours. To exact more   5
Of his dull godhead were sin. I'll implore

Phoebus. No, tend thy cart still: envious day
Shall not give out that I have made thee stay,   ·
And foundered thy hot team, to tune my lay.

Nor will I beg of thee, lord of the vine,   10
To raise my spirits with thy conjuring wine,
In the green circle of thy ivy twine.

Pallas, nor thee I call on, mankind maid,
That at thy birth mad'st the poor smith afraid,
Who with his axe thy father's midwife played.   15

Go, cramp dull Mars, light Venus, when he snorts,
Or with thy tribade trine invent new sports;
Thou nor thy looseness with my making sorts.

Let the old boy, your son, ply his old task,
Turn the stale prologue to some painted masque;   20
His absence in my verse is all I ask.

Chester's *Love's Martyr* in 1601. Carleton Brown argues that the turtle, a type of male
devotion, and the phoenix, a type of female perfection, are intended to represent Sir
John and Lady Salusbury, married in 1586; see his introduction to *Poems by Sir John
Salusbury and Robert Chester*, E.E.T.S. (Oxford and London, 1914). B. H. Newdigate,
finding the inscription 'To L: C: off: B' in a MS. version of *U.V.* 5, suggests that the
phoenix is Lucy, Countess of Bedford: *TLS*, 24 Oct. 1936, p. 862, and introduction to
his edition of *The Phoenix and Turtle* (Oxford, 1937). The difficulties involved in
accepting either of these theories are discussed by W. H. Matchett, *The Phoenix and
Turtle* (The Hague, 1965). Matchett argues plausibly for a return to Grosart's identifi-
cation of the phoenix and the turtle with Elizabeth and Essex. Reprinting the poem in
The Forest, Jonson dropped its original title, 'Praeludium', and changed almost all first
person plural pronouns to first person singular.   4 *his bones are yet sore*: cf. *Epig.*
133. 55.   7 *cart*: the chariot of the sun.   12 *green circle*: Bacchus's crown of leaves is
likened to the magical circle of a necromancer.   13–15 Pallas Athene was born from
Jupiter's head, cleft open by Vulcan.   13 *mankind*: masculine.   17 *tribade trine*: lesbian
trinity: the three graces, often associated with Venus in the dance.   19 *old boy*: cf.
Und. 36. 16, 'The eldest god, yet still a child': a Platonic notion; cf. *Beauty*, 329–31.

Hermes, the cheater, shall not mix with us,
Though he would steal his sisters' Pegasus
And riffle him, or pawn his petasus.

Nor all the ladies of the Thespian lake   25
(Though they were crushed into one form) could make
A beauty of that merit that should take

My muse up by commission: no, I bring
My own true fire. Now my thought takes wing,
And now an epode to deep ears I sing.   30

Proludium

An elegy? No, muse; it asks a strain
Too loose and capering for thy stricter vein.
Thy thoughts did never melt in amorous fire,
Like glass blown up and fashioned by desire.
The skilful mischief of a roving eye   5
Could ne'er make prize of thy white chastity.
Then leave these lighter numbers to light brains
In whom the flame of every beauty reigns,
Such as in lust's wild forest love to range,
Only pursuing constancy in change;   10
Let these in wanton feet dance out their souls.
A farther fury my raised spirit controls,
Which raps me up to the true heaven of love,
And conjures all my faculties to approve
The glories of it. Now our muse takes wing,   15
And now an epode to deep ears we sing.

---

22 *Hermes, the cheater*: Mercury, god of thieves; at his birth he pilfered from many of the gods.   23 *sisters'*: the muses'.   24 *riffle*: raffle, gamble away.   24 *petasus*: Hermes' hat, broad-brimmed or winged.   25 *ladies of the Thespian lake*: the muses.   30 *epode*: a lyric poem, usually of grave character, alternating short and long lines; see Horace's *Odes*, Bk. V.

Proludium. An earlier version of the previous poem, sent to Sir John Salusbury but not printed in *Love's Martyr*. See Matchett's discussion of the two versions, *The Phoenix and Turtle*, pp. 95–6, n. 21. H & S translate the title as 'preliminary canter' (viii. 9).   1 *elegy*: i.e. a poem written in elegiac metre: in Greek and Latin, a dactylic hexameter and pentameter form the elegiac distich.   6 *chastity*: a typical claim by Jonson for his verse (cf. *Epig.* 17. 6, 49. 6), but here especially appropriate to the 'chaste love' about to be celebrated in *For.* 11.   13 *raps*: transports.

11

Epode

Not to know vice at all, and keep true state,
    Is virtue, and not fate;
Next to that virtue, is to know vice well
    And her black spite expel.
Which to effect, since no breast is so sure   5
    Or safe but she'll procure
Some way of entrance, we must plant a guard
    Of thoughts to watch and ward
At the eye and ear, the ports unto the mind,
    That no strange or unkind   10
Object arrive there, but the heart, our spy,
    Give knowledge instantly
To wakeful reason, our affections' king;
    Who, in the examining,
Will quickly taste the treason and commit   15
    Close the close cause of it.
'Tis the securest policy we have
    To make our sense our slave.
But this true course is not embraced by many—
    By many? Scarce by any.   20
For either our affections do rebel,
    Or else the sentinel,
That should ring larum to the heart, doth sleep,
    Or some great thought doth keep
Back the intelligence, and falsely swears   25
    They're base and idle fears

---

11. *Epode*: see *For.* 10. 30 n. Reprinting the poem in *The Forest*, Jonson reversed the procedure he had followed with *For.* 10, and changed almost all first person singular pronouns to first person plural. The discussion of the roles of reason and passion in the early part of the poem is similar to that of Thomas Wright, *The Passions of the Mind in General*, 1601; Jonson was probably acquainted with Wright's ideas before 1601: see Theodore Stroud, *ELH*, xiv (1947), 274–82.   1–4 Plato, *Gorgias*, 478E: 'Happiest therefore is he who has no vice in his soul. . . . Next after him, I take it, is he who is relieved of it'; and cf. Seneca, *Hercules Furens*, 1098–9, *Hippolytus*, 140–1.   2 *virtue, and not fate*: see *Epig.* 63. 2–3 n., 76. 15 n.   13 *reason, our affections' king*: cf. *E.M.I.* (Q), II. ii. 12 ff. The extended figure is developed from classical commonplaces: cf. Plato, *Republic*, iv. 441–2, *Timaeus*, 69–70; Aristotle, *Politics*, I. v.   15 *taste*: detect.   15–16 *commit | Close . . . close*: imprison securely . . . immediate.   18 *sense*: senses.

Whereof the loyal conscience so complains.
    Thus, by these subtle trains,
Do several passions still invade the mind
    And strike our reason blind.   30
Of which usurping rank some have thought love
    The first, as prone to move
Most frequent tumults, horrors and unrests
    In our inflamed breasts;
But this doth from their cloud of error grow   35
    Which thus we overblow:
The thing they here call love is blind desire,
    Armed with bow, shafts and fire;
Inconstant like the sea, of whence 'tis born,
    Rough, swelling, like a storm;   40
With whom who sails, rides on the surge of fear,
    And boils as if he were
In a continual tempest. Now true love
    No such effects doth prove;
That is an essence far more gentle, fine,   45
    Pure, perfect, nay divine;
It is a golden chain let down from heaven,
    Whose links are bright and even,
That falls like sleep on lovers, and combines
    The soft and sweetest minds   50
In equal knots. This bears no brands nor darts
    To murther different hearts,
But in a calm and god-like unity
    Preserves community.
Oh, who is he that, in this peace, enjoys   55
    The elixir of all joys?
A form more fresh than are the Eden bowers,
    And lasting as her flowers;
Richer than time, and as time's virtue rare;
    Sober as saddest care;   60
A fixed thought, an eye untaught to glance.
    Who, blest with such high chance,

---

36 *overblow*: blow away.   37 ff. Cf. *In Praise of Demosthenes*, 13*, attributed to Lucian.   47 *golden chain*: cf. *Hym.* 320, and Jonson's note on allegorical interpretations of Zeus's golden chain leading from heaven to earth, *Iliad*, viii. 19.   50 *soft*: softest.   56 *elixir*: quintessence.   59 *time's virtue*: to reveal truth, proverbially the daughter of time.

Would at suggestion of a steep desire
    Cast himself from the spire
Of all his happiness? But soft, I hear   65
    Some vicious fool draw near
That cries, we dream, and swears there's no such thing
    As this chaste love we sing.
Peace, luxury, thou art like one of those
    Who, being at sea, suppose   70
Because they move, the continent doth so;
    No, vice, we let thee know,
Though thy wild thoughts with sparrows' wings do fly,
    Turtles can chastely die.
And yet, in this to express ourselves more clear,   75
    We do not number here
Such spirits as are only continent
    Because lust's means are spent;
Or those who doubt the common mouth of fame,
    And for their place and name   80
Cannot so safely sin: their chastity
    Is mere necessity.
Nor mean we those whom vows and conscience
    Have filled with abstinence:
Though we acknowledge, who can so abstain   85
    Makes a most blessed gain.
He that for love of goodness hateth ill
    Is more crown-worthy still
Than he which for sin's penalty forbears;
    His heart sins, though he fears.   90
But we propose a person like our dove,
    Graced with a phoenix' love;
A beauty of that clear and sparkling light
    Would make a day of night,
And turn the blackest sorrows to bright joys;   95
    Whose odorous breath destroys
All taste of bitterness, and makes the air
    As sweet as she is fair.

63–5 Hunter suggests a memory of Satan's tempting of Christ, Luke 4, Matt. 4.   69
luxury: lust.   69–71 Cf. Montaigne, 'Of Judging of Others' Death', *Essays*, trs. John
Florio, ed. J. I. M. Stewart (London, 1931), i. 694.   73 *sparrows' wings*: the sparrow is
a symbol of lechery.   79 *doubt*: fear.   80 *for*: because of.   87–90 Horace, *Epist.* I.
xvi. 52–3*, Ovid, *Amores*, III. iv. 3–4*. G. B. Johnston, *Ben Jonson: Poet*, p. 151,
suggests a possible recall of Matt. 5 : 28.

A body so harmoniously composed
    As if nature disclosed   100
All her best symmetry in that one feature!
    Oh, so divine a creature
Who could be false to? chiefly, when he knows
    How only she bestows
The wealthy treasure of her love on him,   105
    Making his fortunes swim
In the full flood of her admired perfection?
    What savage, brute affection
Would not be fearful to offend a dame
    Of this excelling fame?   110
Much more a noble and right generous mind,
    To virtuous moods inclined,
That knows the weight of guilt: he will refrain
    From thoughts of such a strain,
And to his sense object this sentence ever: 115
Man may securely sin, but safely, never.

12

Epistle
To Elizabeth, Countess of Rutland

MADAM,
Whilst that for which all virtue now is sold
    And almost every vice, almighty gold;
That which, to boot with hell, is thought worth heaven,
    And for it life, conscience, yea souls, are given;

---

101 *feature*: form.   115 *object*: place before.   116 This line is attributed to Jonson in Robert Allot's *England's Parnassus*, published in 1600, the year before *Love's Martyr*.   116 *securely*: carelessly. Jonson's wordplay is not original: see *OED*, 'secure', *a*. I. l. a, and examples. Possibly Jonson is remembering, and reversing, Seneca, *Hippolytus*, 163–4: 'Some women have sinned with safety, but none with peace of soul' (*scelus aliqua tutum, nulla securum tulit*).

12. *Elizabeth, Countess of Rutland*: see *Epig.* 79 n. She had married early in 1599; this poem was sent to her on New Year's Day, 1600. It was a contemporary custom to send New Year poems (cf. *Und.* 79 and e.g. *The Poems of Thomas Carew*, ed. R. Dunlap (Oxford, 1949), pp. 32–3, 89–90), but Jonson may have been specifically remembering Tibullus, III. i, and Horace, *Odes*, IV. iii. Jackson I. Cope, *Eng. Misc.* x (1959), 61–6, suggests the poem owes a general debt to Spenser's *Complaints*, but overlooks other sources and commonplaces.   2–4 Cf. *King's Ent.* 743, *Volp.* I. i. 24–5.

Toils by grave custom up and down the court 5
  To every squire or groom that will report
Well or ill only, all the following year,
  Just to the weight their this day's presents bear;
While it makes ushers serviceable men,
  And someone apteth to be trusted then, 10
Though never after; whiles it gains the voice
  Of some grand peer, whose air doth make rejoice
The fool that gave it, who will want and weep
  When his proud patron's favours are asleep;
While thus it buys great grace and hunts poor fame, 15
  Runs between man and man, 'tween dame and dame;
Solders cracked friendship, makes love last a day
  Or perhaps less: whilst gold bears all this sway,
I, that have none to send you, send you verse.
  A present which, if elder writs rehearse 20
The truth of times, was once of more esteem
  Than this, our gilt, nor golden age can deem;
When gold was made no weapon to cut throats
  Or put to flight Astraea, when her ingots
Were yet unfound, and better placed in earth 25
  Than here, to give pride fame, and peasants birth.
But let this dross carry what price it will
  With noble ignorants, and let them still
Turn upon scorned verse their quarter-face;
  With you, I know, my offering will find grace. 30
For what a sin 'gainst your great father's spirit
  Were it to think that you should not inherit
His love unto the muses, when his skill
  Almost you have, or may have, when you will?
Wherein wise nature you a dowry gave 35
  Worth an estate treble to that you have.
Beauty, I know, is good, and blood is more;
  Riches thought most. But, madam, think what store

9 *serviceable*: in this derogatory sense, cf. Edgar on Oswald, 'a serviceable villain', *King Lear*, IV. vi. 254. 10 *apteth*: makes fit. 22 Than this our age, which is not golden but merely gilded, can judge. The Folio spelling, 'guilt', perhaps enlarges the meaning. Cf. *Epig.* 64. 3–4, and n. 24 *Astraea*: see *Epig.* 74. 9 n. 25 *better placed*: Horace, *Odes*, III. iii. 49–52. 29 *quarter-face*: *Sej.* V. 389; cf. 'half-fac'd fellowship', *I Henry IV*, I. iii. 208. 31 *great father's*: Sir Philip Sidney's. Jonson reckoned Elizabeth 'nothing inferior to her father' in poetry (*Conv. Dr.* 213). 32 *to think*: thinkable.

The world hath seen, which all these had in trust
    And now lie lost in their forgotten dust.   40
It is the muse alone can raise to heaven,
    And, at her strong arm's end, hold up and even
The souls she loves. Those other glorious notes,
    Inscribed in touch or marble, or the coats
Painted or carved upon our great men's tombs,   45
    Or in their windows, do but prove the wombs
That bred them graves; when they were born, they died
    That had no muse to make their fame abide.
How many equal with the Argive Queen
    Have beauty known, yet none so famous seen?   50
Achilles was not first that valiant was,
    Or, in an army's head, that, locked in brass,
Gave killing strokes. There were brave men before
    Ajax, or Idomen, or all the store
That Homer brought to Troy; yet none so live,   55
    Because they lacked the sacred pen could give
Like life unto 'em. Who heaved Hercules
    Unto the stars? or the Tyndarides?
Who placed Jason's Argo in the sky,
    Or set bright Ariadne's crown so high?   60
Who made a lamp of Berenice's hair,
    Or lifted Cassiopeia in her chair,
But only poets, rapt with rage divine?
    And such, or my hopes fail, shall make you shine.

---

43 ff. Cf. Seneca, *Ad Polybium de Consolatione*, xviii. 2*, Horace, *Odes*, IV. viii. 13–34, IV. ix. 13–28. 43 *glorious*: vainglorious, boastful. 44 *touch*: black marble. 45–7 *tombs . . . wombs . . . graves*: a traditional collocation: see G. A. E. Parfitt, *RMS*, xv (1971), 23–33. 49 *the Argive Queen*: Helen. 52 *in an army's head*: at the head of an army. 52 *locked in brass*: Vulcan fashioned for Achilles an impenetrable suit of armour (*Iliad*, xviii), in which Achilles subsequently killed many Trojans. 54 *Idomen*: Idomeneus, King of Crete, who accompanied the Greeks to Troy with 90 ships. 57 ff. After his death, *Hercules* was immortalized in the heavens as the sun, his twelve labours being represented in the twelve signs of the zodiac. The *Tyndarides* are Castor and Pollux, twin sons of Tyndareus; Jupiter made them into the constellation Gemini. *Argo*, the famous ship in which Jason recovered the golden fleece, was sent by Minerva (Hyginus, *Fab.* xiv. 13) to the sky, and is the southern constellation, the Ship. *Ariadne's crown* of seven stars, given to her after she had been forsaken by Theseus, was similarly dispatched by Bacchus after her death (Ovid, *Metam.* viii. 177–82). *Berenice's hair*, donated to Venus and subsequently lost, was changed by Venus (or by Jupiter) into the constellation *Coma Berenices* (Catullus, *Carm.* lxvi. 59–68). *Cassiopeia*, mother of Andromeda, was made into a northern constellation of 13 stars.

You, and that other star, that purest light   65
    Of all Lucina's train, Lucy the bright;
Than which a nobler, heaven itself knows not.
    Who, though she have a better verser got
(Or 'poet', in the court account) than I,
    And who doth me, though I not him, envy   70
Yet, for the timely favours she hath done
    To my less sanguine muse, wherein she hath won
My grateful soul, the subject of her powers,
    I have already used some happy hours
To her remembrance; which, when time shall bring   75
    To curious light, to notes I then shall sing,
Will prove old Orpheus' act no tale to be;
    For I shall move stocks, stones no less than he.
Then all that have but done my muse least grace
    Shall thronging come, and boast the happy place   80
They hold in my strange poems, which, as yet,
    Had not their form touched by an English wit.
There like a rich and golden pyramid,
    Borne up by statues shall I rear your head
Above your under-carved ornaments,   85
    And show how, to the life, my soul presents
Your form impressed there: not with tinkling rhymes,
    Or commonplaces, filched, that take these times,
But high and noble matter, such as flies
    From brains entranced and filled with ecstasies,   90

66 *Lucina* ... *Lucy the bright*: Queen Elizabeth, and Lucy, Countess of Bedford, respectively; playing in each case simply on etymology: *lux*, *lucis* = light. Cf. *Epig.* 76. 8 n., 94.   68–9 *verser* ... *'poet'*: cf. *Disc.* 2448–9, *C.R.* II. i. 48–9. The distinction is an ancient one. R. W. Short, *RES*, xv (1939), 315–17, argues that the allusion is not to Daniel ('no poet', *Conv. Dr.* 24) but to Michael Drayton.   72 *less sanguine*: less bloody: than e.g. Drayton's *Piers Gaveston* (or Daniel's *Civil Wars*).   75 *which*: the fruits of the hours: the poems to Lucy.   75–6 *bring | To curious light*: Jonson had planned to write an account of the worthy ladies of the kingdom, hoping it would be 'helped to light' by the favour of the queen (*Queens*, 666–9). It did not appear.   76 *to notes*: variant reading; H & S: 'the notes'.   79 ff. cf. the emphasis of *Epig.* Ded. 14–21.   81 *strange*: probably in the sense of un-English, based on classical models; cf. *Epig.* 18.   83–5 Cf. the descriptions of the House of Fame in *Queens*, 360–3, 683–8. 87 *tinkling rhymes*: variant reading; H & S: 'tickling'. For 'tinkling rhymes', cf. *Fort. Is.* 291 ('fine tinkling rime!'), *Disc.* 283, Sidney, *Apology for Poetry*, ed. G. Shepherd (London, 1965), p. 133. With the animus against rhyme expressed here, cf. *Und.* 29.   88 *commonplaces*: for a different attitude towards their use in poetry, see *Disc.* 2281.

Moods which the god-like Sidney oft did prove,
   And your brave friend and mine so well did love.
Who, wheresoe'er he be. . . ,

                              The rest is lost

[Who, wheresoe'er he be, on what dear coast,
Now thinking on you, though to England lost,
For that firm grace he holds in your regard,   95
I, that am grateful for him, have prepared
This hasty sacrifice; wherein I rear
A vow as new and ominous as the year:
Before his swift and circled race be run,
My best of wishes, may you bear a son.]   100

                        13

                      Epistle
            To Katherine, Lady Aubigny

'Tis grown almost a danger to speak true
   Of any good mind now, there are so few.
The bad, by number, are so fortified,
   As, what they've lost to expect, they dare deride.
So both the praised and praisers suffer: yet,   5
   For others' ill, ought none their good forget.
I, therefore, who profess myself in love
   With every virtue, wheresoe'er it move,
And howsoever; as I am at feud
   With sin and vice, though with a throne endued,   10
And, in this name, am given out dangerous
   By arts and practice of the vicious,
Such as suspect themselves, and think it fit
   For their own capital crimes to indict my wit;
I, that have suffered this, and though forsook   15
   Of fortune, have not altered yet my look,

93 ff. This part of the poem was not lost, but cancelled after it was revealed that the
Earl of Rutland was impotent, and the poem's final wish consequently unrealizable.
98 *ominous*: auspicious.

13. *Katherine, Lady Aubigny*: daughter of Sir Gervase Clifton (later Baron Clifton) of
Nottinghamshire; married Esmé Stuart, Seigneur d'Aubigny (*Epig.* 127 n.), 1609; d.
1627. For the poem's approximate date, see note to ll. 94 ff. below.

Or so myself abandoned, as, because
   Men are not just, or keep no holy laws
Of nature and society, I should faint,
   Or fear to draw true lines, 'cause others paint: 20
I, madam, am become your praiser. Where,
   If it may stand with your soft blush to hear
Yourself but told unto yourself, and see
   In my character what your features be,
You will not from the paper slightly pass;   25
   No lady but, at some time, loves her glass.
And this shall be no false one, but as much
   Removed, as you from need to have it such.
Look then, and see yourself. I will not say
   Your beauty, for you see that every day,   30
And so do many more. All which can call
   It perfect, proper, pure and natural,
Not taken up o'the doctors, but as well
   As I can say, and see, it doth excel.
That asks but to be censured by the eyes,   35
   And in those outward forms all fools are wise.
Nor that your beauty wanted not a dower
   Do I reflect. Some alderman has power,
Or cozening farmer of the customs, so
   To advance his doubtful issue, and o'erflow   40
A prince's fortune: these are gifts of chance,
   And raise not virtue; they may vice enhance.
My mirror is more subtle, clear, refined,
   And takes and gives the beauties of the mind;
Though it reject not those of fortune, such   45
   As blood and match. Wherein, how more than much
Are you engaged to your happy fate
   For such a lot! that mixed you with a state
Of so great title, birth, but virtue most,
   Without which all the rest were sounds, or lost.   50
'Tis only that can time and chance defeat,
   For he that once is good is ever great.

---

25 *slightly*: with slight attention.   27–8 'The glass he presents (i.e. his poem) is as *much remov'd* from *giving* a flattering and false likeness, as the *lady* from *wanting* any such assistance' (Whalley).   33 *taken up o'*: acquired from.   35 *censured*: judged.   39 *farmer of the customs*: tax-collector.   52 *good . . . great*: see *Epig.* Ded. 15–16 n.

Wherewith, then, madam, can you better pay
  This blessing of your stars, than by that way
Of virtue, which you tread? What if alone, 55
  Without companions? 'Tis safe to have none.
In single paths dangers with ease are watched;
  Contagion in the press is soonest catched.
This makes, that wisely you decline your life
  Far from the maze of custom, error, strife, 60
And keep an even and unaltered gait,
  Not looking by, or back (like those that wait
Times and occasions, to start forth and seem);
  Which though the turning world may disesteem,
Because that studies spectacles and shows, 65
  And after varied, as fresh, objects goes,
Giddy with change, and therefore cannot see
  Right the right way; yet must your comfort be
Your conscience; and not wonder if none asks
  For truth's complexion, where they all wear masks. 70
Let who will, follow fashions and attires;
  Maintain their liegers forth; for foreign wires
Melt down their husbands' land, to pour away
  On the close groom and page on New Year's Day
And almost all days after, while they live 75
  (They find it both so witty and safe to give).
Let 'em on powders, oils, and paintings spend
  Till that no usurer nor his bawds dare lend
Them or their officers; and no man know
  Whether it be a face they wear, or no. 80
Let 'em waste body and state, and after all,
  When their own parasites laugh at their fall,
May they have nothing left, whereof they can
  Boast, but how oft they have gone wrong to man,
And call it their brave sin. For such there be 85
  That do sin only for the infamy,

59 *decline*: turn aside. 63 *seem*: come into view. 64 *the turning world*: Jonas A. Barish detects a glancing allusion to the *machina versatilis* of the masques; 'Jonson and the Loathed Stage', in *A Celebration of Ben Jonson*, ed. W. Blissett et al. (Toronto, 1974), p. 39. 72 *wires*: frames to support hair. 72 *liegers*: agents. 73–4 Cf. *For.* 8. 29–30, *S.W.* II. ii. 109–11. On New Year gifts, cf. *For.* 12. 1–19. 78 *bawds*: henchmen. 80 Cf. *Cat.* II. 62–3. 84 *gone wrong to man*: cf. the idiom of *Epig.* 90. 3. 85–6 Cf. Seneca, *Epist.* cxxii. 18: 'They are unwilling to be wicked in the conventional way, because notoriety is the reward of their sort of wickedness.'

And never think how vice doth every hour
  Eat on her clients and some one devour.
You, madam, young have learned to shun these shelves,
  Whereon the most of mankind wrack themselves,   90
And, keeping a just course, have early put
  Into your harbour, and all passage shut
'Gainst storms, or pirates, that might charge your peace;
  For which you worthy are the glad increase
Of your blessed womb, made fruitful from above,   95
  To pay your lord the pledges of chaste love,
And raise a noble stem, to give the fame
  To Clifton's blood that is denied their name.
Grow, grow fair tree, and as thy branches shoot,
  Hear what the muses sing about thy root,   100
By me, their priest (if they can aught divine):
  Before the moons have filled their triple trine,
To crown the burthen which you go withal,
  It shall a ripe and timely issue fall,
To expect the honours of great Aubigny,   105
  And greater rites, yet writ in mystery,
But which the fates forbid me to reveal.
  Only thus much, out of a ravished zeal
Unto your name and goodness of your life,
  They speak; since you are truly that rare wife   110
Other great wives may blush at, when they see
  What your tried manners are, what theirs should be.
How you love one, and him you should; how still
  You are depending on his word and will;
Not fashioned for the court, or strangers' eyes,   115
  But to please him, who is the dearer prize
Unto himself, by being dear to you.
  This makes, that your affections still be new,
And that your souls conspire, as they were gone
  Each into other, and had now made one.   120

Live that one still; and as long years do pass,
  Madam, be bold to use this truest glass,
Wherein your form you still the same shall find,
  Because nor it can change, nor such a mind.

14

Ode
To Sir William Sidney, on His Birthday

Now that the hearth is crowned with smiling fire,
  And some do drink, and some do dance,
      Some ring,
      Some sing,
  And all do strive to advance   5
The gladness higher;
    Wherefore should I
    Stand silent by,
      Who not the least
      Both love the cause and authors of the feast?   10

Give me my cup, but from the Thespian well,
  That I may tell to Sidney what
      This day
      Doth say,
  And he may think on that   15
Which I do tell;
    When all the noise
    Of these forced joys
      Are fled and gone,
      And he with his best genius left alone.   20

---

14. *Sir William Sidney*: 1590–1612, eldest son of Robert Sidney, Lord Lisle. The poem was evidently written in November 1611, for his twenty-first birthday. His career so far had been unpromising: see L. C. John, *MLR*, lii (1957), 168–76. 11 *Thespian well*: of poetic inspiration; cf. *For.* 10. 25.

This day says, then, the number of glad years
  Are justly summed, that make you man;
      Your vow
      Must now
  Strive all right ways it can   25
To outstrip your peers:
    Since he doth lack
    Of going back
      Little, whose will
      Doth urge him to run wrong, or to stand still.   30

Nor can a little of the common store
  Of nobles' virtue show in you;
      Your blood
      So good
  And great must seek for new,   35
And study more;
    Not, weary, rest
    On what's deceased.
      For they that swell
      With dust of ancestors, in graves but dwell.   40

'Twill be exacted of your name, whose son,
  Whose nephew, whose grandchild you are;
      And men
      Will then
  Say you have followed far,   45
When well begun;
    Which must be now:
    They teach you how.
      And he that stays
      To live until tomorrow hath lost two days.   50

---

21 *number*: twenty-one; a fact appropriately introduced at this line of the poem.   34–5
good | And great: see *Epig.* Ded. 15–16 n.   42 *Whose nephew, whose grandchild*: Sir
Philip Sidney's, and Sir Henry Sidney's, respectively.   49–50 Proverbial: 'One today
is worth two tomorrows', Tilley, T370.

So may you live in honour as in name,
    If with this truth you be inspired;
        So may
        This day
Be more, and long desired;   55
And with the flame
    Of love be bright,
    As with the light
        Of bonfires. Then
        The birthday shines, when logs not burn, but men. 60

15

To Heaven

Good and great God, can I not think of thee,
    But it must straight my melancholy be?
Is it interpreted in me disease
    That, laden with my sins, I seek for ease?
Oh, be thou witness, that the reins dost know   5
    And hearts of all, if I be sad for show;
And judge me after, if I dare pretend
    To aught but grace, or aim at other end.
As thou art all, so be thou all to me,
    First, midst, and last; converted one and three;   10
My faith, my hope, my love; and in this state,
    My judge, my witness, and my advocate.
Where have I been this while exiled from thee?
    And whither rapt, now thou but stoop'st to me?

15. 1 ff. '. . . the meaning is not—"Can I not think of God without its making me
melancholy?" but "Can I not think of God without its being imputed or set down by
others to a fit of dejection?" ' (A. C. Swinburne, *A Study of Ben Jonson* (London,
1889), p. 103). It was thought that melancholy gave rise to specious piety: see Burton,
The Anatomy of Melancholy, III. iv. and L. Babb, *The Elizabethan Malady* (East
Lansing, 1951), pp. 47–54, 177–8. As W. Kerrigan argues, however, Jonson's primary
concern throughout the poem is with the judgement not of his fellows but of God:
SLI, vi (1973), 199–217. 5–6 *reins . . . hearts*: Ps. 7 : 9: 'the righteous God trieth the
hearts and reins'; cf. Jer. 17 : 10. 10 *converted*: perhaps in the sense of a mathematical
conversion, in which different numbers may be taken as equivalents. 11 *My faith, my
hope, my love*: 1 Cor. 13 : 13. 12 *judge . . . witness . . . advocate*: respectively: Acts
10 : 42, Rev. 1 : 5, 1 John 2 : 1; and elsewhere.

Dwell, dwell here still: Oh, being everywhere,   15
 How can I doubt to find thee ever here?
I know my state, both full of shame and scorn,
 Conceived in sin, and unto labour born,
Standing with fear, and must with horror fall,
 And destined unto judgement, after all.   20
I feel my griefs too, and there scarce is ground
 Upon my flesh to inflict another wound.
Yet dare I not complain, or wish for death
 With holy Paul, lest it be thought the breath
Of discontent; or that these prayers be   25
 For weariness of life, not love of thee.

17 *scorn*: matter for scorn (*OED*, 3. † a: cf. Milton, *Samson Agonistes*, 34).   21–2
Adapted from a very different context, exiled Ovid complaining that the gods are
conspiring to destroy him: '. . . so am I wounded by the steady blows of fate until now
I have scarce space upon me for a new wound', *Epist. Ex Ponto*, II. vii. 41–2.   24 *holy
Paul*: 'O wretched man that I am! who shall deliver me from the body of this death?',
Rom. 7 : 24; cf. Phil. 1 : 23, 2 Cor. 5 : 1. Kerrigan (pp. 210–12) pertinently compares
Calvin's commentary on these passages.

THE UNDERWOOD

[*Title-page*] *Under-woods*: the plural form, also found in some of the headlines in the 1640 Folio, is probably a misprint: see B. H. Newdigate, *TLS*, 7 Feb. 1935, p. 76, W. W. Greg, *RES*, xviii (1942), 159–60 and n.

Cineri, gloria sera venit: Martial, *Epig.* I. xxv. 8, urging Faustinus to publish: 'to the ashes of the dead, glory comes too late'.

UNDER-WOODS

CONSISTING OF
DIVERS
POEMS.

By

BEN. IOHNSON.

Martial——*Cineri, gloria sera venit.*

LONDON.
Printed M. DC. XL.

TO THE READER

With the same leave, the ancients called that kind of body *sylva*, or
῞Υλη, in which there were works of diverse nature and matter con-
gested, as the multitude call timber-trees, promiscuously growing, a
wood or forest; so am I bold to entitle these lesser poems of later
growth by this of *Underwood*, out of the analogy they hold to *The
Forest* in my former book, and no otherwise.

Ben Jonson

To the Reader: partly a translation of Caspar Gavartius's Latin note in his edition of
Statius's *Silvae* in 1616; the Latin note is also prefixed to Jonson's *Timber, or
Discoveries*. On the arboreal metaphor, see H. H. Hudson, *The Epigram in the English
Renaissance* (Princeton, 1947), p. 26.

THE UNDERWOOD

I

Poems of Devotion

i

The Sinner's Sacrifice
To the Holy Trinity

1. O holy, blessed, glorious Trinity
Of persons, still one God in unity,
The faithful man's believed mystery,
    Help, help to lift

2. Myself up to thee, harrowed, torn, and bruised   5
By sin and Satan; and my flesh misused,
As my heart lies in pieces, all confused,
    Oh, take my gift.

3. All-gracious God, the sinner's sacrifice,
A broken heart thou wert not wont despise,   10
But 'bove the fat of rams, or bulls, to prize
    An offering meet

4. For thy acceptance. Oh, behold me right,
And take compassion on my grievous plight.
What odour can be, than a heart contrite,   15
    To thee more sweet?

5. Eternal Father, God who didst create
This all of nothing, gavest it form and fate,
And breath'st into it life and light, with state
    To worship thee;   20

---

1. *Poems of Devotion*: examined by P. M. Cubeta, *JEGP*, lxii (1963), 96–110. H & S suggest the poems are the work of Jonson's closing years; for a further clue to dating, see 1. ii n., below.

1. i. 10 Ps. 51 : 16–17: 'For thou desirest not sacrifice; else would I give it; thou delightest not in burnt offering. The sacrifices of God are a broken spirit: a broken and a contrite heart, O God, thou wilt not despise.' Cf. Herbert, 'The Altar'. 11 1 Sam. 15 : 22: 'Behold, to obey is better than sacrifice, and to hearken than the fat of rams.' 19 *with state*: meet, ready.

6. Eternal God the Son, who not denied'st
To take our nature, becam'st man, and died'st
To pay our debts, upon thy cross, and cried'st
      All's done in me;

7. Eternal Spirit, God from both proceeding,   25
Father and Son, the comforter in breeding
Pure thoughts in man; with fiery zeal them feeding
      For acts of grace:

8. Increase those acts, O glorious Trinity
Of persons, still one God in unity,   30
Till I attain the longed-for mystery
      Of seeing your face,

9. Beholding one in three, and three in one,
A Trinity, to shine in union;
The gladdest light dark man can think upon,   35
      O grant it me!

10. Father, and Son, and Holy Ghost, you three
All coeternal in your majesty,
Distinct in persons, yet in unity
      One God to see,   40

11. My Maker, Saviour, and my Sanctifier,
To hear, to meditate, sweeten my desire
With grace, with love, with cherishing entire;
      O, then how blessed,

12. Among thy saints elected to abide,   45
And with thy angels placed side by side,
But in thy presence truly glorified,
      Shall I there rest!

---

24 *All's done*: 'It is finished', John 19 : 30.   32 *seeing your face*: Rev. 22 : 4.   42
meditate: 1640 Folio; H & S (following Gifford), 'mediate'. But the sense is, 'sweeten
my desire to hear and to meditate'; see Cubeta, pp. 105–6.

I. *ii*

A Hymn to God the Father

Hear me, O God!   5
  A broken heart
  Is my best part;
Use still thy rod,
  That I may prove
  Therein thy love.

                  10

If thou hadst not
  Been stern to me,
  But left me free,
I had forgot
  Myself and thee.

                  15

For sin's so sweet,
  As minds ill bent
  Rarely repent,
Until they meet
  Their punishment.

                  20

Who more can crave
  Than thou hast done?
  That gav'st a Son,
To free a slave,
  First made of nought;
  With all since bought.

Sin, death, and hell
  His glorious name
  Quite overcame,   25
Yet I rebel,
  And slight the same.

I. ii. Daniel O'Connor, *N & Q*, xii (1965), 379–80, finds borrowings from this hymn in John Cosin's *A Collection of Private Devotions*, compiled late 1626/early 1627 for Anglican ladies at Charles I's court (Stationers' Register, 1 March 1627). Herrick imitates Jonson's poem in 'An Ode, or Psalm, to God', Herrick, p. 363.

But I'll come in,
Before my loss
Me farther toss,   30
As sure to win
Under his cross.

1. iii
A Hymn
On the Nativity of My Saviour

I sing the birth was born tonight,
The author both of life and light;
      The angels so did sound it,
And like, the ravished shepherds said,
Who saw the light and were afraid,   5
      Yet searched, and true they found it.

The Son of God, the Eternal King,
That did us all salvation bring,
      And freed the soul from danger;
He whom the whole world could not take,   10
The Word, which heaven and earth did make,
      Was now laid in a manger.

The Father's wisdom willed it so,
The Son's obedience knew no No,
      Both wills were in one stature;   15
And as that wisdom had decreed,
The Word was now made flesh indeed,
      And took on him our nature.

What comfort by him do we win,
Who made himself the price of sin,   20
      To make us heirs of glory!
To see the babe, all innocence,
A martyr born in our defence,
      Can man forget this story?

1. iii. 10 *take*: contain; a Latinism.   15 *stature*: state, condition.

2

A Celebration of Charis in Ten Lyric Pieces

i

His Excuse for Loving

Let it not your wonder move,
Less your laughter, that I love.
Though I now write fifty years,
I have had, and have, my peers;
Poets, though divine, are men: 5
Some have loved as old again.
And it is not always face,
Clothes, or fortune gives the grace,
Or the feature, or the youth;
But the language, and the truth,   10
With the ardour and the passion,
Gives the lover weight and fashion.
If you will then read the story,
First prepare you to be sorry
That you never knew till now   15
Either whom to love, or how;
But be glad as soon with me,
When you know that this is she,
Of whose beauty it was sung,
She shall make the old man young,   20

2. *A Celebration of Charis*: Charis is the name of one of Vulcan's wives (see note to *Und*. 2. v. 37–8, below), and also the singular form of *Charites*, the graces. Fleay (i. 324–5) believed that Charis was Elizabeth, Lady Hatton, and guessed that she was also addressed in *Epig*. 124 and *Und*. 19, and that she was Venus in *Haddington*—a role which would in fact have been played by a boy actor—and the model for Mrs. Fitzdottrell in *D. is A.* (1616). Parallel passages are noted below; in themselves, they hardly add up to evidence of a long-continued real-life love-affair. H & S consider the first poem of the sequence was written in 1622 or 1623 (a literal reading of 'fifty years', l. 3), subsequent poems between 1612 and 1616. For further clues to dating, see notes to *Und*. 2. vi. 28 and 2. vii, below. The sequence is discussed by P. M. Cubeta, *ELH*, xxv (1958), 163–80, by Trimpi, pp. 209–28, and by R. S. Peterson, *SLI*, vi (1973), 219–68.

2. i. 3 *fifty years*: perhaps remembering Horace's appeal to Venus not to trouble him 'at his fiftieth year', *Odes*, IV. i: see Jonson's trs., *Und*. 86. Cf. Jonson's references to his age in *For*. 1. 10–12, *Und*. 20. 5.   4 *my peers*: e.g. Horace, Anacreon, Ronsard. 9 *feature*: proportions, shape of body.   19 *it was sung*: not identified. H & S compare *Gyp. Met*. 484–5: a long shot.

Keep the middle age at stay,
And let nothing high decay,
Till she be the reason why
All the world for love may die.

2. *ii*
How He Saw Her

I beheld her, on a day,
When her look out-flourished May,
And her dressing did outbrave
All the pride the fields then have;
Far I was from being stupid,   5
For I ran and called on Cupid:
Love, if thou wilt ever see
Mark of glory, come with me.
Where's thy quiver? Bend thy bow!
Here's a shaft—thou art too slow!   10
And (withal) I did untie
Every cloud about his eye;
But he had not gained his sight
Sooner than he lost his might
Or his courage; for away   15
Straight he ran, and durst not stay,
Letting bow and arrow fall,
Nor for any threat or call
Could be brought once back to look.
I, foolhardy, there up-took   20
Both the arrow he had quit
And the bow, with thought to hit
This my object. But she threw
Such a lightning, as I drew,
At my face, that took my sight   25
And my motion from me quite;

22 *high*: in both temporal and qualitative senses: well advanced, exalted (*OED*, 11,
6).   23–4 Cubeta (p. 166) suspects burlesque; but these lines are also addressed,
without apparent mockery, to Lady Purbeck, *Gyp. Met.* 540–1. Trimpi (pp. 215–16)
compares *N.I.* III. ii. 96–8, 105, and Ficino, *Commentary on Plato's 'Symposium'*, II.
viii.

2. ii.   12 *cloud*: perhaps = clout, cloth.

So that there I stood a stone,
Mocked of all, and called of one
(Which with grief and wrath I heard)
Cupid's statue with a beard,   30
Or else one that played his ape
In a Hercules's shape.

2. iii
What He Suffered

After many scorns like these,
Which the prouder beauties please,
She content was to restore
Eyes and limbs, to hurt me more.
And would, on conditions, be   5
Reconciled to Love and me.
First, that I must kneeling yield
Both the bow and shaft I held
Unto her; which Love might take
At her hand, with oath, to make   10
Me the scope of his next draught,
Aimed with that self-same shaft.
He no sooner heard the law,
But the arrow home did draw,
And, to gain her by his art,   15
Left it sticking in my heart;
Which when she beheld to bleed,
She repented of the deed,
And would fain have changed the fate,
But the pity comes too late.   20
Loser-like, now, all my wreak
Is that I have leave to speak,
And in either prose, or song,
To revenge me with my tongue;
Which how dexterously I do,   25
Hear, and make example too.

32 *Hercules's shape*: cf. Odysseus's sight of Hercules in Hades, *Odyssey*, ii. 606–8: 'like dark night, with his bow bare and with arrow on the string, he glared about him terribly, like one ready to shoot'. But the reference may be to Hercules during his period of infatuation with Omphale, Queen of Lydia.

2. iii. 11 *scope*: target, aim.   11 *draught*: shot (literally, drawing of the bowstring). 21 *Loser-like*: proverbial: 'Give losers leave to speak', Tilley, L458; cf. *S.W.* II. iv. 39–40.   21 *wreak*: revenge.   26 *make example*: take example; learn from this.

2. iv

Her Triumph

See the chariot at hand here of Love,
    Wherein my lady rideth!
Each that draws is a swan or a dove,
    And well the car Love guideth.
As she goes, all hearts do duty   5
        Unto her beauty;
And enamoured, do wish, so they might
        But enjoy such a sight,
    That they still were to run by her side,
Through swords, through seas, whither she would ride.   10

Do but look on her eyes, they do light
    All that Love's world compriseth!
Do but look on her hair, it is bright
    As Love's star when it riseth!
Do but mark, her forehead's smoother   15
        Than words that soothe her!
And from her arched brows, such a grace
        Sheds itself through the face,
    As alone there triumphs to the life
All the gain, all the good, of the elements' strife.   20

---

2. iv. *Triumph*: i.e. victory procession, as well as victory; ironically following on the promise of 'revenge', *Und*. 2. iii. 24. 3 *a swan or a dove*: birds sacred to Venus; in *Haddington*, 44, they draw her chariot, in which the *Charites* or graces are seated. Cf. Ovid, *Amores*, I. ii. 23 ff. 10 Trimpi (p. 219) compares Spenser, 'An Hymn in Honour of Love', 228, where lovers follow Cupid 'Through seas, through flames, through thousand swords and spears', and sees a source in Plato, *Symposium*, 178–80; cf. Castiglione, *The Book of the Courtier*, trs. Sir T. Hoby (London, 1928), p. 234. Marie Borroff's suggested reading, *swards*, is unconvincing in view of these parallels (*ELN*, viii (1971), 257–9). 10 *whither*: wherever. 11–30 Wittipol woos Mrs Fitzdottrell with these lines, *D. is A*. II. vi. 94– 113. For Suckling's parody, 'A Song to a Lute', see *Non-Dramatic Works*, ed. T. Clayton (Oxford, 1971), pp. 29–30. H & S (viii. 135–6) also print an alternative version of ll. 21–30, and two additional stanzas, from a seventeenth-century manuscript. 14 *Love's star*: Venus. In *Haddington*, 46, Venus sits in her chariot 'crowned with her star'. 17–18 Deplored by Swinburne, *A Study of Ben Jonson*, p. 104. Trimpi (pp. 219–20) and Peterson (p. 227) refer to the neoplatonic association of the light which emanates from God and from a lady's eyes.

Have you seen but a bright lily grow,
   Before rude hands have touched it?
Have you marked but the fall o' the snow,
   Before the soil hath smutched it?
Have you felt the wool o' the beaver?   25
      Or swan's down ever?
Or have smelled o' the bud o' the briar?
      Or the nard i' the fire?
Or have tasted the bag o' the bee?
O so white! O so soft! O so sweet is she!   30

2. *v*
His Discourse with Cupid

Noblest Charis, you that are
Both my fortune and my star!
And do govern more my blood
Than the various moon the flood!
Hear what late discourse of you   5
Love and I have had, and true.
'Mongst my muses finding me,
Where he chanced your name to see
Set, and to this softer strain:
Sure, said he, if I have brain,   10
This, here sung, can be no other
By description but my mother!
So hath Homer praised her hair,
So Anacreon drawn the air
Of her face, and made to rise,   15
Just above her sparkling eyes,
Both her brows, bent like my bow.
By her looks I do her know,
Which you call my shafts. And see!
Such my mother's blushes be,   20

21–4 Cf. Martial, *Epig.* V. xxxvii. 4–6*, on Erotion; and Ovid, *Metam.* xiii. 789 ff.   24
smutched: blackened, smudged.

2. v. 10 ff. Cf. Cupid's confusion in *Songs*, 4. 10–11, and in Marlowe's *Hero and
Leander*, i. 39–40.   13 *Homer*: see *Iliad*, xvii. 51.   14 *Anacreon*: see the advice to a
painter concerning a lady, *Anacreontea*, xvi.   17 *like my bow*: cf. *Chall. Tilt*, 45 ff.,
Sidney, *Astrophil and Stella*, xvii. 10.

As the bath your verse discloses
In her cheeks, of milk and roses;
Such as oft I wanton in!
And, above her even chin,
Have you placed the bank of kisses,   25
Where, you say, men gather blisses,
Ripened with a breath more sweet,
Than when flowers and west winds meet.
Nay, her white and polished neck,
With the lace that doth it deck,   30
Is my mother's! Hearts of slain
Lovers, made into a chain!
And between each rising breast
Lies the valley called my nest,
Where I sit and proyne my wings   35
After flight, and put new stings
To my shafts! Her very name
With my mother's is the same.
I confess all, I replied,
And the glass hangs by her side,   40
And the girdle 'bout her waist:
All is Venus, save unchaste.
But alas, thou seest the least
Of her good, who is the best
Of her sex; but could'st thou, Love,   45
Call to mind the forms that strove
For the apple, and those three
Make in one, the same were she.

21–6 Cf. *Und.* 19. 7–10: Cubeta (pp. 171–2) thinks Jonson is allowing Cupid to make fun of this poem; but cf. the version of these lines addressed to Lady Purbeck in *Gyp. Met.* 534 ff. See also *C.R.* V. iv. 439–40, and Wittipol to Mrs. Fitzdottrell, *D. is A.* II. vi. 82–7.   31–2 The chain of lovers' hearts is usually associated with Cupid himself: see E. Panofsky, *Studies in Iconology* (New York, 1972), pp. 115 ff.   35 *proyne*: preen.   37–8 A double confusion of identities: for one of Vulcan's wives, in Lemnos, is also called Charis, while the other, in heaven, is Venus: *Iliad*, xviii. 382, *Odyssey*, viii. 266–366.   40 *the glass hangs by her side*: a contemporary fashion; but a large looking-glass was also associated with Venus, as an emblem of pride, in Renaissance art: see Peterson, p. 237.   41 *girdle*: which aroused desire; Venus lent it to Hera for her seduction of Zeus, *Iliad*, xiv. 214. Cf. *Chall. Tilt*, 50–2, and Jonson's note to *Hym.* 407.   47 *the apple*: thrown down by the goddess of discord, it led to the dispute between Venus, Juno, and Minerva, concerning their relative beauty; Paris judged in favour of Venus.

For this beauty yet doth hide
Something more than thou hast spied.   50
Outward grace weak love beguiles;
She is Venus, when she smiles,
But she's Juno, when she walks,
And Minerva, when she talks.

2. *vi*
Claiming a Second Kiss by Desert

Charis, guess, and do not miss,
Since I drew a morning kiss
From your lips, and sucked an air
Thence as sweet as you are fair,
What my muse and I have done: 5
Whether we have lost, or won,
If by us the odds were laid
That the bride, allowed a maid,
Looked not half so fresh and fair,
With the advantage of her hair   10
And her jewels, to the view
Of the assembly, as did you!
Or, that did you sit, or walk,
You were more the eye and talk
Of the court, today, than all   15
Else that glistered in Whitehall;
So as those that had your sight
Wished the bride were changed tonight,
And did think such rites were due
To no other grace but you!   20
Or, if you did move tonight
In the dances, with what spite
Of your peers you were beheld,
That at every motion swelled

51 Cf. *Und.* 2. i. 7 ff., above. 52 ff. From Angerianus, 'Erotopaegnion'; for Juno's stately walk, cf. *The Tempest*, IV. i. 102: from Virgil, *Aeneid*, i. 46 (and see T. W. Baldwin, *Shakspere's Small Latine* (Urbana, 1944), ii. 481). 'But' (l. 53) appears to come too early for the contrast proposed.

2. vi. 10 *her hair*: a bride traditionally wore her hair loose: cf. *Und.* 75. 45, *Hym.* 57.

So to see a lady tread,   25
As might all the graces lead,
And was worthy, being so seen,
To be envied of the queen.
Or if you would yet have stayed,
Whether any would upbraid   30
To himself his loss of time,
Or have charged his sight of crime
To have left all sight for you.
Guess of these which is the true:
And if such a verse as this   35
May not claim another kiss.

2. vii
Begging Another, on Colour of Mending the Former

For Love's sake, kiss me once again,
I long, and should not beg in vain,
Here's none to spy, or see:
    Why do you doubt, or stay?
I'll taste as lightly as the bee,   5
That doth but touch his flower, and flies away.
Once more, and (faith) I will be gone;
Can he that loves ask less than one?
    Nay, you may err in this,
      And all your bounty wrong: 10
This could be called but half a kiss.
What we are but once to do, we should do long.
I will but mend the last, and tell
Where, how, it would have relished well;
    Join lip to lip, and try;   15
      Each suck the other's breath.
And whilst our tongues perplexed lie,
Let who will think us dead, or wish our death.

26 *all the graces lead*: as Venus does in the dance.   28 *the queen*: Queen Anne of Denmark was famed for her dancing, and had performed in Jonson's masques. She died on 2 March 1619.

2. vii. Amongst 'the most commonplace of his repetition' when Jonson visited Drummond in 1618–19; see *Conv. Dr.* 89–101.   17 *perplexed*: intertwined.

2. viii
Urging Her of a Promise

Charis one day in discourse
Had of Love and of his force,
Lightly promised she would tell
What a man she could love well;
And that promise set on fire   5
All that heard her, with desire.
With the rest I long expected
When the work would be effected;
But we find that cold delay
And excuse spun every day,   10
As, until she tell her one,
We all fear she loveth none.
Therefore, Charis, you must do it,
For I will so urge you to it,
You shall neither eat, nor sleep,   15
No, nor forth your window peep
With your emissary eye
To fetch in the forms go by,
And pronounce which band or lace
Better fits him than his face;   20
Nay, I will not let you sit
'Fore your idol glass a whit,
To say over every purl
There or to reform a curl;
Or with secretary Cis   25
To consult, if fucus this
Be as good as was the last:
All your sweet of life is past,
Make accompt, unless you can
(And that quickly) speak your man.   30

2. viii. 17 *emissary eye*: spying eye; from Plautus, *Aulularia*, 41. A new word in
English; see *S. of N.* I. ii. 47 ff.   18 i.e. assess the passing men.   19 *band*: collar or
ruff.   23 *say over*: try on.   23 *purl*: lace, frill.   25 *secretary*: confidante. The original
name of Prue, the chambermaid in *N.I.*, was Secretary Cis.   26 *fucus*: cosmetic.

2. ix

Her Man Described by Her Own Dictamen

Of your trouble, Ben, to ease me,
I will tell what man would please me.
I would have him, if I could,
Noble, or of greater blood;
Titles, I confess, do take me,   5
And a woman God did make me;
French to boot, at least in fashion,
And his manners of that nation.

     Young I'd have him too, and fair,
Yet a man; with crisped hair   10
Cast in thousand snares and rings
For Love's fingers and his wings:
Chestnut colour, or more slack
Gold, upon a ground of black.
Venus' and Minerva's eyes,   15
For he must look wanton-wise.

     Eyebrows bent like Cupid's bow,
Front an ample field of snow;
Even nose, and cheek (withal)
Smooth as is the billiard ball;   20
Chin as woolly as the peach,
And his lip should kissing teach,
Till he cherished too much beard,
And make Love or me afeared.

     He would have a hand as soft   25
As the down, and show it oft;
Skin as smooth as any rush,
And so thin to see a blush
Rising through it ere it came;
All his blood should be a flame   30
Quickly fired as in beginners
In Love's school, and yet no sinners.

---

2. ix. *Dictamen*: pronouncement. Peterson (pp. 362–3) detects a play on 'dictamen'/
'dictamnum'/'dittany', a herb eaten—according to Pliny—by wounded deer to make
their arrows drop out; Renaissance commentators contrasted the plight of lovers irre-
trievably wounded by Cupid. Hence 'cure for love' is possibly an ironical secondary
meaning.   7–8 See Jonson's attack on French fashions in *Epig.* 88.   9 ff. Cf. the
advice to a painter on painting a boy, *Anacreontea*, xvii.   10–11 Cf. Wittipol to Mrs.
Fitzdottrell, *D. is A.* II. vi. 80–2.   13 *slack*: pale, dull.   20 *billiard ball*: *D. is A.* II. vi.
85.   23 *cherished*: cultivated.   26 *show it oft*: cf. *S.W.* I. i. 109.

'Twere too long to speak of all;
What we harmony do call
In a body should be there.   35
Well he should his clothes, too, wear,
Yet no tailor help to make him;
Dressed, you still for man should take him,
And not think he'd ate a stake
Or were set up in a brake.   40
  Valiant he should be as fire,
Showing danger more than ire.
Bounteous as the clouds to earth,
And as honest as his birth.
All his actions to be such,   45
As to do no thing too much.
Nor o'erpraise, nor yet condemn,
Nor out-value, nor contemn;
Nor do wrongs, nor wrongs receive;
Nor tie knots, nor knots unweave;   50
And from baseness to be free,
As he durst love truth and me.
  Such a man, with every part,
I could give my very heart;
But of one, if short he came,   55
I can rest me where I am.

2. x
Another Lady's Exception, Present at the Hearing

For his mind I do not care,
That's a toy that I could spare;
Let his title be but great,
His clothes rich, and band sit neat,
Himself young, and face be good,   5
All I wish is understood.
What you please you parts may call,
'Tis one good part I'd lie withal.

37–8 Playing on the proverb, 'the tailor makes the man', Tilley, T17; cf. *S. of N.* I. ii.
111, *T. of T.* I. vii. 25–7.   39 Proverbial: Tilley, S810.   40 *brake*: a framework to
hold something steady, e.g. a horse's foot while being shod; cf. *S.W.* IV. vi. 28.   42
danger: power.   50 Cf. Mosca's praise of the lawyer Voltore's ability to 'make knots,
and undo them', *Volp.* I. iii. 57.

3

The Musical Strife; in a Pastoral Dialogue

She

Come, with our voices let us war,
    And challenge all the spheres,
Till each of us be made a star
    And all the world turn ears.

He

At such a call what beast or fowl   5
    Of reason empty is?
What tree or stone doth want a soul?
    What man but must lose his?

She

Mix then your notes, that we may prove
    To stay the running floods,   10
To make the mountain quarries move,
    And call the walking woods.

He

What need of me? Do you but sing,
    Sleep and the grave will wake;
No tunes are sweet, nor words have sting,   15
    But what those lips do make.

She

They say the angels mark each deed
    And exercise below,
And out of inward pleasure feed
    On what they viewing know.   20

He

O sing not you then, lest the best
    Of angels should be driven
To fall again, at such a feast
    Mistaking earth for heaven.

3. A version of this poem existed before 1618; Jonson quoted it to Drummond (*Conv. Dr.* 89–91). The poem originally had six stanzas, and was a musical duet between two women; for details, see H & S, viii. 11. The present version is examined by John Hollander in *The Untuning of the Sky* (Princeton, 1961), pp. 338–40.   5 ff. Alluding to the legendary effects of Orpheus's music; see *Epig.* 130.   24 Echoed by Dryden, *A Song for St. Cecilia's Day*, 53–4; cf. *Alexander's Feast*, 161–70. See also *Und.* 67. 20–1.

She

Nay, rather both our souls be strained 25
   To meet their high desire;
So they in state of grace retained
   May wish us of their choir.

4

A Song

Oh, do not wanton with those eyes
   Lest I be sick with seeing;
Nor cast them down, but let them rise,
   Lest shame destroy their being.

Oh, be not angry with those fires, 5
   For then their threats will kill me;
Nor look too kind on my desires,
   For then my hopes will spill me.

Oh, do not steep them in thy tears,
   For so will sorrow slay me; 10
Nor spread them as distract with fears,
   Mine own enough betray me.

5

*In the Person of Womankind
A Song Apologetic*

Men, if you love us, play no more
   The fools or tyrants with your friends,
To make us still sing o'er and o'er
   Our own false praises, for your ends;
     We have both wits and fancies too, 5
     And if we must, let's sing of you.

---

4. 8 *spill*: destroy, kill.   11 *spread*: open wide.

Nor do we doubt but that we can,
    If we would search with care and pain,
Find some one good in some one man;
So going thorough all your strain   10
        We shall, at last, of parcels make
        One good enough for a song's sake.

And as a cunning painter takes
    In any curious piece you see
More pleasure while the thing he makes   15
    Than when 'tis made, why so will we.
        And having pleased our art, we'll try
        To make a new, and hang that by.

6

Another: in Defence of Their Inconstancy
A Song

Hang up those dull and envious fools
    That talk abroad of woman's change,
We were not bred to sit on stools,
    Our proper virtue is to range;
        Take that away, you take our lives,   5
        We are no women then, but wives.

Such as in valour would excel
    Do change, though man, and often fight,
Which we in love must do as well,
    If ever we will love aright.   10
        The frequent varying of the deed
        Is that which doth perfection breed.

Nor is't inconstancy to change
    For what is better, or to make,
By searching, what before was strange   15
    Familiar, for the use's sake;
        The good from bad is not descried
        But as 'tis often vexed and tried.

5. 13–16 Seneca, *Epist.* ix. 7; cf. *Disc.* 1719–21, and *Dubia*, 1. 57–8.

6. 1 *Hang up*: ignore.   4 *proper virtue*: natural function; but the words also hint playfully at a moral duty.   18 *vexed*: severely tested.

And this profession of a store
   In love, doth not alone help forth   20
Our pleasure, but preserves us more
   From being forsaken than doth worth;
      For were the worthiest woman cursed
      To love one man, he'd leave her first.

7

A Nymph's Passion

I love and he loves me again,
   Yet dare I not tell who;
For if the nymphs should know my swain,
   I fear they'd love him too;
      Yet if it be not known   5
      The pleasure is as good as none,
For that's a narrow joy is but our own.

I'll tell, that if they be not glad,
   They yet may envy me;
But then if I grow jealous mad   10
   And of them pitied be,
      It were a plague 'bove scorn;
      And yet it cannot be forborne,
Unless my heart would as my thought be torn.

He is, if they can find him, fair,   15
   And fresh and fragrant too
As summer's sky or purged air,
   And looks as lilies do
      That were this morning blown;
      Yet, yet I doubt he is not known,   20
And fear much more that more of him be shown.

7. For Coleridge's reworking of this poem, see his *Complete Poetical Works*, ed. E. H.
Coleridge (Oxford, 1912), ii. 1118–19.

But he hath eyes so round and bright
    As make away my doubt,
Where Love may all his torches light
    Though hate had put them out;   25
        But then, to increase my fears,
        What nymph soe'er his voice but hears
Will be my rival, though she have but ears.

I'll tell no more, and yet I love,
    And he loves me; yet no   30
One unbecoming thought doth move
    From either heart, I know;
        But so exempt from blame,
        As it would be to each a fame,
If love, or fear, would let me tell his name.   35

8

The Hour-Glass

Do but consider this small dust
    Here running in the glass,
        By atoms moved:
Could you believe that this
    The body ever was   5
        Of one that loved?
And in his mistress' flame, playing like a fly,
    Turned to cinders by her eye?
Yes; and in death, as life, unblessed,
    To have 't expressed,   10
Even ashes of lovers find no rest.

24 *torches*: Tibullus, III. viii; cf. *Und.* 19. 2, *C.R.* V. iv. 441, *Chall. Tilt*, 44-5.

8. Borrowed from the Italian poet Girolamo Amaltei's *Horologium Pulverum*. The conceit is varied by Herrick, 'The Hour-glass', Herrick, p. 44; and Jonson's poem is reworked by Coleridge: *Complete Poetical Works*, ed. E. H. Coleridge, ii. 1119-20. Ferrabosco's musical setting of the poem is discussed by E. Doughtie, *RQ*, xxii (1969), 148-50.   1 *small*: fine.

9

My Picture Left in Scotland

I now think Love is rather deaf than blind,
　　For else it could not be
　　　　That she
Whom I adore so much should so slight me,
　　And cast my love behind;   5
I'm sure my language to her was as sweet,
　　　　And every close did meet
　　　　In sentence of as subtle feet,
　　　　　　As hath the youngest he
　　That sits in shadow of Apollo's tree.   10

　　Oh, but my conscious fears
　　　　That fly my thoughts between,
　　　　Tell me that she hath seen
　　　　My hundred of grey hairs,
　　　　Told seven-and-forty years,   15
　　Read so much waste, as she cannot embrace
　　　　My mountain belly, and my rocky face;
And all these through her eyes have stopped her ears.

10

Against Jealousy

Wretched and foolish jealousy
How cam'st thou thus to enter me?
　　I ne'er was of thy kind,
　　Nor have I yet the narrow mind
　　　　To vent that poor desire   5
That others should not warm them at my fire;
　　I wish the sun should shine
On all men's fruit and flowers, as well as mine.

9. This poem, 'which is (as he said) a picture of himself', was sent to William
Drummond in a slightly different form: *Conv. Dr.* 660–78. The present version dates
from 1619.　8 *sentence*: thought.　11 *conscious*: guilty.　16 *waste*: a pun.　17 *mountain belly*: echoed ironically by Dryden, of Shadwell: *MacFlecknoe*, 193–4.

But under the disguise of love
Thou say'st thou only cam'st to prove　　　10
　　What my affections were.
　　Think'st thou that love is helped by fear?
　　Go, get thee quickly forth,
Love's sickness and his noted want of worth,
　　Seek doubting men to please;　　　15
I ne'er will owe my health to a disease.

I I

The Dream

Or scorn, or pity on me take,
I must the true relation make:
　　I am undone tonight;
　　Love in a subtle dream disguised
　　Hath both my heart and me surprised,　　　5
Whom never yet he durst attempt awake;
Nor will he tell me for whose sake
　　He did me the delight,
　　　　Or spite,
　　But leaves me to inquire,　　　10
　　In all my wild desire
　　　Of sleep again, who was his aid;
　　　And sleep so guilty and afraid
As, since, he dares not come within my sight.

I2

An Epitaph on Master Vincent Corbett

I have my piety too, which could
It vent itself but as it would,
　　Would say as much as both have done
　　Before me here, the friend and son;

---

12. *Vincent Corbett*: father of Richard Corbett, Bishop of Oxford and Norwich, and Caroline poet, whose elegy to his father is referred to in ll. 3–4 below; see *Poems*, ed. J. A. W. Bennett and H. R. Trevor-Roper (Oxford, 1955), pp. 67–9. (The identity of the

For I both lost a friend and father,   5
Of him whose bones this grave doth gather:
   Dear Vincent Corbett, who so long
   Had wrestled with diseases strong
That though they did possess each limb,
Yet he broke them, ere they could him,   10
   With the just canon of his life;
   A life that knew nor noise nor strife,
But was, by sweetening so his will,
All order and disposure still.
   His mind as pure, and neatly kept,   15
   As were his nurseries, and swept
So of uncleanness or offence,
That never came ill odour thence;
   And add his actions unto these,
   They were as specious as his trees.   20
'Tis true, he could not reprehend;
His very manners taught to amend,
   They were so even, grave, and holy;
   No stubbornness so stiff, nor folly
To licence ever was so light   25
As twice to trespass in his sight;
   His looks would so correct it, when
   It chid the vice, yet not the men.
Much from him I profess I won,
And more and more I should have done,   30
   But that I understood him scant.
   Now I conceive him by my want,
And pray, who shall my sorrows read,
That they for me their tears will shed;
   For truly, since he left to be,   35
   I feel I'm rather dead than he!

Reader, whose life and name did e'er become
   An epitaph, deserved a tomb;
Nor wants it here, through penury or sloth;
   Who makes the one, so it be first, makes both.   40

'friend' of l. 4 is not known.) The father, a nurseryman of Whitton in the parish of
Twickenham, died on 29 April 1619. 6 *Of*: in. 11 *canon*: a pun. 20 *specious*:
lovely. 28 Martial, *Epig.* X. xxxiii. 10. 37 ff. Richard Corbett also concludes his
elegy with a four-line address to the reader. 37 *whose*: he whose. 40 *first*: perhaps
'first-rate' (Hunter).

13

An Epistle to Sir Edward Sackville, now Earl of Dorset

If, Sackville, all that have the power to do
Great and good turns, as well could time them too,
And knew their how and where, we should have then
Less list of proud, hard, or ungrateful men.
For benefits are owed with the same mind   5
As they are done, and such returns they find.
You then whose will not only, but desire
To succour my necessities took fire,
Not at my prayers, but your sense, which laid
The way to meet what others would upbraid,   10
And in the act did so my blush prevent,
As I did feel it done as soon as meant;
You cannot doubt but I, who freely know
This good from you, as freely will it owe.
And though my fortune humble me to take   15
The smallest courtesies with thanks, I make
Yet choice from whom I take them, and would shame
To have such do me good I durst not name.
They are the noblest benefits, and sink
Deepest in man, of which, when he doth think,   20
The memory delights him more from whom
Than what he hath received. Gifts stink from some,
They are so long a-coming, and so hard;
Where any deed is forced, the grace is marred.
    Can I owe thanks for courtesies received   25
Against his will that does 'em; that hath weaved
Excuses or delays; or done 'em scant,
That they have more oppressed me than my want?

13. Sir Edward Sackville: 1591–1652; succeeded his brother Richard as Earl of Dorset,
1624; Lord Chamberlain, 1628; Commissioner for Planting Virginia, 1631, 1634; Lord
Privy Seal, 1644. See *Und.* 26 and n. Jonson's extensive borrowings in this poem from
Seneca's *De Beneficiis* are traced by W. D. Briggs, *MP*, x (1913), 573–85. 1–4
Seneca, *De Beneficiis*, I. i. 1, 1–2. 2 *Great and good*: see l. 124 below, and *Epig.* Ded.
15–16 n. 5–6 Seneca, *De Ben.* I. i. 3, 8; cf. Tilley, G97, 'A gift is valued by the mind
of the giver'. 5 *owed*: acknowledged. 9–12 *De Ben.* II. i. 3. 11 *prevent*: come
before. 16–17 *De Ben.* II. xviii. 2. 18–22 *De Ben.* I. xv. 4. 24 *De Ben.* I. i. 7.
25–8, 30–2 *De Ben.* I. i. 6.

Or if he did it not to succour me
But by mere chance, for interest, or to free   30
Himself of farther trouble, or the weight
Of pressure, like one taken in a strait?
All this corrupts the thanks; less hath he won
That puts it in his debt-book ere it be done;
Or that doth sound a trumpet, and doth call   35
His grooms to witness; or else lets it fall
In that proud manner, as a good so gained
Must make me sad for what I have obtained.

No! Gifts and thanks should have one cheerful face,
So each that's done and ta'en becomes a brace.   40
He neither gives, nor does, that doth delay
A benefit, or that doth throw it away;
No more than he doth thank that will receive
Nought but in corners, and is loath to leave
Least air or print, but flies it: such men would   45
Run from the conscience of it, if they could.

As I have seen some infants of the sword,
Well known and practised borrowers on their word,
Give thanks by stealth, and whispering in the ear,
For what they straight would to the world forswear;   50
And speaking worst of those from whom they went
But then, fist-filled, to put me off the scent:
Now, damn me, sir, if you should not command
My sword ('tis but a poor sword, understand)
As far as any poor sword in the land.   55
Then turning unto him is next at hand,
Damns whom he damned to, as the veriest gull
Has feathers, and will serve a man to pull.

Are they not worthy to be answered so,
That to such natures let their full hands flow,   60
And seek not wants to succour, but enquire,
Like money-brokers, after names, and hire
Their bounties forth to him that last was made,
Or stands to be, in commission of the blade?
Still, still the hunters of false fame apply   65
Their thoughts and means to making loud the cry;

33 *corrupts the thanks*: Seneca's phrase, *gratiam omnem corrupimus: De Ben*. I. i. 4.   34
De Ben. I. ii. 3.   37–8 *De Ben*. II. xiii. 1.   40 Cf. Tilley, G125: 'He that gives quickly
gives twice'.   45 *air or print*: i.e. spoken or written thanks.   47–9 *De Ben*. II. xxiii. 1–
2.   57 *De Ben*. II. xxiv. 1.   64 *commission of the blade*: knighthood. Cf. *Epig*. 19. 1 n.

But one is bitten by the dog he fed,
And, hurt, seeks cure: the surgeon bids take bread
And sponge-like with it dry up the blood quite,
Then give it to the hound that did him bite.   70
Pardon, says he, that were a way to see
All the town curs take each their snatch at me.
Oh, is it so? Knows he so much? And will
Feed those at whom the table points at still?
I not deny it, but to help the need   75
Of any is a great and generous deed:
Yea, of the ungrateful; and he forth must tell
Many a pound and piece, will place one well.
But these men ever want: their very trade
Is borrowing; that but stopped, they do invade   80
All as their prize, turn pirates here at land,
Have their Bermudas, and their straits i' the Strand;
Man out their boats to the Temple; and not shift
Now, but command, make tribute what was gift;
And it is paid 'em with a trembling zeal,   85
And superstition I dare scarce reveal
If it were clear; but being so in cloud
Carried and wrapped, I only am allowed
My wonder why the taking a clown's purse,
Or robbing the poor market-folks should nurse   90
Such a religious horror in the breasts
Of our town gallantry! Or why there rests
Such worship due to kicking of a punk,
Or swaggering with the watch, or drawer, drunk,
Or feats of darkness acted in mid-sun,   95
And told of with more licence than they were done!
Sure there is mystery in it I not know,
That men such reverence to such actions show!
And almost deify the authors: make
Loud sacrifice of drink for their health's sake,   100
Rere-suppers in their names, and spend whole nights
Unto their praise in certain swearing rites!

67–72 From a fable of Phaedrus, *Fab*. II. iii.   77–8 Based on a couplet attacked by Seneca, *De Ben*. I. ii. 1: 'To shower bounties on the mob should you delight, / Full many must you lose, for one you place aright'.   82 *Bermudas . . . straits*: a notorious maze of lanes and alleyways north of the Strand, near Covent Garden.   101 *Rere-suppers*: lavish late-night suppers.

Cannot a man be reckoned in the state
Of valour, but at this idolatrous rate?
I thought that fortitude had been a mean   105
'Twixt fear and rashness; not a lust obscene,
Or appetite of offending, but a skill
Or science of discerning good and ill.
And you, sir, know it well, to whom I write,
That with these mixtures we put out her light.   110
Her ends are honesty and public good,
And where they want, she is not understood.
No more are these of us, let them then go;
I have the list of mine own faults to know,
Look to, and cure. He's not a man hath none,   115
But like to be, that every day mends one
And feels it; else he tarries by the beast.
Can I discern how shadows are decreased
Or grown, by height or lowness of the sun,
And can I less of substance? When I run,   120
Ride, sail, am coached, know I how far I have gone,
And my mind's motion not? Or have I none?
No! he must feel and know that will advance.
Men have been great, but never good, by chance
Or on the sudden. It were strange that he   125
Who was this morning such a one should be
Sidney ere night! Or that did go to bed
Coryate should rise the most sufficient head
Of Christendom! And neither of these know,
Were the rack offered them, how they came so;   130
'Tis by degrees that men arrive at glad
Profit in aught; each day some little add,
In time 'twill be a heap; this is not true
Alone in money, but in manners too.
Yet we must more than move still, or go on,   135
We must accomplish: 'tis the last keystone

105–8 Cf. Lovel's speech on true valour, *N.I.* IV. iv. 40 ff.; from Seneca, *Epist.* lxxxv.
28.   114–15 Seneca, *De Vita Beata*, xviii. 3: 'It is enough for me if every day I reduce
the number of my vices, and blame my mistakes'.   118–23 Cf. Plutarch, *How a Man
May Become Aware of His Progress in Virtue*, iii.   124–30 Plutarch, ibid., i.   128
Coryate: Thomas Coryate, the traveller: see *U.V.* 10, n.   136 ff. Seneca, *Epist.* cxi. 3,
on the true philosopher's not needing to walk on tip-toe.

That makes the arch. The rest that there were put
Are nothing till that comes to bind and shut.
Then stands it a triumphal mark! Then men
Observe the strength, the height, the why, and when   140
It was erected; and still walking under
Meet some new matter to look up and wonder!
Such notes are virtuous men: they live as fast
As they are high; are rooted, and will last.
They need no stilts, nor rise upon their toes.   145
As if they would belie their stature; those
Are dwarfs of honour, and have neither weight
Nor fashion; if they chance aspire to height,
'Tis like light canes, that first rise big and brave,
Shoot forth in smooth and comely spaces, have   150
But few and fair divisions; but being got
Aloft, grow less and straitened, full of knot,
And last, go out in nothing; you that see
Their difference cannot choose which you will be.
You know (without my flattering you) too much   155
For me to be your indice. Keep you such,
That I may love your person (as I do)
Without your gift, though I can rate that too,
By thanking thus the courtesy to life,
Which you will bury; but therein the strife   160
May grow so great to be example, when
(As their true rule or lesson) either men,
Donors or donees, to their practice shall
Find you to reckon nothing, me owe all.

14

An Epistle to Master John Selden

I know to whom I write. Here, I am sure,
Though I am short, I cannot be obscure;
Less shall I for the art or dressing care,
Truth and the graces best when naked are.

156 *indice*: indicator.

14. *John Selden*: 1584–1654, the jurist: 'the law-book of the judges of England, the bravest man in all languages' (*Conv. Dr.* 604–5). This poem is prefixed to Selden's *Titles of Honour*, 1614; in his preface, Selden testifies to Jonson's learning and friendship.   1–2 Horace warns that brevity can lead to obscurity, *Ars Poetica*, 25–6; Jonson

Your book, my Selden, I have read, and much   5
Was trusted, that you thought my judgement such
To ask it; though in most of works it be
A penance, where a man may not be free,
Rather than office, when it doth or may
Chance that the friend's affection proves allay   10
Unto the censure. Yours all need doth fly
Of this so vicious humanity.
Than which there is not unto study a more
Pernicious enemy; we see before
A many of books, even good judgements wound   15
Themselves through favouring what is there not found.
But I on yours far otherwise shall do,
Not fly the crime, but the suspicion too;
Though I confess (as every muse hath erred,
And mine not least) I have too oft preferred   20
Men past their terms, and praised some names too much;
But 'twas with purpose to have made them such.
Since, being deceived, I turn a sharper eye
Upon myself, and ask to whom, and why,
And what I write? And vex it many days   25
Before men get a verse, much less a praise;
So that my reader is assured I now
Mean what I speak, and still will keep that vow.
Stand forth my object, then, you that have been
Ever at home, yet have all countries seen;   30
And like a compass keeping one foot still
Upon your centre, do your circle fill
Of general knowledge; watched men, manners too,
Heard what times past have said, seen what ours do.
Which grace shall I make love to first: your skill,   35
Or faith in things? Or is 't your wealth and will
To instruct and teach, or your unwearied pain
Of gathering, bounty in pouring out again?

suggests that his own brevity cannot be misunderstood by such a friend as
Selden.   7 ff. i.e. with most works given one for an opinion, one is not performing
a good office so much as undergoing a penance; a man may not be free when he
must temper his criticism out of friendship.   19–20 For Jonson's awareness of par-
ticular instances of over-praise, see *Epig.* 65 and 132. 6 n.   21 *terms*: limits, capacities.
24 Horace, *Epist.* I. xviii. 68, 76.   25 *vex*: test, sift.   30 *at home*: *For.* 4. 68, *Und.*
78. 8.   31–2 Cf. *Und.* 47. 60, and Donne, 'A Valediction: Forbidding Mourning',
25–35.   33 *men, manners*: *Und.* 50. 22, 77. 18.

What fables have you vexed, what truth redeemed,
Antiquities searched, opinions disesteemed,   40
Impostures branded, and authorities urged!
What blots and errors have you watched and purged
Records and authors of! How rectified
Times, manners, customs! Innovations spied!
Sought out the fountains, sources, creeks, paths, ways,   45
And noted the beginnings and decays!
Where is that nominal mark, or real rite,
Form, art, or ensign that hath 'scaped your sight?
How are traditions there examined, how
Conjectures retrieved! And a story now   50
And then of times, besides the bare conduct
Of what it tells us, weaved in to instruct!
I wondered at the richness, but am lost
To see the workmanship so exceed the cost;
To mark the excellent seasoning of your style,   55
And manly elocution, not one while
With horror rough, then rioting with wit:
But to the subject still the colours fit
In sharpness of all search, wisdom of choice,
Newness of sense, antiquity of voice!   60
    I yield, I yield, the matter of your praise
Flows in upon me, and I cannot raise
A bank against it. Nothing but the round
Large clasp of nature such a wit can bound.
Monarch in letters! 'mongst thy titles shown   65
Of others' honours, thus enjoy thine own.
I first salute thee so, and gratulate,
With that thy style, thy keeping of thy state,
In offering this thy work to no great name
That would, perhaps, have praised and thanked the same, 70

---

39 *vexed*: see note to l. 25, above.   47 *nominal mark, or real rite*: title of address or
royal ceremony: Selden's own terms in *Titles of Honour*.   54 Ovid, *Metam.* ii. 5: 'the
workmanship was more beautiful than the material'.   56 *manly*: *Disc.* 797, *Und.* 46.
13.   58 *colours*: of rhetoric.   59–60 'a masterly summary of the intellectual qualities
Jonson most admired. The newness of sense is the Baconian insistence upon the
observation of particulars and all that this implies in the investigation of the real world;
antiquity of voice is the use of recorded history which gives chronological and ethical
coherence to the fragments of individual experience' (Trimpi, p. 146).

But nought beyond. He thou hast given it to,
Thy learned chamber-fellow, knows to do
It true respects. He will not only love,
Embrace, and cherish, but he can approve
And estimate thy pains, as having wrought   75
In the same mines of knowledge, and thence brought
Humanity enough to be a friend,
And strength to be a champion and defend
Thy gift 'gainst envy. O how I do count
Among my comings-in, and see it mount,   80
The gain of your two friendships! Hayward and
Selden: two names that so much understand;
On whom I could take up, and ne'er abuse
The credit, what would furnish a tenth muse!
But here's no time, nor place, my wealth to tell;   85
You both are modest: so am I. Farewell.

15

An Epistle to a Friend, to Persuade Him to the Wars

Wake, friend, from forth thy lethargy; the drum
Beats brave and loud in Europe, and bids come
All that dare rouse, or are not loath to quit
Their vicious ease and be o'erwhelmed with it.
It is a call to keep the spirits alive   5
That gasp for action, and would yet revive

---

72 *chamber-fellow*: Edward Hayward (d. 1658), of the Inner Temple.   80 *comings-in*:
gains, income.   83 *take up*: borrow.

15. The friend is named as Colby in l. 176; Newdigate conjectures he may be Sir
Huntingdon Colby of Suffolk, or another member of that family. Judith K. Gardiner
(*N & Q*, xxii, July 1975) identifies him with the 'Colbie XXV yeres' who went with
Suckling to join the English volunteers serving under the Earl of Wimbledon in the
Dutch Wars in Oct. 1629 (P.R.O., E.157/14, f. 40: see Herbert Berry, *Sir John
Suckling's Poems and Letters from Manuscript* (London, Ontario, 1960), pp. 49–50).
But the poem was probably written, as H & S suggest, near the beginning of the
Thirty Years' War, e.g. in 1620, when English volunteers were recruited to aid
Frederick of the Palatinate.

Man's buried honour in his sleepy life,
Quickening dead nature to her noblest strife.
All other acts of worldlings are but toil
In dreams, begun in hope, and end in spoil.   10
Look on the ambitious man, and see him nurse
His unjust hopes with praises begged, or (worse)
Bought flatteries, the issue of his purse,
Till he become both their and his own curse!
Look on the false and cunning man, that loves   15
No person, nor is loved; what ways he proves
To gain upon his belly, and at last
Crushed in the snaky brakes that he had passed!
See the grave, sour, and supercilious sir—
In outward face, but, inward, light as fur   20
Or feathers—lay his fortune out to show,
Till envy wound or maim it at a blow!
See him, that's called and thought the happiest man,
Honoured at once and envied (if it can
Be honour is so mixed) by such as would,   25
For all their spite, be like him if they could.
No part or corner man can look upon,
But there are objects bid him to be gone
As far as he can fly, or follow day,
Rather than here, so bogged in vices, stay.   30
The whole world here, leavened with madness, swells,
And being a thing blown out of nought, rebels
Against his Maker; high alone with weeds
And impious rankness of all sects and seeds;
Not to be checked or frighted now with fate,   35
But more licentious made, and desperate!
Our delicacies are grown capital,
And even our sports are dangers; what we call
Friendship is now masked hatred; justice fled,
And shamefastness together; all laws dead   40
That kept man living; pleasures only sought!
Honour and honesty as poor things thought
As they are made; pride and stiff clownage mixed
To make up greatness! And man's whole good fixed

10 *spoil*: destruction.   16 *proves*: tries.   26 Cf. Seneca on flatterers, *De Ben.*, I. ix.
2.   34 *sects*: a pun: religious factions; cuttings from plants.   37 *capital*: fatal.   43
clownage: rudeness.

In bravery or gluttony, or coin,   45
All which he makes the servants of the groin:
Thither it flows! How much did Stallion spend
To have his court-bred filly there commend
His lace and starch, and fall upon her back
In admiration, stretched upon the rack   50
Of lust, to his rich suit and title, lord?
Aye, that's a charm and half! She must afford
That all respect; she must lie down—nay, more,
'Tis there civility to be a whore.
He's one of blood and fashion! and with these   55
The bravery makes; she can no honour leese.
To do 't with cloth, or stuffs, lust's name might merit;
With velvet, plush, and tissues, it is spirit.
Oh, these so ignorant monsters! light, as proud;
Who can behold their manners and not cloud-   60
Like upon them lighten? If nature could
Not make a verse, anger or laughter would,
To see 'em aye discoursing with their glass
How they may make someone that day an ass;
Planting their purls, and curls spread forth like net,   65
And every dressing for a pitfall set
To catch the flesh in, and to pound a prick.
Be at their visits: see 'em squeamish, sick,
Ready to cast, at one whose band sits ill,
And then leap mad on a neat piccadill,   70
And if a breeze were gotten in their tail;
And firk and jerk, and for the coachman rail,
And jealous each of other, yet think long
To be abroad chanting some bawdy song,
And laugh, and measure thighs, then squeak, spring, itch,   75
Do all the tricks of a salt lady bitch;

51 *lord*: see *Epig.* 11. 5 and n.   55 ff. Cf. the association of sex and clothes in *Und.* 42. 25 ff., and in Jonson's late plays; and see E. B. Partridge, *JEGP*, lvi (1957), 396–409.   55 *blood*: good stock; also hinting at a secondary sense, sexual appetite.   56 *leese*: lose.   57–8 Cf. Juvenal, *Sat.* xi. 176–8*.   57 *stuffs*: woven fabrics; with a sexual innuendo (cf. Nick and Pinnacia Stuff in *N.I.*).   58 *spirit*: style; but also sexual energy: cf. l. 82 below, and Shakespeare's Sonnet 129.   60–1 Cf. *B.F.* V. ii. 5–6.   61–2 Cf. Juvenal, *Sat.* i. 79: 'Though nature say me nay, indignation will prompt my verse'.   65 *purls*: laces, frills.   65 *like net*: cf. *D. is A.* II. ii. 111–14.   66 *pitfall*: also the name of the serving-woman in *D. is A.*   67 *pound a prick*: for the *double entendre*, cf. Middleton, *The Changeling*, III. iii. 9–10.   69 *cast*: vomit.   69 *band*: collar or ruff.   70 *piccadill*: frame to hold band upright.   72 *firk*: move briskly.   76 *salt*: on heat.

For t'other pound of sweetmeats, he shall feel
That pays, or what he will: the dame is steel.
For these with her young company she'll enter
Where Pitts, or Wright, or Modet would not venter,   80
And comes by these degrees the style to inherit
Of woman of fashion, and a lady of spirit;
Nor is the title questioned with our proud,
Great, brave, and fashioned folk; these are allowed
Adulteries, now, are not so hid, or strange: 85
They're grown commodity upon exchange.
He that will follow but another's wife
Is loved, though he let out his own for life;
The husband now's called churlish, or a poor
Nature, that will not let his wife be a whore;   90
Or use all arts, or haunt all companies
That may corrupt her, even in his eyes.
The brother trades a sister, and the friend
Lives to the lord, but to the lady's end.
Less must not be thought on than mistress, or,   95
If it be thought, killed like her embryons; for,
Whom no great mistress hath as yet infamed,
A fellow of coarse lechery is named;
The servant of the serving-woman, in scorn,
Ne'er came to taste the plenteous marriage-horn.   100
    Thus they do talk. And are these objects fit
For man to spend his money on? His wit,
His time, health, soul? Will he for these go throw
Those thousands on his back, shall after blow
His body to the Counters, or the Fleet?   105
Is it for these that Fine-man meets the street

77–8 he who pays may feel her, or do what he will.   78 *steel*: i.e. in her moral insensibility, not (like Milton's Lady, *Comus*, l. 421) in moral invincibility.   80 *Pitts, or Wright, or Modet*: notorious women of the time.   80 *venter*: venture.   85 ff. Cf. Seneca on adultery, *De Ben.*, I. ix. 3–4, I. x. 2–3.   86 *commodity*: see *Epig.* 12. 13n. 92 *in his eyes*: 'before his face' (Hunter); or, in his esteem.   94 *to the lady's end*: to seduce his lady.   96 *like her embryons*: see *Epig.* 62, and n.   97 ff. for he who has not yet made a great mistress infamous is scornfully named a fellow of coarse lechery, the servant of the serving-woman, one who has never tasted the plenteous delights of marriage. (The 'marriage-horn' is at once the cornucopia of marriage and the cuckold's emblem; 'plenteous' refers ironically to lovers rather than offspring.) Cf. *E.M.I.* III. vi. 23–5.   104 *his back*: i.e. in clothing.   105 *Counters*: the city prisons for debtors.   105 *Fleet*: the prison for Star Chamber and Chancery Court offenders.

Coached, or on foot-cloth, thrice changed every day,
To teach each suit he has the ready way
From Hyde Park to the stage, where at the last
His dear and borrowed bravery he must cast?   110
When not his combs, his curling irons, his glass,
Sweet bags, sweet powders, nor sweet words will pass
For less security? O God, for these
Is it that man pulls on himself disease,
Surfeit, and quarrel; drinks the tother health,   115
Or by damnation voids it, or by stealth?
What fury of late is crept into our feasts!
What honour given to the drunkenest guests!
What reputation to bear one glass more,
When oft the bearer is borne out of door!   120
This hath our ill-used freedom and soft peace
Brought on us, and will every hour increase.
Our vices do not tarry in a place,
But being in motion still, or rather in race,
Tilt one upon another, and now bear   125
This way, now that, as if their number were
More than themselves, or than our lives, could take,
But both fell pressed under the load they make.
    I'll bid thee look no more, but flee, flee, friend,
This precipice and rocks that have no end   130
Or side, but threatens ruin. The whole day
Is not enough now, but the night's to play;
And whilst our states, strength, body, and mind we waste,
Go make ourselves the usurer's at a cast.
He that no more for age, cramps, palsies can   135
Now use the bones, we see doth hire a man
To take the box up for him, and pursues
The dice with glassen eyes to the glad views
Of what he throws: like lechers grown content
To be beholders, when their powers are spent.   140
        Can we not leave this worm? Or will we not?
Is that the truer excuse, or have we got

---

107 *on foot-cloth*: on a richly caparisoned horse.   110 *cast*: abandon.   113 *God*: the
Folio leaves a blank.   116 *voids*: abstains from.   119–20 Cf. Herbert, 'The Church
Porch', 25 ff.   121 *soft peace*: cf. *U.V.* 3, 'Peace'.   128 *fell*: hard.   135–9 Horace,
Sat. II. vii. 15–18*.   141 *this worm*: *Epig.* 15.1.

In this, and like, an itch of vanity,
That scratching now's our best felicity?
Well, let it go. Yet this is better than   145
To lose the forms and dignities of men,
To flatter my good lord, and cry his bowl
Runs sweetly as it had his lordship's soul;
Although perhaps it has: what's that to me,
That may stand by and hold my peace? Will he,   150
When I am hoarse with praising his each cast,
Give me but that again, that I must waste
In sugar candied or in buttered beer,
For the recovery of my voice? No, there
Pardon his lordship. Flattery's grown so cheap   155
With him, for he is followed with that heap
That watch and catch at what they may applaud,
As a poor single flatterer, without bawd,
Is nothing; such scarce meat and drink he'll give;
But he that's both, and slave to boot, shall live   160
And be beloved, while the whores last. O times!
Friend, flee from hence, and let these kindled rhymes
Light thee from hell on earth; where flatterers, spies,
Informers, masters both of arts and lies,
Lewd slanderers, soft whisperers that let blood   165
The life and fame-veins (yet not understood
Of the poor sufferers); where the envious, proud,
Ambitious, factious, superstitious, loud
Boasters, and perjured, with the infinite more
Prevaricators swarm. Of which the store   170
(Because they are everywhere amongst mankind
Spread through the world) is easier far to find
Than once to number, or bring forth to hand,
Though thou wert muster-master of the land.
    Go, quit 'em all. And take along with thee   175
Thy true friend's wishes, Colby, which shall be

143–4 Seneca, *De Tranquillitate Animi*, ii, 11–12. 153 *buttered beer*: 'a beverage composed of sugar, cinnamon, butter and beer brewed without hops' (*OED*). 161 *O times!*: Cicero, *In Catilinam*, I. i. 2: *O tempora!* Cf. *Und.* 64. 17, and Justice Overdo in *B.F.* II. ii. 113. 165–6 *let blood* | *The life and fame-veins*: bleed the lives and reputations of others.

That thine be just and honest; that thy deeds
Not wound thy conscience, when thy body bleeds;
That thou dost all things more for truth than glory,
And never but for doing wrong be sorry;   180
That by commanding first thyself, thou mak'st
Thy person fit for any charge thou tak'st;
That fortune never make thee to complain,
But what she gives thou dar'st give her again;
That whatsoever face thy fate puts on,   185
Thou shrink or start not, but be always one;
That thou think nothing great but what is good,
And from that thought strive to be understood.
So, 'live or dead, thou wilt preserve a fame
Still precious with the odour of thy name.   190
And last, blaspheme not; we did never hear
Man thought the valianter 'cause he durst swear,
No more than we should think a lord had had
More honour in him 'cause we have known him mad:
These take, and now go seek thy peace in war;   195
Who falls for love of God shall rise a star.

16

An Epitaph on Master Philip Gray

Reader, stay,
And if I had no more to say
But here doth lie, till the last day,
All that is left of Philip Gray,
It might thy patience richly pay: 5
For if such men as he could die,
What surety of life have thou, and I?

181 *commanding first thyself*: *Disc.* 1005–7.   183–4 *Disc.* 1–7.   187 *great ... good*: *Epig.* Ded. 15–16 n.

16. *Philip Gray*: perhaps the eldest son of Sir Edward Gray of Morpeth Castle; d. *c.* 1625/6.

17

Epistle to a Friend

They are not, sir, worst owers, that do pay
   Debts when they can; good men may break their day,
And yet the noble nature never grudge;
   'Tis then a crime, when the usurer is judge,
And he is not in friendship. Nothing there   5
   Is done for gain; if't be, 'tis not sincere.
Nor should I at this time protested be,
   But that some greater names have broke with me,
And their words too, where I but break my band.
   I add that 'but' because I understand   10
That as the lesser breach; for he that takes
   Simply my band, his trust in me forsakes
And looks unto the forfeit. If you be
   Now so much friend as you would trust in me,
Venture a longer time, and willingly;   15
   All is not barren land doth fallow lie.
Some grounds are made the richer for the rest,
   And I will bring a crop, if not the best.

18

An Elegy

Can beauty that did prompt me first to write,
   Now threaten with those means she did invite?
Did her perfections call me on to gaze,
   Then like, then love, and now would they amaze?
Or was she gracious afar off, but near   5
   A terror? Or is all this my fear?

17. 5 *there*: in friendship.   7 *protested*: declared bankrupt.   9 *where*: whereas.
9 *band*: bond.

18. 4 *amaze*: bewilder.

That as the water makes things put in 't straight,
  Crooked appear, so that doth my conceit;
I can help that with boldness; and love sware,
  And fortune once, to assist the spirits that dare.   10
But which shall lead me on? Both these are blind;
  Such guides men use not, who their way would find,
Except the way be error to those ends,
  And then the best are, still, the blindest friends!
Oh how a lover may mistake! To think   15
  Or love or fortune blind, when they but wink
To see men fear; or else, for truth and state,
  Because they would free justice imitate,
Veil their own eyes, and would impartially
  Be brought by us to meet our destiny.   20
If it be thus, come love, and fortune go;
  I'll lead you on; or if my fate will so
That I must send one first, my choice assigns
  Love to my heart, and fortune to my lines.

19

An Elegy

By those bright eyes, at whose immortal fires
  Love lights his torches to inflame desires;
By that fair stand, your forehead, whence he bends
  His double bow, and round his arrows sends;
By that tall grove, your hair, whose globy rings   5
  He flying curls and crispeth with his wings;
By those pure baths your either cheek discloses,
  Where he doth steep himself in milk and roses;
And lastly by your lips, the bank of kisses,
  Where men at once may plant and gather blisses: 10

7–8 Cf. Seneca, *Epist.* lxxi. 24.   8 *so that doth my conceit*: i.e. so fear acts on my imagination.   10 Ovid, *Ars Amatoria*, i. 608: 'Fortune and Venus help the brave'; cf. Tilley, F601, H302.   13 *error*: winding.   16 *wink*: close their eyes.   21 *come love, and fortune go*: parallel constructions; both are summoned.

19. 1–10 For parallels, see *Und.* 2. v. 21–6 n.; *C.R.* V. iv. 439–42; *D. is A.* II. vi. 78–87; and, for the *torches* of l. 2, *Und.* 7. 24, and n.   3 *stand*: place from which to shoot.

Tell me, my loved friend, do you love, or no,
  So well as I may tell in verse, 'tis so?
You blush, but do not; friends are either none,
  Though they may number bodies, or but one.
I'll therefore ask no more, but bid you love;   15
  And so, that either may example prove
Unto the other, and live patterns how
  Others in time may love, as we do now.
Slip no occasion; as time stands not still,
  I know no beauty, nor no youth that will.   20
To use the present, then, is not abuse,
  You have a husband is the just excuse
Of all that can be done him; such a one
  As would make shift to make himself, alone,
That which we can; who both in you, his wife,   25
  His issue, and all circumstance of life,
As in his place, because he would not vary,
  Is constant to be extraordinary.

20

A Satirical Shrub

A woman's friendship! God whom I trust in,
  Forgive me this one foolish deadly sin,
Amongst my many other, that I may
  No more (I am sorry for so fond cause) say
At fifty years, almost, to value it   5
  That ne'er was known to last above a fit!
Or have the least of good, but what it must
  Put on for fashion, and take up on trust.
Knew I all this afore? Had I perceived
  That their whole life was wickedness, though weaved   10

17–18 Cf. Donne, 'The Canonization'.   22–3 Cf. Wittipol's plea to Mrs. Fitzdottrell, *D. is A.* II. vi. 64–6.

20. *Shrub*: playing on the title of the collection, *The Underwood*; and in *Und.* 21 picking up a secondary sense, a mean or insignificant person. Neither victim in these two poems has been identified. Probably written in 1621 or 1622 ('fifty years, almost', l. 5).

Of many colours; outward, fresh from spots,
  But their whole inside full of ends and knots?
Knew I that all their dialogues and discourse
  Were such as I will now relate, or worse?
      Here something is wanting.
        *   *   *   *   *

        *   *   *   *   *

Knew I this woman? Yes; and you do see   15
  How penitent I am, or I should be!
Do not you ask to know her; she is worse
  Than all the ingredients made into one curse.
And that poured out upon mankind, can be!
  Think but the sin of all her sex, 'tis she!   20
I could forgive her being proud, a whore,
  Perjured, and painted, if she were no more:
But she is such, as she might yet forestall
  The devil, and be the damning of us all.

21

A Little Shrub Growing By

Ask not to know this man. If fame should speak
  His name in any metal, it would break.
Two letters were enough the plague to tear
  Out of his grave, and poison every ear.
A parcel of court dirt, a heap and mass   5
  Of all vice hurled together; there he was
Proud, false, and treacherous, vindictive, all
  That thought can add: unthankful, the lay-stall
Of putrid flesh alive; of blood the sink;
  And so I leave to stir him, lest he stink.   10

---

21. See *Und.* 20 and n. Jonson's reluctance to name his victim is characteristic: see
Epig. Ded. 15–16 n., *Epig.* 30. 4, 38. 8, 77.   8 *lay-stall*: dung-heap.   10 Proverbial:
'The more you stir the more you stink' (Tilley, S862).

22

An Elegy

Though beauty be the mark of praise,
    And yours of whom I sing be such
    As not the world can praise too much,
Yet is't your virtue now I raise.

A virtue, like alloy, so gone   5
    Throughout your form, as though that move,
    And draw, and conquer all men's love,
This subjects you to love of one.

Wherein you triumph yet, because
    'Tis of yourself, and that you use   10
    The noblest freedom, not to choose
Against or faith, or honour's laws.

But who should less expect from you,
    In whom alone Love lives again?
    By whom he is restored to men,   15
And kept, and bred, and brought up true.

His falling temples you have reared,
    The withered garlands ta'en away,
    His altars kept from the decay
That envy wished, and nature feared.   20

And on them burn so chaste a flame
    With so much loyalty's expense,
    As Love, to acquit such excellence,
Is gone himself into your name.

---

22. 5 *alloy*: i.e. a tempering (rather than a devaluing) additive. 6 *that*: beauty. 8 *This*: virtue. 23 *acquit*: requite. 23–4 From this hint Fleay deduces (i. 326) that the elegy is addressed to Lady Covell, of *Und.* 56.

And you are he, the deity   25
   To whom all lovers are designed
   That would their better objects find;
Among which faithful troop am I.

Who, as an offering at your shrine,
   Have sung this hymn, and here entreat   30
   One spark of your diviner heat
To light upon a love of mine.

Which if it kindle not, but scant
   Appear, and that to shortest view,
   Yet give me leave to adore in you   35
What I, in her, am grieved to want.

23

An Ode. To Himself

Where dost thou careless lie,
   Buried in ease and sloth?
Knowledge that sleeps doth die;
   And this security,
     It is the common moth   5
That eats on wits and arts, and oft destroys them both.

Are all the Aonian springs
   Dried up? Lies Thespia waste?
Doth Clarius' harp want strings,
   That not a nymph now sings?   10
     Or droop they, as disgraced
To see their seats and bowers by chattering pies defaced?

---

23. Cf. *Songs*, 14, on the failure of *The New Inn*. 1–2 Cf. Ovid, *Amores*, I. xv. 1–2, trs. by Jonson in *Poet*. I. i. 43–4: 'Envy, why twit'st thou me, my time's spent ill? / And call'st my verse fruits of an idle quill?' 4 *security*: carelessness. 7 *Aonian springs*: sacred to the muses. 8 *Thespia*: town at the foot of Mt. Helicon, also sacred to the muses. 9 *Clarius'*: Apollo's (worshipped in Claros, in Ionia). 12 *chattering pies*: cf. Pindar, *Olymp.* ii. 87; Jonson's birds are not only noisy but dirty (Carol Maddison, *Apollo and the Nine* (London, 1960), p. 298).

If hence thy silence be,
        As 'tis too just a cause,
Let this thought quicken thee: 15
Minds that are great and free,
        Should not on fortune pause;
'Tis crown enough to virtue still, her own applause.

What though the greedy fry
        Be taken with false baits   20
Of worded balladry,
And think it poesie?
        They die with their conceits,
And only piteous scorn upon their folly waits.

Then take in hand thy lyre,   25
        Strike in thy proper strain;
With Japhet's line, aspire
Sol's chariot for new fire
        To give the world again;
Who aided him, will thee, the issue of Jove's brain.   30

And since our dainty age
        Cannot endure reproof,
Make not thyself a page
To that strumpet, the stage;
        But sing high and aloof,   35
Safe from the wolf's black jaw, and the dull ass's hoof.

17–18 For the contrast of fortune and virtue, see *Epig.* 63. 2–3, and n.   18 A commonplace; see e.g. Seneca, *De Vita Beata*, ix. 4, *De Clementia*, i. 1.   21 *worded balladry*: Fleay (i. 327) suspects a gibe at Daniel—'no poet' (*Conv. Dr.* 24).   27 *Japhet's line*: Prometheus; see Horace, *Odes*, I. iii. 25–8; cf. *Epig.* 110. 17.   27 *aspire*: aspire to.   30 *issue of Jove's brain*: Minerva; see *For.* 10. 13–15 n.   35–6 These lines are also spoken by the Author in *Poet. Apol. Dial.* 238–9; 'dull ass's hoof' is echoed by W. B. Yeats, *Collected Poems* (London, 1963), p. 143.

The frontispiece to Sir Walter Raleigh's *History of the World*, 1614, engraved by Renold Elstracke from a design by Raleigh.

24

The Mind of the Frontispiece to a Book

From death and dark oblivion (near the same)
    The mistress of man's life, grave history,
Raising the world to good or evil fame
    Doth vindicate it to eternity.
Wise providence would so, that nor the good   5
    Might be defrauded, nor the great secured,
But both might know their ways were understood,
    When vice alike in time with virtue dured.
Which makes that, lighted by the beamy hand
    Of truth that searcheth the most hidden springs,   10
And guided by experience, whose straight wand
    Doth mete, whose line doth sound the depth of things,
She cheerfully supporteth what she rears,
    Assisted by no strengths but are her own;
Some note of which each varied pillar bears;   15
    By which, as proper titles, she is known
Time's witness, herald of antiquity,
    The light of truth, and life of memory.

25

An Ode to James, Earl of Desmond.
Writ in Queen Elizabeth's Time,
Since Lost, and Recovered

Where art thou, genius? I should use
    Thy present aid; arise invention,
Wake, and put on the wings of Pindar's muse,
    To tower with my intention

24. *a Book*: Sir Walter Raleigh's *History of the World*, 1614 (see p. 169). Jonson's poem describes its frontispiece; A. H. Gilbert suspects Jonson may have helped with its design (*The Symbolic Persons in the Masques of Ben Jonson* (Durham, N. Carolina, 1948), pp. 121–2). Jonson had helped in other ways with Raleigh's work: see *Conv. Dr.* 200–1. With 'The Mind', cf. the body/soul distinction of Jonson's masques: see *U.V.* 34. 50 n.   6 *secured*: protected.   8 *dured*: lasted.   17–18 Cicero's definition of history, *De Oratore*, II. ix. 36*.

25. *James, Earl of Desmond*: James Fitzgerald (?1570–1601), son of the attainted 'Rebel Earl', Gerald Fitzgerald. Imprisoned in Dublin Castle and the Tower of London from 1579 to 1600; restored as Earl of Desmond, 1 Oct. 1600, to defeat the pretensions of a cousin. The poem was probably written shortly before this date.

High as his mind, that doth advance 5
Her upright head above the reach of chance,
    Or the time's envy;
    Cinthius, I apply
My bolder numbers to thy golden lyre:
    O, then inspire 10
Thy priest in this strange rapture; heat my brain
    With Delphic fire,
That I may sing my thoughts in some unvulgar strain.

    Rich beam of honour, shed your light
    On these dark rhymes, that my affection 15
May shine through every chink, to every sight,
    Graced by your reflection!
    Then shall my verses, like strong charms,
Break the knit circle of her stony arms
    That holds your spirit, 20
    And keeps your merit
Locked in her cold embraces, from the view
    Of eyes more true,
Who would with judgement search, searching conclude
    (As proved in you) 25
True noblesse. Palm grows straight, though handled ne'er so
                     rude.

    Nor think yourself unfortunate,
    If subject to the jealous errors
Of politic pretext, that wries a state;
    Sink not beneath these terrors, 30
    But whisper, O glad innocence,
Where only a man's birth is his offence;
    Or the disfavour,
    Of such as savour
Nothing, but practise upon honour's thrall. 35
    O virtue's fall!
When her dead essence, like the anatomy
    In Surgeons' Hall,
Is but a statist's theme, to read phlebotomy.

---

8 *Cinthius*: Apollo, born on Mt. Cinthus.   19 *her stony arms*: the Tower.   26 *Palm*: an emblem of patience and fortitude; said to flourish even when beaten; see Alciati, *Emblemata* (Paris, 1602), pp. 231–4.   29 *wries*: distorts, convulves.   38 *Surgeons' Hall*: or Barber-Surgeons' Hall, in Monkswell St., near Cripplegate; here criminals' bodies ('anatomies') were dissected.

Let Brontes and black Steropes   40
    Sweat at the forge, their hammers beating;
Pyracmon's hour will come to give them ease,
    Though but while metal's heating;
    And after all the Aetnean ire
Gold that is perfect will outlive the fire.   45
        For fury wasteth,
        As patience lasteth.
No armour to the mind! He is shot-free
        From injury
That is not hurt, not he that is not hit;   50
        So fools, we see,
Oft scape an imputation more through luck than wit.

    But to yourself, most loyal lord,
        Whose heart in that bright sphere flames clearest,
Though many gems be in your bosom stored,   55
        Unknown which is the dearest,
        If I auspiciously divine,
As my hope tells, that our fair Phoebe's shine
        Shall light those places
        With lustrous graces,   60
Where darkness with her gloomy-sceptred hand
        Doth now command;
O then, my best-best loved, let me importune,
        That you will stand
As far from all revolt, as you are now from fortune.   65

40–2 *Brontes ... Steropes ... Pyracmon*: the three Cyclopes who assist Vulcan at his forge on Mt. Aetna; they make the shield of Aeneas, *Aeneid*, viii. 424 ff.   45 Proverbial, of stoical sufferers: see Tilley, G284; cf. *Und.* 47. 3–4.   48–50 Seneca, *De Constantia*, iii. 3: 'The invulnerable thing is not that which is not struck, but that which is not hurt'; cf. *Poet.* Apol. Dial. 38–9, *N.I.* IV. iv. 204–5.   58 *fair Phoebe's*: Queen Elizabeth's.

26

An Ode

High-spirited friend,
I send nor balms nor corsives to your wound;
Your fate hath found
A gentler and more agile hand to tend
The cure of that, which is but corporal,   5
And doubtful days (which were named critical)
Have made their fairest flight,
And now are out of sight.
Yet doth some wholesome physic for the mind
Wrapped in this paper lie,   10
Which in the taking, if you misapply,
You are unkind.

Your covetous hand,
Happy in that fair honour it hath gained,
Must now be reined.   15
True valour doth her own renown command
In one full action; nor have you now more
To do than be a husband of that store.
Think but how dear you bought
This fame which you have caught;   20
Such thoughts will make you more in love with truth.
'Tis wisdom, and that high,
For men to use their fortune reverently,
Even in youth.

---

26. Trimpi (pp. 197–8, 278–9) and Alvaro Ribeiro (*RES*, xxiv (1973), 164, n. 1) suspect that this poem is addressed to Sir John Roe; see *Epig.* 27 and n., and the duels referred to in *Epig.* 32. But the 'high-spirited friend' might equally be Sir Edward Sackville (see *Und.* 13 and n.), who killed Lord Bruce, and was himself seriously wounded, in a duel in 1613; Sidney Lee (*DNB*) guessed that Venetia Stanley (see *Und.* 78 and n.) was the cause of the quarrel. On the duel and on Sackville's qualities, see Clarendon, *History of the Rebellion*, i. § 129–30. Sackville had married in 1612; the reference in ll. 3–4 may be to his wife. 2 *corsives*: corrosives, sharp medicines. Cf. Plutarch, *How to Tell a Flatterer*, xi. 10 *this paper*: i.e. the poem itself; playing on the fact that drugs were wrapped in paper. 20 *fame*: Newdigate's emendation of Folio 'same'. 23 From Ausonius, *Epig.* ii. 7–8; cf. *Sej.* II. 137, *Volp.* III. vii. 88–9.

27

An Ode

Helen, did Homer never see
Thy beauties, yet could write of thee?
Did Sappho, on her seven-tongued lute,
So speak, as yet it is not mute,
Of Phaon's form? Or doth the boy   5
In whom Anacreon once did joy,
Lie drawn to life in his soft verse,
As he whom Maro did rehearse?
Was Lesbia sung by learned Catullus,
Or Delia's graces by Tibullus?   10
Doth Cinthia in Propertius' song
Shine more than she the stars among?
Is Horace his each love so high,
Rapt from the earth, as not to die;
With bright Lycoris, Gallus' choice,   15
Whose fame hath an eternal voice?
Or hath Corinna, by the name
Her Ovid gave her, dimmed the fame
Of Caesar's daughter, and the line
Which all the world then styled divine?   20
Hath Petrarch since his Laura raised
Equal with her; or Ronsard praised
His new Cassandra 'bove the old
Which all the fate of Troy foretold?
Hath our great Sidney Stella set,   25
Where never star shone brighter yet;

27. 5 *Phaon*: handsome boatman of Mitylene; for love of him, Sappho is said to have thrown herself into the sea.   5 *the boy*: named Bathyllus; see *Anacreontea*, xvii. Horace speaks of this boy living on in Anacreon's verse, *Odes*, IV. ix. 9–12.   8 *he whom Maro did rehearse*: the child Virgil prophesies in *Eclogue* iv.   11 *Propertius' song*: Propertius himself, in celebrating his Cinthia, names—as Jonson does—other women celebrated by other poets: *Elegies*, ii. 85–94.   15 *Lycoris*: also called Cytheris and Volumnia; a freedwoman, loved by Gallus. See Virgil, *Eclogue* x.   17–20 Jonson's Ovid in *Poet*. I. iii. 37 'veils' Caesar's daughter, Julia, under the name of Corinna; cf. Ovid. *Tristia*, II. 339–40.   23–4 *new Cassandra . . . old*: Cassandre Salviati, to whom Ronsard addressed his sonnet cycle, *Amours*, in 1552; Cassandra the Greek prophetess, daughter of Priam.   25 *Stella*: Lady Rich (née Penelope Devereux), of *Astrophil and Stella*.

Or Constable's ambrosiac muse
Made Dian not his notes refuse?
Have all these done—and yet I miss
The swan that so relished Pancharis—   30
And shall not I my Celia bring
Where men may see whom I do sing?
Though I, in working of my song,
Come short of all this learned throng,
Yet sure my tunes will be the best,   35
So much my subject drowns the rest.

28

A Sonnet
to the Noble Lady, the Lady Mary Worth

I, that have been a lover, and could show it,
    Though not in these, in rhythms not wholly dumb,
    Since I exscribe your sonnets, am become
A better lover, and much better poet.
Nor is my muse, or I ashamed to owe it   5
    To those true numerous graces, whereof some
    But charm the senses, others overcome
Both brains and hearts; and mine now best do know it:
For in your verse all Cupid's armory,
    His flames, his shafts, his quiver, and his bow,   10
    His very eyes are yours to overthrow.
But then his mother's sweets you so apply,
    Her joys, her smiles, her loves, as readers take
    For Venus' ceston every line you make.

27 *Constable's ambrosiac muse*: Henry Constable, 1562–1613, author of *Diana*, 1592.   30 *The swan*: Hugh Holland; see Jonson's poem on his *Pancharis* (1603), *U.V.* 6.   31 *my Celia*: suggesting a real identity for the girl addressed in *For.* 5, 6, and 9.

28. *Lady Mary Worth*: or Wroth; see *Epig.* 103, and n.   3 *exscribe*: copy out.   3 *your sonnets*: 'Pamphilia to Amphilanthus', appended to Lady Mary Wroth's *Urania* (London, 1621).   6 *numerous*: poetic, musical.   14 *Venus' ceston*: the girdle which conferred beauty on the wearer and excited love in the beholder; see *Und.* 2. v. 41.

29

A Fit of Rhyme against Rhyme

Rhyme, the rack of finest wits
That expresseth but by fits
        True conceit;
Spoiling senses of their treasure,
Cozening judgement with a measure   5
        But false weight.
Wresting words from their true calling,
Propping verse for fear of falling
        To the ground.
Jointing syllabes, drowning letters,   10
Fastening vowels, as with fetters
        They were bound!
Soon as lazy thou wert known,
All good poetry hence was flown,
        And art banished.   15
For a thousand years together
All Parnassus' green did wither,
        And wit vanished.
Pegasus did fly away,
At the well no muse did stay,   20
        But bewailed
So to see the fountain dry,
And Apollo's music die,
        All light failed!

---

29. In 1602 Campion published his *Observations in the Art of English Poesy*, attacking the 'childish titillation' of rhyme; in 1603 Daniel replied with *A Defence of Rhyme*. Jonson told Drummond that 'he had written a discourse of poesy both against Campion and Daniel, especially this last, where he proves couplets to be the bravest sort of verses'; 'he detesteth all other rhymes' (*Conv. Dr.* 1–11). G. B. Johnston, *Ben Jonson: Poet*, p. 7, points out that every poem in the Folio is rhymed. 2 *fits*: a pun: spasms, sections of poems. 5 *measure*: a pun. 10 *Jointing syllabes*: i.e. breaking a word on a syllabic unit to achieve a rhyme, as in *Und.* 70. 92–3; on the practice, see K. A. McEuen, *Classical Influence on the Tribe of Ben*, pp. 166, 277. 'Syllabe' is Jonson's usual spelling, e.g. in *Eng. Gram.* and *Und.* 70. 63. 17 *Parnassus*: the mountain of the muses. 19 *Pegasus*: the muses' winged horse. 20 *well*: Hippocrene, sacred to the muses: it was produced by Pegasus striking the ground with his hoof.

Starveling rhymes did fill the stage,   25
Not a poet in an age
            Worth a-crowning.
Not a work deserving bays,
Nor a line deserving praise,
            Pallas frowning.   30
Greek was free from rhyme's infection,
Happy Greek by this protection,
            Was not spoiled.
Whilst the Latin, queen of tongues,
Is not yet free from rhyme's wrongs,   35
            But rests foiled.
Scarce the hill again doth flourish,
Scarce the world a wit doth nourish,
            To restore
Phoebus to his crown again,   40
And the muses to their brain,
            As before.
Vulgar languages that want
Words and sweetness, and be scant
            Of true measure,   45
Tyrant rhyme hath so abused,
That they long since have refused
            Other caesure.
He that first invented thee,
May his joints tormented be,   50
            Cramped for ever;
Still may syllabes jar with time,
Still may reason war with rhyme,
            Resting never.
May his sense, when it would meet   55
The cold tumor in his feet,
            Grow unsounder.
And his title be long fool,
That in rearing such a school,
            Was the founder.   60

---

25 *starveling rhymes did fill the stage*: cf. Marlowe's gibe in the prologue to *Tamburlaine the Great*, Part I.   34–5 Contemporary Latin verse was sometimes rhymed.   48 *caesure*: caesura, metrical pause.   56 *feet*: a pun: poetic feet.

30

*An Epigram
on William, Lord Burghley,
Lord High Treasurer of England*

If thou wouldst know the virtues of mankind,
  Read here in one, what thou in all canst find,
And go no farther; let this circle be
  Thy universe, though his epitome.
Cecil, the grave, the wise, the great, the good: 5
  What is there more that can ennoble blood?
The orphan's pillar, the true subject's shield,
  The poor's full store-house, and just servant's field.
The only faithful watchman for the realm,
  That in all tempests never quit the helm,   10
But stood unshaken in his deeds and name,
  And laboured in the work, not with the fame;
That still was good for goodness' sake, nor thought
  Upon reward, till the reward him sought.
Whose offices and honours did surprise   15
  Rather than meet him; and before his eyes
Closed to their peace, he saw his branches shoot,
  And in the noblest families took root
Of all the land. Who now at such a rate
  Of divine blessing, would not serve a state?   20

30. 'Presented upon a plate of gold to his son, Robert, Earl of Salisbury, when he was also Treasurer' (Jonson's note). William Cecil, Lord Burghley (1521–98), was Elizabeth's senior minister; Baron of Burghley, 1571, Knight of the Garter and Lord High Treasurer, 1572. For Robert Cecil, Earl of Salisbury, see *Epig.* 43, n. A version of ll. 1–10 is inscribed on the underside of a silver-gilt paten (hall-mark 1609) in the chapel of Burghley House, with the concluding lines: 'Whose worthy sonne besides his owne high graces / Inheritts all his vertues, all his places. / —To the memorye of W Lo: Burleigh, / late high Treasurer of Englande.' C. C. Oman, who describes the paten, doubts if a 'plate of gold' ever existed: *English Church Plate, 597–1830* (London, 1957), pp. 233–4. The poem may be a delayed tribute to father and son after the latter's preferment to the Treasurership in May 1608, see *Epig.* 64. 16–19 Five of Burghley's seven children actually predeceased him. Several of his children and grandchildren married into other noble families.

31

An Epigram
to Thomas, Lord Ellesmere,
the Last Term He Sat Chancellor

So, justest lord, may all your judgements be
    Laws, and no change e'er come to one decree;
So may the king proclaim your conscience is
    Law to his law, and think your enemies his;
So from all sickness may you rise to health,   5
    The care and wish still of the public wealth;
So may the gentler muses, and good fame
    Still fly about the odour of your name:
As, with the safety and honour of the laws,
    You favour truth, and me, in this man's cause.   10

32

Another to Him

The judge his favour timely then extends
    When a good cause is destitute of friends,
Without the pomp of counsel, or more aid
    Than to make falsehood blush, and fraud afraid,
When those good few that her defenders be   5
    Are there for charity, and not for fee.
Such shall you hear today, and find great foes,
    Both armed with wealth and slander to oppose,
Who, thus long safe, would gain upon the times
    A right by the prosperity of their crimes;   10
Who, though their guilt and perjury they know,
    Think—yea, and boast—that they have done it so,
As, though the court pursues them on the scent,
    They will come off, and scape the punishment.
When this appears, just lord, to your sharp sight,   15
    He does you wrong that craves you to do right.

---

31. 'For a poor man' (Jonson's note). Jonson's intercession may be related to a suit in a letter to Ellesmere's secretary, printed in H & S, i. 201. For Ellesmere, see *Epig.* 74 n. Hilary term, 1617, was his last as Chancellor.

32. 'For the same' (Jonson's note).

33

An Epigram to the Counsellor that Pleaded
and Carried the Cause

That I, hereafter, do not think the Bar
    The seat made of a more than civil war,
Or the Great Hall at Westminster the field
    Where mutual frauds are fought, and no side yield;
That, henceforth, I believe nor books nor men   5
    Who 'gainst the law weave calumnies, my Benn,
But when I read or hear the names so rife
    Of hirelings, wranglers, stitchers-to of strife,
Hook-handed harpies, gowned vultures, put
    Upon the reverend pleaders; do now shut   10
All mouths that dare entitle them (from hence)
    To the wolf's study, or dog's eloquence:
Thou art my cause; whose manners since I knew,
    Have made me to conceive a lawyer new.
So dost thou study matter, men, and times,   15
    Mak'st it religion to grow rich by crimes;
Dar'st not abuse thy wisdom in the laws,
    Or skill, to carry out an evil cause,
But first dost vex and search it. If not sound,
    Thou prov'st the gentler ways to cleanse the wound,   20
And make the scar fair; if that will not be,
    Thou hast the brave scorn to put back the fee.
But in a business that will bide the touch,
    What use, what strength of reason! and how much
Of books, of precedents hast thou at hand!   25
    As if the general store thou didst command
Of argument, still drawing forth the best,
    And not being borrowed by thee, but possessed.

33. *the Counsellor*: the rhyme-word in l. 6 is missing in Folio; Whalley conjectured 'Benn', i.e. Sir Anthony Benn, Recorder of London, d. 1618.   9 *gowned vultures*: cf. the lawyer Voltore (= vulture) in *Volp.* The association is traditional.   12 *dog's eloquence*: from Quintilian, XII. ix. 9: 'For it is a dog's eloquence, as Appius says, to undertake the task of abusing one's opponent'.   19 *vex*: try.   21 *the scar fair*: cf. *Und.* 38. 49–52, and n.   23 *touch*: test.

So com'st thou like a chief into the court,
    Armed at all pieces, as to keep a fort   30
Against a multitude, and (with thy style
    So brightly brandished) wound'st, defend'st—the while
Thy adversaries fall, as not a word
    They had, but were a reed unto thy sword.
Then com'st thou off with victory and palm,   35
    Thy hearers' nectar, and thy client's balm,
The court's just honour, and thy judge's love.
    And (which doth all achievements get above)
Thy sincere practice breeds not thee a fame
    Alone, but all thy rank a reverend name.   40

34

An Epigram to the Smallpox

Envious and foul disease, could there not be
    One beauty in an age, and free from thee?
What did she worth thy spite? Were there not store
    Of those that set by their false faces more
Than this did by her true? She never sought   5
    Quarrel with nature, or in balance brought
Art, her false servant; nor, for Sir Hugh Platt
    Was drawn to practise other hue than that
Her own blood gave her; she ne'er had, nor hath
    Any belief in Madam Bawd-be's bath,   10
Or Turner's oil of talc; nor ever got
    Spanish receipt to make her teeth to rot.
What was the cause, then? Thought'st thou in disgrace
    Of beauty so to nullify a face
That heaven should make no more; or should amiss   15
    Make all hereafter, hadst thou ruined this?
Aye, that thy aim was: but her fate prevailed;
    And, scorned, thou hast shown thy malice, but hast failed.

30 *at all pieces*: at all points. 31 *style*: stylus, pen.

34. Cf. *For.* 8, 'To Sickness', and *Conv. Dr.* 348–9, on Sir Philip Sidney's mother's smallpox. 7 *Sir Hugh Platt*: or Plat (1552–1608); his *Jewell House of Art and Nature* (1594) and *Delights for Ladies* (1602) include advice on cosmetics. 10 *Madam Bawd-be's bath*: probably a sweating-bath to improve the complexion (recommended by Platt). Bath-houses enjoyed a doubtful reputation: see *Epig.* 7 and n. 11 *Turner's oil of talc*: see *For.* 8. 33 n. H & S take 'Turner' to be Anne Turner, the poisoner of Overbury. Her husband, George, had been a doctor of physic. 12 Several popular dentifrices of the time had this calamitous effect.

35

An Epitaph on Elizabeth Chute

What beauty would have lovely styled,
What manners pretty, nature mild,
What wonder perfect, all were filed,
Upon record, in this blessed child.
    And till the coming of the soul   5
    To fetch the flesh, we keep the roll.

36

A Song

Lover

Come, let us here enjoy the shade,
For love in shadow best is made.
Though envy oft his shadow be,
None brooks the sunlight worse than he.

Mistress

Where love doth shine, there needs no sun,   5
All lights into his one doth run;
Without which all the world were dark,
Yet he himself is but a spark.

Arbiter

A spark to set whole worlds afire,
Who more they burn, they more desire,   10
And have their being their waste to see,
And waste still, that they still might be.

---

35. *Elizabeth Chute*: d. 18 May 1627, aged 3½ years. The epitaph is inscribed on her memorial tablet in Sonning church, near Reading. (The present text is from the Folio.) 3–6 *filed, | Upon record . . . the roll*: cf. *Und.* 70. 126, 75. 155–6.

Chorus

Such are his powers, whom time hath styled
Now swift, now slow, now tame, now wild;
Now hot, now cold, now fierce, now mild;   15
The eldest god, yet still a child.

37

An Epistle to a Friend

Sir, I am thankful, first to heaven for you;
   Next to yourself, for making your love true;
   Then to your love and gift. And all's but due.

You have unto my store added a book,
   On which with profit I shall never look   5
   But must confess from whom what gift I took.

Not like your country neighbours, that commit
   Their vice of loving for a Christmas fit,
   Which is indeed but friendship of the spit;

But as a friend, which name yourself receive,   10
   And which you, being the worthier, gave me leave
   In letters, that mix spirits, thus to weave.

Which, how most sacred I will ever keep,
   So may the fruitful vine my temples steep,
   And fame wake for me, when I yield to sleep.   15

Though you sometimes proclaim me too severe,
   Rigid, and harsh, which is a drug austere
   In friendship, I confess: but, dear friend, hear:

36. 16. *For*. 10. 19 and n.

37. 9 *friendship of the spit*: as in *Und*. 45. 8; cf. the proverbial 'trencher-friendship',
Tilley, F762.   12 *mix spirits*: cf. Donne's 'To Sir Henry Wotton', l. 1: 'Sir, more than
kisses, letters mingle souls' (*Satires*, p. 71).

Little know they that profess amity,
  And seek to scant her comely liberty,   20
  How much they lame her in her property.

And less they know, who being free to use
  That friendship which no chance, but love, did choose,
  Will unto licence that fair leave abuse.

It is an act of tyranny, not love,   25
  In practised friendship wholly to reprove,
  As flattery with friends' humours still to move.

From each of which I labour to be free;
  Yet if with either's vice I tainted be,
  Forgive it as my frailty, and not me.   30

For no man lives so out of passion's sway,
  But shall sometimes be tempted to obey
  Her fury, yet no friendship to betray.

19–33 These lines are repeated in *U.V.* 49. 12–26.   20 *scant*: restrict, do small justice
to.   27 As it is an act of flattery always to fall in with a friend's whims.

38

An Elegy

'Tis true, I'm broke! Vows, oaths, and all I had
   Of credit lost. And I am now run mad,
Or do upon myself some desperate ill;
   This sadness makes no approaches but to kill.
It is a darkness hath blocked up my sense,   5
   And drives it in to eat on my offence,
Or there to starve it. Help, O you that may
   Alone lend succours, and this fury stay,
Offended mistress; you are yet so fair,
   As light breaks from you that affrights despair,   10
And fills my powers with persuading joy
   That you should be too noble to destroy.
There may some face or menace of a storm
   Look forth, but cannot last in such a form.
If there be nothing worthy you can see   15
   Of graces, or your mercy here in me,
Spare your own goodness yet, and be not great
   In will and power, only to defeat.
God, and the good, know to forgive and save;
   The ignorant and fools no pity have.   20
I will not stand to justify my fault,
   Or lay the excuse upon the vintner's vault,
Or in confessing of the crime be nice,
   Or go about to countenance the vice,
By naming in what company 'twas in,   25
   As I would urge authority for sin.
No, I will stand arraigned and cast, to be
   The subject of your grace in pardoning me,
And, styled your mercy's creature, will live more
   Your honour now than your disgrace before.   30

---

38. The authorship of this poem and of *Und.* 39, 40, and 41 has been the subject of doubt. *Und.* 39 was printed with Donne's *Poems* in 1633, and all four poems have from time to time been thought to be Donne's. Evelyn Simpson, *RES*, xv (1939), 274–82, argues persuasively for the ascription of *Und.* 38, 40, and 41 to Jonson. Helen Gardner, agreeing with Mrs. Simpson that *Und.* 39 is not Jonson's, suspects it may have been written not by Donne but by Sir Thomas Roe: see *Elegies*, pp. xxxv–xxxviii, 224–5. 5 *blocked up*: cf. *Und.* 71. 10. 17 *goodness . . . great*: see *Epig.* Ded. 15–16 n. 27 *cast*: found guilty. 30 Seneca, *De Clementia*, I. xxi. 2.

Think it was frailty, mistress, think me man,
　Think that yourself, like heaven, forgive me can;
Where weakness doth offend, and virtue grieve,
　There greatness takes a glory to relieve.
Think that I once was yours, or may be now; 35
　Nothing is vile that is a part of you.
Error and folly in me may have crossed
　Your just commands, yet those, not I, be lost.
I am regenerate now, become the child
　Of your compassion. Parents should be mild; 40
There is no father that for one demerit,
　Or two, or three, a son will disinherit;
That as the last of punishments is meant:
　No man inflicts that pain till hope be spent.
An ill-affected limb, whate'er it ail, 45
　We cut not off till all cures else do fail;
And then with pause; for severed once, that's gone
　Would live his glory, that could keep it on.
Do not despair my mending; to distrust
　Before you prove a medicine, is unjust. 50
You may so place me, and in such an air,
　As not alone the cure, but scar be fair.
That is, if still your favours you apply,
　And not the bounties you have done, deny.
Could you demand the gifts you gave again? 55
　Why was't? Did e're the clouds ask back their rain?
The sun his heat and light, the air his dew,
　Or winds the spirit by which the flower so grew?
That were to wither all, and make a grave
　Of that wise nature would a cradle have. 60
Her order is to cherish and preserve;
　Consumption's nature to destroy and starve.
But to exact again what once is given
　Is nature's mere obliquity!—as heaven
Should ask the blood and spirits he hath infused 65
　In man, because man hath the flesh abused.
O may your wisdom take example hence:
　God lightens not at man's each frail offence;
He pardons slips, goes by a world of ills,
　And then his thunder frights more than it kills. 70

40–50 Ibid., I. xiv. 1–3*.　49–52 Ibid., I. xvii. 2*; cf. *Und.* 33. 21.　67 ff. *De Clementia*, I. vii. 1–3.

He cannot angry be, but all must quake,
    It shakes even him that all things else doth shake.
And how more fair and lovely looks the world
    In a calm sky, than when the heaven is hurled
About in clouds, and wrapt in raging weather,   75
    As all with storm and tempest ran together.
O imitate that sweet serenity
    That makes us live, not that which calls to die.
In dark and sullen morns, do we not say,
    This looketh like an execution day?   80
And with the vulgar doth it not obtain
    The name of cruel weather, storm, and rain?
Be not affected with these marks too much
    Of cruelty, lest they do make you such.
But view the mildness of your Maker's state,   85
    As I the penitent's here emulate:
He, when he sees a sorrow such as this,
    Straight puts off all his anger, and doth kiss
The contrite soul, who hath no thought to win
    Upon the hope to have another sin   90
Forgiven him. And in that line stand I
    Rather than once displease you more, to die;
To suffer tortures, scorn, and infamy;
    What fools, and all their parasites can apply,
The wit of ale and genius of the malt   95
    Can pump for, or a libel without salt
Produce; though threatening with a coal or chalk
    On every wall, and sung where'er I walk.
I number these as being of the chore
    Of contumely, and urge a good man more   100
Than sword, or fire, or what is of the race
    To carry noble danger in the face:
There is not any punishment, or pain,
    A man should fly from, as he would disdain.
Then mistress, here, here let your rigour end,   105
    And let your mercy make me ashamed to offend.
I will no more abuse my vows to you
    Than I will study falsehood, to be true.

95–6 Cf. *B.F.* I. i. 33–41.   96 *pump for*: work for.   99 *chore*: band, company.   101 *of the race*: of that kind.   102 *danger*: perhaps, disdain (cf. *OED* † 2).

Oh, that you could but by dissection see
    How much you are the better part of me;   110
How all my fibres by your spirit do move;
    And that there is no life in me, but love.
You would be then most confident, that though
    Public affairs command me now to go
Out of your eyes, and be awhile away,   115
    Absence or distance shall not breed decay.
Your form shines here, here fixed in my heart:
    I may dilate myself, but not depart.
Others by common stars their courses run,
    When I see you, then I do see my sun,   120
Till then 'tis all but darkness that I have;
    Rather than want your light, I wish a grave.

[For *Und*. 39, see *Dubia*, 1]

40

An Elegy

That love's a bitter sweet I ne'er conceive
    Till the sour minute comes of taking leave,
And then I taste it. But as men drink up
    In haste the bottom of a medicined cup,   5
And take some syrup after, so do I,
    To put all relish from my memory
Of parting, drown it in the hope to meet
    Shortly again, and make our absence sweet.
This makes me, mistress, that sometime by stealth,   10
    Under another name, I take your health,
And turn the ceremonies of those nights
    I give or owe my friends, into your rites:
But ever without blazon, or least shade
    Of vows so sacred, and in silence made;

114–15 Possibly a masquing expedition out of London, as H & S suggest, though the lines seem to hint at a weightier mission.

40. See *Und*. 38 n.   13 *blazon*: show (literally, of heraldic arms); i.e., divulging of identity.

For though love thrive, and may grow up with cheer   15
    And free society, he's born elsewhere,
And must be bred so to conceal his birth,
    As neither wine do rack it out, or mirth.
Yet should the lover still be airy and light,
    In all his actions rarified to sprite;   20
Not, like a Midas, shut up in himself,
    And turning all he toucheth into pelf,
Keep in, reserved, in his dark-lantern face,
    As if that excellent dullness were love's grace;
No, mistress, no, the open merry man   25
    Moves like a sprightly river, and yet can
Keep secret in his channels what he breeds,
    'Bove all your standing waters, choked with weeds.
They look at best like cream bowls, and you soon
    Shall find their depth: they're sounded with a spoon.   30
They may say grace, and for love's chaplains pass,
    But the grave lover ever was an ass;
Is fixed upon one leg, and dares not come
    Out with the other, for he's still at home;
Like the dull wearied crane that, come on land,   35
    Doth, while he keeps his watch, betray his stand.
Where he that knows will, like a lapwing, fly
    Far from the nest, and so himself belie
To others as he will deserve the trust
    Due to that one that doth believe him just.   40
And such your servant is, who vows to keep
    The jewel of your name as close as sleep
Can lock the sense up, or the heart a thought,
    And never be by time or folly brought,
Weakness of brain, or any charm of wine,   45
    The sin of boast, or other countermine
(Made to blow up love's secrets) to discover
    That article may not become your lover:
Which in assurance to your breast I tell,
    If I had writ no word but Dear, farewell.   50

21 *Not ... shut up in himself*: cf. Truewit's advice in *S.W.* IV. i. 55–66.   29–30 Cf. *Disc.* 716–18.

41

An Elegy

Since you must go, and I must bid farewell,
  Hear, mistress, your departing servant tell
What it is like, and do not think they can
  Be idle words, though of a parting man:
It is as if a night should shade noonday,   5
  Or that the sun was here, but forced away,
And we were left under that hemisphere
  Where we must feel it dark for half a year.
What fate is this, to change men's days and hours,
  To shift their seasons and destroy their powers!   10
Alas, I ha' lost my heat, my blood, my prime,
  Winter is come a quarter ere his time;
My health will leave me; and when you depart
  How shall I do, sweet mistress, for my heart?
You would restore it? No, that's worth a fear   15
  As if it were not worthy to be there:
Oh, keep it still, for it had rather be
  Your sacrifice than here remain with me.
And so I spare it. Come what can become
  Of me, I'll softly tread unto my tomb;   20
Or like a ghost walk silent amongst men,
  Till I may see both it and you again.

42

An Elegy

Let me be what I am: as Virgil cold,
  As Horace fat, or as Anacreon old;
No poet's verses yet did ever move,
  Whose readers did not think he was in love.

---

41. See *Und.* 38 n. Cf. 'His Parting From Her', formerly attributed to Donne: 'Since
she must go, and I must mourn, come night'.

42. Written about 1624. For criticism, see Barbara Hutchison, *ELN*, ii (1965), 185–
90.   1–2 Suetonius, *Vita Virgili*, 10–11, *Vita Horatii*.   3–4 A classical commonplace;
e.g. Cicero, *De Oratore*, II. xlv.

Who shall forbid me then in rhythm to be   5
   As light and active as the youngest he
That from the muses' fountains doth endorse
   His lines, and hourly sits the poet's horse?
Put on my ivy garland; let me see
   Who frowns, who jealous is, who taxeth me.   10
Fathers and husbands, I do claim a right
   In all that is called lovely: take my sight
Sooner than my affection from the fair.
   No face, no hand, proportion, line, or air
Of beauty, but the muse hath interest in;   15
   There is not worn that lace, purl, knot, or pin,
But is the poet's matter; and he must,
   When he is furious, love, although not lust.
But then consent, your daughters and your wives,
   If they be fair and worth it, have their lives   20
Made longer by our praises. Or, if not,
   Wish you had foul ones and deformed got,
Cursed in their cradles, or there changed by elves,
   So to be sure you do enjoy yourselves.
Yet keep those up in sackcloth too, or leather,   25
   For silk will draw some sneaking songster thither.
It is a rhyming age, and verses swarm
   At every stall; the city cap's a charm.
But I who live, and have lived, twenty year
   Where I may handle silk as free and near   30
As any mercer, or the whale-bone man
   That quilts those bodies, I have leave to span;
Have eaten with the beauties and the wits
   And braveries of court, and felt their fits
Of love and hate, and came so nigh to know   35
   Whether their faces were their own or no;
It is not likely I should now look down
   Upon a velvet petticoat or a gown,
Whose like I have known the tailor's wife put on
   To do her husband's rites in, ere 'twere gone   40

9 *ivy garland*: of Bacchus, seen as god of inspiration.   11–12 Distantly following Ovid, *Amores*, II. iv.   16 *purl*: lace, frill.   28 *city cap*: cf. Crispinus's verses on the velvet cap of a jeweller's wife, *Poet*. III. i.   29 ff. Jonson alludes to his involvement with masquing at court.   31 *whale-bone*: for farthingales, to hold out skirts of kirtles.   32 *span*: measure with outstretched hand.   36 Cf. *Sej*. I. 307.   39–42 An idea more fully developed in *N.I*. IV. iii. 63–73.

Home to the customer; his lechery
   Being, the best clothes still to preoccupy.
Put a coach-mare in tissue, must I horse
   Her presently? or leap thy wife of force,
When by thy sordid bounty she hath on   45
   A gown of that was the caparison?
So I might dote upon thy chairs and stools
   That are like clothed: must I be of those fools
Of race accounted, that no passion have
   But when thy wife, as thou conceiv'st, is brave?   50
Then ope thy wardrobe, think me that poor groom
   That from the footman, when he was become
An officer there, did make most solemn love
   To every petticoat he brushed, and glove
He did lay up, and would adore the shoe   55
   Or slipper was left off, and kiss it too;
Court every hanging gown, and after that
   Lift up some one and do I'll tell not what.
Thou didst tell me, and wert o'erjoyed to peep
   In at a hole, and see these actions creep   60
From the poor wretch, which, though he played in prose,
   He would have done in verse with any of those
Wrung on the withers by Lord Love's despite,
   Had he'd the faculty to read and write!
Such songsters there are store of: witness he   65
   That chanced the lace laid on a smock to see
And straightway spent a sonnet; with that other
   That (in pure madrigal) unto his mother
Commended the French hood and scarlet gown
   The Lady Mayoress passed in through the town   70
Unto the Spittle sermon. Oh, what strange
   Variety of silks were on the Exchange,

---

42 *preoccupy*: have first use of; the word also had a bawdy sense. Cf. *N.I.* IV. iii. 79–
80.   50 *brave*: finely dressed.   63 *withers*: Hutchison (p. 189) sees a secondary refer-
ence to George Wither. ('Withers' was in fact an alternative form of the poet's name;
he was attacked by Jonson in *Time Vind.* in 1623.)   67 *spent*: with a play on the sexual
sense.   69 *French hood*: with a round front, framing the face; out-of-date with all but
citizens' wives by the early seventeenth century.   71 *Spittle sermon*: preached in
Easter week near St. Mary Spittle of Bishopsgate Without, and attended by civic
dignitaries.   72 *Exchange*: the new Exchange on the south side of the Strand; oc-
cupied by milliners' shops; a fashionable resort for ladies.

Or in Moorfields this other night! sings one;
    Another answers, 'las, those silks are none,
In smiling l'envoy, as he would deride   75
    Any comparison had with his Cheapside.
And vouches both the pageant and the day,
    When not the shops but windows do display
The stuffs, the velvets, plushes, fringes, lace,
    And all the original riots of the place.   80
Let the poor fools enjoy their follies, love
    A goat in velvet, or some block could move
Under that cover, an old midwife's hat,
    Or a close-stool so cased, or any fat
Bawd in a velvet scabbard! I envy   85
    None of their pleasures, nor will ask thee why
Thou art jealous of thy wife's or daughter's case:
    More than of either's manners, wit, or face.

43

An Execration upon Vulcan

And why to me this, thou lame lord of fire,
    What had I done that might call on thine ire?
Or urge thy greedy flame thus to devour
    So many my years' labours in an hour?
I ne'er attempted, Vulcan, 'gainst thy life,   5
    Nor made least line of love to thy loose wife,
Or in remembrance of thy affront and scorn,
    With clowns and tradesmen, kept thee closed in horn.
'Twas Jupiter that hurled thee headlong down,
    And Mars that gave thee a lanthorn for a crown.   10

---

73 *Moorfields*: flat marshy ground just north of the city; laundresses dried their clothes here.   75 *l'envoy*: after-thought; literally, a concluding stanza.   76 *Cheapside*: at the east end of which were mercers' shops.   82 *block*: wooden dummy for hat; blockhead: for the play, cf. *S. of N.* I. ii. 133.   85 *scabbard*: i.e. dress.   87 *case*: clothing.

43. On the fire of Nov. 1623, which destroyed Jonson's library and many of his unpublished writings. Chapman's 'Invective written against Mr. Ben Jonson' (*Poems*, ed. P. B. Bartlett, pp. 374–8) was evidently prompted by this poem; see R. B. Sharpe, *SP*, xlii (1945), 555–63, and H & S, x. 692–7.   1 *lame*: see note to ll. 111–17 below.   8 *in horn*: in a lantern. Lanterns were made of horn; hence, by popular etymology, the spelling 'lanthorn': in l. 10, a punning allusion to Vulcan's cuckolding.

Was it because thou wert of old denied
   By Jove to have Minerva for thy bride
That, since, thou tak'st all envious care and pain
   To ruin any issue of the brain?
Had I wrote treason there, or heresy,   15
   Imposture, witchcraft, charms, or blasphemy,
I had deserved, then, thy consuming looks;
   Perhaps to have been burned with my books.
But, on thy malice, tell me, didst thou spy
   Any least loose or scurrile paper lie   20
Concealed or kept there, that was fit to be,
   By thy own vote, a sacrifice to thee?
Did I there wound the honour of the crown,
   Or tax the glories of the church and gown,
Itch to defame the state, or brand the times,   25
   And myself most, in some self-boasting rhymes?
If none of these, then why this fire? Or find
   A cause before, or leave me one behind.
Had I compiled from Amadis de Gaul,
   The Esplandians, Arthurs, Palmerins, and all   30
The learned library of Don Quixote,
   And so some goodlier monster had begot;
Or spun out riddles, and weaved fifty tomes
   Of logogriphs, and curious palindromes;
Or pumped for those hard trifles, anagrams,   35
   Or eteostics, or those finer flams

11–12 Jove agreed that Vulcan might marry Minerva, but privately persuaded
Minerva to refuse him.   14 *issue of the brain*: Minerva sprang from Jove's head, cleft
open by Vulcan (see *For.* 10. 13–15). Jonson wryly contrasts the fate of his own
brain-children. 20 *scurrile*: scurrilous. 29 ff. *Amadis de Gaul*, early sixteenth-
century romance by Garcia de Montalvo, from older originals. Its fifth book concerns
Esplandian, son of Amadis; separate continuations of his adventures were also pub-
lished. *Palmerin d'Oliva* and *Palmerin of England*, romances translated into English by
Antony Munday (ridiculed in *C. is A.*). *The Adventures of Splandian* and *Palmerin
d'Oliva* join the bonfire of romances from Don Quixote's library made by Cervantes's
curate and barber, who nevertheless spare *Amadis de Gaul* and *Palmerin of England*
(*Don Quixote*, I. vi).   34 ff. Some of these 'courtly trifles' are described by Thomas
Puttenham in *The Art of English Poesie* (1589), ed. G. D. Willcock and A. Walker
(Cambridge, 1936), Bk. II, ch. xi.   34 *logogriphs*: 'A kind of enigma, in which a certain
word, and other words that can be formed out of all or any of its letters, are to be
guessed from synonyms of them introduced into a set of verses' (*OED*).   34 *palin-
dromes*: words, phrases, or verses that read the same backwards as forwards: e.g. 'A
man, a plan, a canal: Panama'.   35 *pumped*: laboured.   35 *anagrams*: See Puttenham,
ed. cit., pp. 108 ff. Jonson expresses his contempt for anagrams again in *Conv.*

Of eggs, and halberds, cradles and a hearse,
    A pair of scissors and a comb in verse,
Acrostics and telestichs on jump names,
    Thou then hadst had some colour for thy flames   40
On such my serious follies. But, thou'lt say,
    There were some pieces of as base allay,
And as false stamp there: parcels of a play,
    Fitter to see the fire-light than the day,
Adulterate moneys, such as might not go;   45
    Thou shouldst have stayed till public fame said so.
She is the judge, thou executioner;
    Or if thou needs wouldst trench upon her power,
Thou mightst have yet enjoyed thy cruelty
    With some more thrift and more variety: 50
Thou mightst have had me perish piece by piece,
    To light tobacco, or save roasted geese,
Singe capons, or poor pigs, dropping their eyes;
    Condemned me to the ovens with the pies,
And so have kept me dying a whole age,   55
    Not ravished all hence in a minute's rage.
But that's a mark whereof thy rites do boast,
    To make consumption ever, where thou goest.
Had I foreknown of this thy least desire
    To have held a triumph or a feast of fire,   60
Especially in paper, that that steam
    Had tickled your large nostril, many a ream
To redeem mine I had sent in: Enough!
    Thou should'st have cried, and all been proper stuff.

Dr. 437–9, but employs them in *Hym.* 232–3 ('Iuno'/'Unio'), *U.V.* 40. 23–4 ('Celia'/'Alice'), and elsewhere.   36 *eteostics*: or chronograms, in which certain letters indicate a date or a numerical value: e.g. 'LorD haVe MerCIe Vpon Vs', for a day of national humiliation in 1666 (the date being also the sum of the capitalized letters) (*OED*).   36 *flams*: fanciful compositions, conceits.   37–8 See Puttenham, ed. cit. pp. 91 ff., for illustrations of verses in similarly fanciful shapes.   39 *Acrostics*: a form Jonson occasionally in fact used, e.g. *Epig.* 40.   39 *telestichs*: a kind of acrostic, in which the final (rather than the initial) letters of each line of verse spell a word or phrase.   39 *jump*: coinciding, exactly equivalent.   43 *parcels*: parts; probably of *The Staple of News*: see G. B. Johnston, *MLN*, xlvi (1931), 150–3.   45 *go*: pass as current.   52 *light tobacco*: cf. *Poet.* Apol. Dial. 171–2.   54 *with the pies*: cf. Dryden, *Mac Flecknoe*, l. 101 ('Martyrs of pies, and relics of the bum'), Addison, *The Spectator*, no. 85, Fielding, *Tom Jones*, IV. i, etc.   55 *a whole age*: the pies being continually rejected, and continually warmed up for resale: cf. *Epig.* 133. 149 ff.

The Talmud and the Alcoran had come,   65
  With pieces of the *Legend*; the whole sum
Of errant knighthood, with the dames and dwarfs,
  The charmed boats, and the enchanted wharves;
The Tristrams, Lancelots, Turpins and the Peers,
  All the mad Rolands, and sweet Olivers,   70
To Merlin's marvels and his cabal's loss,
  With the chimera of the Rosy Cross,
Their seals, their characters, hermetic rings,
  Their gem of riches, and bright stone that brings
Invisibility, and strength, and tongues;   75
  The Art of Kindling the True Coal, by Lungs:
With Nicholas Pasquill's *Meddle With Your Match*,
  And the strong lines, that so the time do catch;
Or Captain Pamphlet's horse and foot, that sally
  Upon the Exchange, still, out of Pope's Head Alley;   80
The weekly *Courants*, with Paul's seal, and all
  The admired discourses of the prophet Ball:
These, hadst thou pleased either to dine or sup,
  Had made a meal for Vulcan to lick up.

---

66 *the* Legend: or *Golden Legend*, a medieval manual of ecclesiastical lore.   68 *charmed boats*: such as Guingelot, the magic boat of Wade.   69–70 *Tristrams, Lancelots*: of Arthurian legend.   69 *Turpins*: Turpin, eighth-century A.D. archbishop of Rheims; thought to have been at Roncesvalles, and to have chronicled the deeds of Charlemagne. 69 *Peers*: or Paladins, who accompanied Charlemagne. Roland, the most famous of the Peers, is celebrated by Boiardo and (mad) by Ariosto; Oliver is his companion.   71 *his cabal's loss*: the loss of Merlin's secret art after his infatuation with Niviene; see *The Works of Sir Thomas Malory*, ed. E. Vinaver (Oxford, 1967), i. 125–32.   72 ff. *chimera of the Rosy Cross*: Rosicrucianism, first heard of in England around 1614, though allegedly founded much earlier by (the probably mythical) Christian Rosenkreuz (1378–1484). In 1652 Thomas Vaughan was to defend 'that admirable chimera' of Rosicrucianism, and to describe some of its mysteries (e.g. the properties of the 'bright stone' of l. 74): *Works*, ed. A. E. Waite (London, 1919), pp. 343–76. Members of the fraternity were bound by seals of secrecy, and devised a magic writing, based on divine 'characters' which they found in nature and the bible.   76 *Lungs*: one who blows an alchemist's fire: see *Alch*. II. i. 27.   77 Probably a lost pamphlet by Nicholas Breton (?1545–?1626), who is addressed in *U.V.* 2. The title is proverbial: Tilley, M747; cf. *E.M.I.* III. v. 121, *B.F.* I. iv. 102.   78 *strong lines*: for the term, see George Williamson, *Seventeenth-Century Contexts* (London, 1960), pp. 120–31.   79–80 The 'Letters of News' of Captain Thomas Gainford (see *Epig*. 107 n.), ostensibly written from foreign parts but in fact from Pope's Head Alley, home of printers and booksellers.   81 *weekly* Courants: the news-sheets of the printer and journalist Nathaniel Butter, published from St. Paul's Churchyard.   82 *prophet Ball*: a mad tailor who prophesied that King James would become Pope.   84 *Vulcan to lick up*: cf. Pope, *Dunciad* (A), iii. 73: 'From shelves to shelves see greedy Vulcan roll'.

But in my desk what was there to accite   85
    So ravenous and vast an appetite?
I dare not say a body, but some parts
    There were of search, and mastery in the arts.
All the old Venusine in poetry,
    And lighted by the Stagirite, could spy   90
Was there made English; with a Grammar too,
    To teach some that their nurses could not do,
The purity of language; and among
    The rest, my journey into Scotland sung,
With all the adventures; three books not afraid   95
    To speak the fate of the Sicilian maid
To our own ladies; and in story there
    Of our fifth Henry, eight of his nine year;
Wherein was oil, beside the succour, spent
    Which noble Carew, Cotton, Selden lent;   100
And twice twelve years' stored-up humanity,
    With humble gleanings in divinity,
After the fathers, and those wiser guides
    Whom faction had not drawn to study sides.
How in these ruins, Vulcan, thou dost lurk,   105
    All soot and embers, odious as thy work!
I now begin to doubt if ever grace
    Or goddess could be patient of thy face.
Thou woo Minerva! or to wit aspire!
    'Cause thou canst halt, with us, in arts and fire!   110

---

85 *accite*: arouse, excite. 88 *of search*: of an introspective and reflective nature. 88 *mastery in the arts*: a reference to Jonson's Oxford M.A. of 1619? See the marginal note to the MS. of Chapman's 'Invective', *Poems*, ed. Bartlett, p. 478. 89–91 Horace's *Ars Poetica*, translated and commented upon in the light of the *Poetics*, from which it was thought to derive. (Venusia was Horace's birthplace, Stagira Aristotle's.) The commentary was read to Drummond in 1619: *Conv. Dr.* 82–5. Two versions of Jonson's translation of *Ars Poetica* were published in 1640. 91 *a Grammar*: later rewritten, though not finished. 94 *journey into Scotland*: by foot, 1618–19. Drummond speaks of Jonson's intention of writing this account, *Conv. Dr.* 406–8. 96 *the Sicilian maid*: John Barclay's *Argenis* (1621), a Latin romance which alluded to recent political events in Europe; James I invited Jonson to translate it in 1622. 100 *Carew, Cotton, Selden*: Richard Carew (1555–1620) and Sir Robert Cotton (1571–1631), the antiquaries, and Sir John Selden (1584–1654), the jurist (see *Und.* 14 n.). Some books about Henry V, borrowed from Cotton's library, are known to have perished in Jonson's fire. 101 *stored-up humanity*: evidently a commonplace-book, perhaps along the lines of *Discoveries*. 102 *gleanings in divinity*: probably written after 1610, when Jonson abandoned Catholicism.

Son of the wind—for so thy mother, gone
    With lust, conceived thee; father thou hadst none;
When thou wert born and that thou look'st at best,
    She durst not kiss, but flung thee from her breast.
And so did Jove, who ne'er meant thee his cup: 115
    No mar'l the clowns of Lemnos took thee up,
For none but smiths would have made thee a god.
    Some alchemist there may be yet, or odd
Squire of the squibs, against the pageant day
    May to thy name a *Vulcanale* say,   120
And for it lose his eyes with gunpowder,
    As the other may his brains with quicksilver.
Well fare the wise men yet, on the Bankside,
    My friends the watermen! They could provide
Against thy fury, when to serve their needs   125
    They made a Vulcan of a sheaf of reeds,
Whom they durst handle in their holiday coats,
    And safely trust to dress, not burn, their boats.
But, O those reeds! Thy mere disdain of them
    Made thee beget that cruel stratagem   130
(Which some are pleased to style but thy mad prank)
    Against the Globe, the glory of the Bank.
Which, though it were the fort of the whole parish,
    Flanked with a ditch and forced out of a marish,
I saw with two poor chambers taken in   135
    And razed, ere thought could urge, This might have been!

111–17 For the tradition that Vulcan had no father, see Apollodorus, *Bibliotheca*, I. iii. 5, Hyginus, *Fabulae*, preface; Juno's impregnation by the wind is Jonson's invention. Frightened by Vulcan's appearance, Juno is said to have thrown the child from heaven (*Iliad*, xviii. 394–8). According to other traditions, Jove rejected Vulcan as his cup-bearer ('his cup', l. 115), and threw him from heaven when he tried to intervene in a dispute on behalf of his mother. He landed on Lemnos, was lamed by his fall, and was cared for by peasants.   119 *Squire of the squibs*: John Squire, deviser of the Lord Mayor's show of 30 Oct. 1620 (Fleay, i. 329; J. Nichols, *Progresses . . . of King James I* (London, 1828), iv. 619–27).   120 Vulcanale: hymn to Vulcan.   123 *Bankside*: on the south bank of the Thames, between St. Saviour's Church and the modern Blackfriars Bridge.   126 *a Vulcan of a sheaf of reeds*: i.e. a torch or beacon, made from the reeds which then grew plentifully along the Thames. Boats were sometimes adorned in this way on festive occasions, e.g. for the Lord Mayor's procession: see F. W. Fairholt, *Lord Mayors' Pageants* (London, 1843), p. 11.   132 *the Globe*: the famous theatre was burnt to the ground on 29 June 1613, during a performance of *Henry VIII*. It was rebuilt the following year.   134 The land on which the Globe stood was formerly a marsh; surrounding ditches were for drainage and sewage. See J. C. Adams, *The Globe Playhouse* (London, 1961), p. 13.   135 *chambers*: pieces of ordnance, which caused the fire.

See the world's ruins, nothing but the piles
   Left! and wit since to cover it with tiles.
The brethren, they straight noised it out for news,
   'Twas verily some relic of the stews   140
And this a sparkle of that fire let loose
   That was raked up in the Winchestrian goose
Bred on the Bank, in time of Popery,
   When Venus there maintained the mystery.
But others fell with that conceit by the ears,   145
   And cried, it was a threatening to the bears,
And that accursed ground, the Parish Garden;
   Nay, sighed a sister, 'twas the nun, Kate Arden,
Kindled the fire! But then, did one return,
   No fool would his own harvest spoil or burn!   150
If that were so, thou rather wouldst advance
   The place that was thy wife's inheritance.
O no! cried all, Fortune, for being a whore,
   Scaped not his justice any jot the more;
He burnt that idol of the revels too: 155
   Nay, let Whitehall with revels have to do,
Though but in dances, it shall know his power;
   There was a judgement shown too in an hour.
He is true Vulcan still! He did not spare
   Troy, though it were so much his Venus' care.   160
Fool, wilt thou let that in example come?
   Did not she save from thence to build a Rome?
And what hast thou done in these petty spites,
   More than advanced the houses and their rites?

---

137 *the world's ruins*: playing on the theatre's name.   138 *tiles*: instead of the original
thatch ('those reeds!', l. 129), in which the fire began.   139 *The brethren*:
Puritans, who saw the burning of the Globe as a divine judgement: see E.N.S.
Thompson, *The Controversy Between the Puritans and the Stage* (New York, 1903), p.
151.   142 *Winchestrian goose*: venereal disease (the brothels of the Bankside were
within the liberty of the Bishop of Winchester).   145 *fell with that conceit by the ears*:
quarrelled with that idea.   147 *Parish Garden*: or Paris Garden, the centre for bull-
and bear-baiting; the place had a bad reputation. Cf. *Epig.* 133. 117.   148 *the nun,
Kate Arden*: the unsavoury lady of *Epig.* 133. 118. Three MSS. read 'Venus' nun':
evidently Jonson is remembering Marlowe's phrase in *Hero and Leander*, i. 45, 'So
lovely fair was Hero, Venus' nun'.   153 *Fortune, for being a whore*: the Fortune theatre
burnt to the ground on 9 Dec. 1621. The phrase was probably proverbial: cf. *N.I.* II.
v. 132, *Hamlet*, II. ii. 232-3, and Webster, *The White Devil*, I. i. 4, 'Fortune's a right
whore'.   156 The banqueting house at Whitehall burnt down on 12 Jan. 1618.
160 *Venus' care*: as evidenced in the *Aeneid*.

I will not argue thee, from those, of guilt,   165
    For they were burnt but to be better built.
'Tis true that in thy wish they were destroyed,
    Which thou hast only vented, not enjoyed.
So wouldst thou have run upon the Rolls by stealth,
    And didst invade part of the commonwealth   170
In those records which, were all chroniclers gone,
    Will be remembered by six clerks to one.
But say, all six good men, what answer ye?
    Lies there no writ out of the Chancelry
Against this Vulcan, no injunction,   175
    No order, no decree? Though we be gone
At common law, methinks in his despite
    A court of Equity should do us right,
But to confine him to the brew-houses,
    The glass-house, dye-vats, and their furnaces;   180
To live in sea-coal and go forth in smoke,
    Or—lest that vapour might the city choke—
Condemn him to the brick-kilns, or some hill-
    Foot (out in Sussex) to an iron mill;
Or in small faggots have him blaze about   185
    Vile taverns, and the drunkards piss him out;
Or in the bellman's lanthorn, like a spy,
    Burn to a snuff, and then stink out and die.
I could invent a sentence yet were worse,
    But I'll conclude all in a civil curse: 190
Pox on your flameship, Vulcan; if it be
    To all as fatal as it hath been to me,
And to Paul's steeple, which was unto us
    'Bove all your fireworks had at Ephesus
Or Alexandria; and though a divine   195
    Loss, remains yet as unrepaired as mine.
Would you had kept your forge at Aetna still,
    And there made swords, bills, glaives, and arms your fill,

169 *the Rolls*: the Six Clerks' Office in Chancery Lane was burnt on 20 Dec. 1621. 187 *bellman's lanthorn*: the bellman was a nightwatchman, who called the hours; he commonly carried a lantern, staff, and bell. 187–8 *like a spy*: cf. *Epig.* 59. 193 *Paul's steeple*: struck by lightning and burnt, 4 June 1561; it was not re-erected. 194 *Ephesus*: where Herostratus burnt down the Temple of Diana in 356 B.C. 195 *Alexandria*: in A.D. 640, after Arabian conquest, it is said that books from the great library of Alexandria were used for six months to feed the furnaces of the public baths. 198 *glaives*: swords, bills, or lances.

Maintained the trade at Bilbo, or elsewhere,
    Struck in at Milan with the cutlers there,   200
Or stayed but where the friar and you first met,
    Who from the Devil's Arse did guns beget,
Or fixed in the Low Countries, where you might
    On both sides do your mischiefs with delight,
Blow up and ruin, mine and countermine,   205
    Make your petards and granats, all your fine
Engines of murder, and receive the praise
    Of massacring mankind so many ways.
We ask your absence here, we all love peace,
    And pray the fruits thereof and the increase;   210
So doth the king, and most of the king's men
    That have good places: therefore once again
Pox on thee, Vulcan, thy Pandora's pox,
    And all the evils that flew out of her box
Light on thee; or if those plagues will not do,   215
    Thy wife's pox on thee, and Bess Broughton's too.

44

A Speech according to Horace

Why yet, my noble hearts, they cannot say
    But we have powder still for the King's Day,
And ordnance too; so much as from the Tower
    To have waked, if sleeping, Spain's ambassador,

199–200 *Bilbo . . . Milan*: Bilboa and Milan were famous for their swords.   201 *the friar*: Roger Bacon, who was (erroneously) thought to have invented gunpowder (H & S); or—more probably—Konstantin Anklitzen, a Franciscan friar and chemist otherwise known as Berthold Schwarz, who was accredited with the invention of the gun, *c.* 1320 (Newdigate).   202 *Devil's Arse*: a cavern near Castleton, in the Peak of Derbyshire.   204 *both sides*: the Dutch and the Spanish..   206 *petards*: small explosive devices used to breach walls, etc.   206 *granats*: grenades.   213 *Pandora's pox*: Pandora and her box of evils were created by Vulcan at Jove's command, to punish mankind for Prometheus's theft of divine fire. Jonson insinuates that Pandora suffered from one of her own evils, the pox.   216 *Bess Broughton's*: a famous courtesan, who died of the pox; her career is described by John Aubrey, *Brief Lives*, ed. O. L. Dick (London, 1949), pp. 40–1.

44. *A Speech according to Horace*: 'Speech' is Jonson's rendering of Horace's *sermo*; as H & S point out, the poem also recalls Horace's attacks on national degeneracy in *Odes*

Old Aesop Gondomar: the French can tell,   5
　For they did see it the last tilting well,
That we have trumpets, armour, and great horse,
　Lances, and men, and some a breaking force.
They saw too store of feathers, and more may,
　If they stay here but till Saint George's Day.   10
All ensigns of a war are not yet dead,
　Nor marks of wealth so from our nation fled
But they may see gold chains and pearl worn then,
　Lent by the London dames to the lords' men;
Withal, the dirty pains those citizens take   15
　To see the pride at court their wives do make;
And the return those thankful courtiers yield
　To have their husbands drawn forth to the field,
And coming home, to tell what acts were done
　Under the auspice of young Swinnerton.   20
What a strong fort old Pimlico had been,
　How it held out, how (last) 'twas taken in!
Well, I say, thrive; thrive, brave Artillery Yard,
　Thou seed-plot of the war, that hast not spared
Powder or paper to bring up the youth   25
　Of London in the military truth

I. xxxv. 33–40, III. i–vi, xxiv. 54–8, etc. Jonson's main target is the growing indiffer-
ence of the nobility to military service: see Lawrence Stone's *The Crisis of the
Aristocracy* (Oxford, 1965), ch. v, esp. p. 239, for background. Amateur military
bodies such as the Artillery Company revived considerably after 1610, and especially
after 1613, with increased fears of a Spanish invasion: see Lindsay Boynton, *The
Elizabethan Militia 1558–1638* (London & Toronto, 1967), chs. vii & viii. Dekker,
amongst others, celebrated this activity, in *The Artillery Garden*, 1616 (rediscovered
1936; facsimile ed. F. P. Wilson (Oxford, 1952)). Jonson's poem was probably written
c. 1626; see ll. 27, 40–2 and notes.   2 *King's Day*: the anniversary of James's acces-
sion, 24 March; it had last been celebrated in 1624 (Fleay, i. 330): hence 'we have
powder still'.   5 *Aesop Gondomar*: Diego Sarmiento d'Acuña, Count of Gondomar,
Spanish ambassador in England 1613–18, 1620–22; highly skilful, highly unpopular.
'Aesop', perhaps, 'as the author of diplomatic fables' (H & S).   9 *feathers*: the
Gentlemen of the Artillery wore a scarlet ostrich feather.   10 *Saint George's Day*: on
23 April, when the annual procession of the Feast of the Garter takes place.   15 *dirty*:
despicable.   16 *pride*: display.   20 *Swinnerton*: Captain John Swinnerton, enrolled in
the Artillery Company in 1614 (Lt.-Col. G. A. Raikes, *The Ancient Vellum Book of the
Hon. Artillery Company* (London, 1890), p. 21).   21 *Pimlico*: the fort was used by the
trained bands.   23 *Artillery Yard*: or Artillery Garden (hence 'seed-plot', l. 24),
where the trained bands exercised.   25–6 *to bring up . . . truth*: echoing *D. is A*. III. ii.
45–6.

These ten years' day, as all may swear that look
  But on thy practice and the posture-book;
He that but saw thy curious captain's drill
  Would think no more of Flushing or the Brill,   30
But give them over to the common ear
  For that unnecessary charge they were.
Well did thy crafty clerk and knight, Sir Hugh,
  Supplant bold Panton, and brought there to view
Translated Aelian's *Tactics* to be read   35
  And the Greek discipline, with the modern, shed
So in that ground, as soon it grew to be
  The City question whether Tilly or he
Were now the greater captain; for they saw
  The Bergen siege, and taking in Breda,   40
So acted to the life, as Maurice might
  And Spinola, have blushed at the sight.
O happy art, and wise epitome
  Of bearing arms, most civil soldiery!
Thou canst draw forth thy forces, and fight dry   45
  The battles of thy aldermanity,
Without the hazard of a drop of blood
  More than the surfeits in thee that day stood.

27 *ten years' day*: i.e. the period since 1616, when special *Orders* were drawn up for training in the Artillery Garden.   28 *posture-book*: the standard, but not the only, work was Jacob de Gheyn's *The Exercise of Arms* (1607). Cf. *D. is A.* III. ii. 38.   29 *curious*: careful, painstaking.   30 *Flushing or the Brill*: towns in the Netherlands handed over to Elizabeth in August 1585, in return for an English military presence; Leicester had a costly and disastrous period of service there until recalled in Nov. 1586.   33–6 Sir Hugh Hammersley (d. 1636) was President of the Artillery Company 1619–33; Captain Edward Panton had command of the Company 1612–18, when he was suspended for neglect of duty. Panton's immediate successor, as Newdigate points out, was not Hammersley but Sir John Bingham, translator of Aelian's *Tactics* (1616) and advocate for the revival of ancient methods of warfare. Jonson confuses Hammersley and Bingham.   40–2 Ambrosio Spinola (1569–1630), general-in-chief of the Spanish army in the Netherlands, failed to take the town of Bergen-op-Zoom in 1622; after a long siege (28 Aug. 1624–5 June 1625) he successfully took Breda. Maurice of Nassau, Prince of Orange (1567–1625), commander of the armed forces of the Netherlands, was especially famed for his knowledge of siege warfare.   46 *aldermanity*: body of aldermen: Jonson probably coined the word. Many aldermen and several Lord Mayors of London were members of the Artillery Company.   48 *surfeits*: for which blood-letting might be prescribed.

Go on, increase in virtue and in fame,
    And keep the glory of the English name   50
Up among nations. In the stead of bold
    Beauchamps, and Nevilles, Cliffords, Audleys old,
Insert thy Hodges, and those newer men,
    As Styles, Dyke, Ditchfield, Millar, Crips, and Fenn;
That keep the war, though now it be grown more tame,   55
    Alive yet in the noise, and still the same;
And could, if our great men would let their sons
    Come to their schools, show 'em the use of guns,
And there instruct the noble English heirs
    In politic and militar affairs.   60
But he that should persuade to have this done
    For education of our lordings, soon
Should he hear of billow, wind, and storm
    From the tempestuous grandlings: Who'll inform
Us in our bearing, that are thus and thus   65
    Born, bred, allied? What's he dare tutor us?
Are we by bookworms to be awed? Must we
    Live by their scale that dare do nothing free?
Why are we rich or great, except to show
    All licence in our lives? What need we know   70
More than to praise a dog, or horse, or speak
    The hawking language; or our day to break
With citizens? Let clowns and tradesmen breed
    Their sons to study arts, the laws, the creed;
We will believe, like men of our own rank   75
    In so much land a year, or such a bank
That turns us so much moneys, at which rate
    Our ancestors imposed on prince and state.
Let poor nobility be virtuous; we,
    Descended in a rope of titles, be   80

51–2 *bold* | *Beauchamps*: 'a stock epithet of the family' (H & S). 53–4 Names of
captains in the Artillery Company (identifiable from its *Ancient Vellum Book*, ed.
Raikes): John Hodges, Thomas Styles, both enrolled 1614; Richard Dyke, Edward
Ditchfield, enrolled 1624; Nicholas Crips (or Crispe), enrolled 1621; Sir Richard Fenn,
enrolled 1614, sheriff, 1626, President, 1632–4, Lord Mayor, 1637. 'Millar' may be
Edward Mullar, enrolled 1625. In the margin Jonson adds (evidently as an after-
thought) the name 'Waller', captain and Treasurer of the Company in 1621. 58 *the
use of guns*: which the nobility regarded with suspicion and distaste. 60 *militar*:
military. 64 *grandlings*: an oxymoronic coinage. 70–3 Cf. *Disc.* 2672–5. 77 *turns
us*: keeps in circulation for us; cf. *Volp.* I. i. 39.

From Guy, or Bevis, Arthur, or from whom
    The herald will. Our blood is now become
Past any need of virtue. Let them care
    That in the cradle of their gentry are,
To serve the state by counsels and by arms;   85
    We neither love the troubles nor the harms.
What love you then? Your whore. What study? Gait,
    Carriage, and dressing. There is up of late
The Academy, where the gallants meet—
    What, to make legs? Yes, and to smell most sweet: 90
All that they do at plays. Oh, but first here
    They learn and study, and then practise there.
But why are all these irons in the fire
    Of several makings? Helps, helps, to attire
His lordship. That is for his band, his hair   95
    This, and that box his beauty to repair,
This other for his eyebrows—hence, away!
    I may no longer on these pictures stay:
These carcases of honour, tailors' blocks
    Covered with tissue, whose prosperity mocks   100
The fate of things; whilst tottered virtue holds
    Her broken arms up to their empty moulds.

45

An Epistle to Master Arthur Squib

What I am not, and what I fain would be,
    Whilst I inform myself, I would teach thee,
My gentle Arthur, that it might be said
    One lesson we have both learned and well read.
I neither am, nor art thou, one of those   5
    That hearkens to a jack's pulse, when it goes;

---

81 *Guy, or Bevis*: Guy of Warwick, Bevis of Hampton, heroes of romance. 89
Academy: cf. *D. is A.* II. viii. 19–22: it taught 'postures' of a quite unmilitary
kind. 98 *these pictures*: cf. Justice Clement's dismissal of Bobadill—'this picture'—
E.M.I. V. ii. 27; and *U.V.* 34. 25 ff. 99 *tailors' blocks*: see *Und.* 42. 82 n. 102
moulds: tailors' dummies or frames.

45. *Arthur Squib*: a teller in the Exchequer; addressed again in *Und.* 54. 6 *jack's
pulse*: a jack or jack-of-the-clock is a metal figure which strikes the hours; hence, a
would-be friend, chiming in mechanically with one's mood.

Nor ever trusted to that friendship yet
    Was issue of the tavern or the spit:
Much less a name would we bring up or nurse
    That could but claim a kindred from the purse.   10
Those are poor ties depend on those false ends,
    'Tis virtue alone, or nothing, that knits friends.
And as within your office you do take
    No piece of money, but you know or make
Inquiry of the worth, so must we do: 15
    First weigh a friend, then touch, and try him too;
For there are many slips and counterfeits.
    Deceit is fruitful. Men have masks and nets,
But these with wearing will themselves unfold:
    They cannot last. No lie grew ever old.   20
Turn him, and see his threads; look if he be
    Friend to himself, that would be friend to thee.
For that is first required, a man be his own.
    But he that's too much that is friend of none.
Then rest, and a friend's value understand: 25
    It is a richer purchase than of land.

46

An Epigram
on Sir Edward Coke,
When He Was Lord Chief Justice of England

He that should search all glories of the gown,
    And steps of all raised servants of the crown,
He could not find, than thee, of all that store,
    Whom fortune aided less, or virtue more.

---

8 *the spit*: see *Und.* 37. 9 n., and cf. Martial, *Epig.* IX. xiv.   13–16 Plutarch advises that friends, like coins, must be well examined before acceptance: *How to Tell a Flatterer*, ii; *Of Having Many Friends*, iii.   17 *slips*: counterfeit coins.   20 *No lie grew ever old*: a classical commonplace, but Jonson may be remembering Seneca, *Epist.* lxxix.   18. Cf. also *Disc.* 542.   22 *Friend to himself*: proverbial: 'Be a friend to thyself and others will be so too' (Tilley, F684); and cf. Seneca, *Epist.* vi. 7.   26 *richer purchase*: proverbial: 'A true friend is better than a rich farm', James Howell, 1659 (see Tilley, F719).

46. *Sir Edward Coke*: 1552–1634; judge and jurist. Solicitor-General, 1592, Speaker of the House of Commons, 1593, Attorney-General, 1594; Chief Justice of Common Pleas, 1606, and of King's Bench, 1613; dismissed from the Chief Justiceship in 1616.   4 *fortune* . . . *virtue*: a favourite antithesis for Jonson: see *Epig.* 63. 2–3 n.

Such Coke, were thy beginnings, when thy good 5
   In others' evil best was understood;
When, being the stranger's help, the poor man's aid,
   Thy just defences made the oppressor afraid.
Such was thy process, when integrity
   And skill in thee now grew authority; 10
That clients strove, in question of the laws,
   More for thy patronage than for their cause,
And that thy strong and manly eloquence
   Stood up thy nation's fame, her crown's defence.
And now such is thy stand; while thou dost deal 15
   Desired justice to the public weal
Like Solon's self, explait'st the knotty laws
   With endless labours, whilst thy learning draws
No less of praise than readers in all kinds
   Of worthiest knowledge that can take men's minds. 20
Such is thy all, that (as I sung before)
   None fortune aided less, or virtue more.
Of if chance must, to each man that doth rise,
   Needs lend an aid, to thine she had her eyes.

47

*An Epistle Answering to One that Asked to be
Sealed of the Tribe of Ben*

Men that are safe and sure in all they do
   Care not what trials they are put unto;
They meet the fire, the test, as martyrs would,
   And though opinion stamp them not, are gold.

---

13 *manly eloquence*: cf. *Und.* 14. 56, *Disc.* 797.   17 *Solon's self*: the Athenian statesman and lawgiver, born *c.* 638 B.C.   17 *explait'st*: unravellest (lit., take out the plaits or pleats: *OED*'s only example of this word).   24 *her eyes*: Fortune is traditionally represented as blind.

47. *Tribe of Ben*: Jonson's followers or 'sons'; glancing at Rev. 7, where the four angels at the earth's corners undertake to seal 'the servants of our God on their foreheads' (v. 3); 'Of the tribe of Benjamin were sealed twelve thousand' (v. 8). Written in 1623. For criticism, see Hugh Maclean in *Essays in English Literature* . . ., ed. M. MacLure and F. W. Watt (Toronto, 1964), pp. 47–51.   3–4 Cf. *Und.* 25. 45, and n.

I could say more of such, but that I fly 5
   To speak myself out too ambitiously,
And showing so weak an act to vulgar eyes,
   Put conscience and my right to compromise.
Let those that merely talk, and never think,
   That live in the wild anarchy of drink, 10
Subject to quarrel only, or else such
   As make it their proficiency how much
They have glutted in and lechered out that week,
   That never yet did friend or friendship seek
But for a sealing: let these men protest. 15
   Or the other on their borders, that will jest
On all souls that are absent, even the dead,
   Like flies or worms which man's corrupt parts fed:
That to speak well, think it above all sin,
   Of any company but that they are in; 20
Call every night to supper in these fits,
   And are received for the covey of wits;
That censure all the town, and all the affairs,
   And know whose ignorance is more than theirs;
Let these men have their ways, and take their times 25
   To vent their libels, and to issue rhymes,
I have no portion in them, nor their deal
   Of news they get to strew out the long meal;
I study other friendships, and more one
   Than these can ever be; or else wish none. 30
What is't to me whether the French design
   Be, or be not, to get the Valtelline?
Or the States' ships sent forth belike to meet
   Some hopes of Spain in their West Indian Fleet?

5–6 *fly | To speak*: i.e., hate to speak. 9 *merely talk*: cf. *Disc.* 338–42. 10 *anarchy of drink*: *Epig.* 115. 12. 13 *glutted in and lechered out*: cf. *Epig.* 118. 16 *on their borders*: like them. 17 *absent*: cf. *Epig.* 115. 16. 19–20 Cf. *Epig.* 115. 7–8. 22 *covey of wits*: cf. *S. of N.* II. iv. 41. 26 *issue rhymes*: cf. *Und.* 70. 103. 28 *strew out the long meal*: cf. *Epig.* 115. 10. 31–2 *Valtelline*: or Valtellina, the upper valley of the Adda in Lombardy; of critical strategic importance, it was held by the Spanish from 1621 to 1623, by the French from 1624 to 1627, and changed hands frequently thereafter. 33–4 Frequent skirmishing between Dutch and Spanish shipping occurred after the resumption of hostilities between the two countries in 1621. The Spanish Mexico fleet was overtaken by storm and badly damaged before it left the West Indies in 1622.

Whether the dispensation yet be sent,   35
  Or that the match from Spain was ever meant?
I wish all well, and pray high heaven conspire
  My prince's safety and my king's desire;
But if, for honour, we must draw the sword,
  And force back that which will not be restored,   40
I have a body yet that spirit draws
  To live, or fall a carcass in the cause.
So far without inquiry what the States,
  Brunsfield, and Mansfeld, do this year, my fates
Shall carry me at call, and I'll be well,   45
  Though I do neither hear these news, nor tell
Or Spain or France, or were not pricked down one
  Of the late mystery of reception,
Although my fame to his not under-hears,
  That guides the motions and directs the bears.   50
But that's a blow by which in time I may
  Lose all my credit with my Christmas clay
And animated porcelain of the court;
  Aye, and for this neglect, the coarser sort
Of earthen jars there may molest me too: 55
  Well, with mine own frail pitcher, what to do
I have decreed; keep it from waves and press,
  Lest it be jostled, cracked, made nought, or less;

35 *the dispensation*: the papal dispensation to allow Prince Charles to marry the Infanta finally arrived in Madrid at the end of April 1623. There were deep misgivings about the match both in Spain and in England. 40 *that which will not be restored*: the Palatinate, invaded by Spain in 1620. Popular sympathy for the defeated Frederick, James's son-in-law, was strong in England. 44 *Brunsfield*: Hunter suggests the reference may be to Christian of Brunswick (1599–1626), who raised an army in aid of Frederick in 1621, and conducted several unsuccessful battles in subsequent years. 44 *Mansfeld*: Ernst, Graf von Mansfeld (1580–1626), commander of Frederick's army. 48 *late mystery of reception*: elaborate preparations for the reception of the Infanta at Southampton, and subsequently in London, were in train during the summer of 1623. Jonson's collaborator and rival Inigo Jones, who is sniped at in the following lines, had been busy in both places. Other 'mysteries' of Jones are referred to in *U.V.* 34. 46. 49 *to his not under-hears*: is not inferior to his. Probably a coinage: cf. *OED* 'hear', 12: 'to be reported or spoken (well or ill) of'. 50 Cf. the gibes in *Epig.* 97. '*Motions*' = puppets. 52 ff. *Christmas clay . . . animated porcelain . . . earthen jars . . . frail pitcher*: 'The danger of supersession at Court has made Jonson unusually modest' (H & S); but the first three of these references are not only to Jonson's masques but (less modestly) to those who watch and take part in them.

Live to that point I will, for which I am man,
   And dwell as in my centre as I can,   60
Still looking to, and ever loving, heaven;
   With reverence using all the gifts thence given.
'Mongst which, if I have any friendships sent,
   Such as are square, well-tagged, and permanent,
Not built with canvas, paper, and false lights,   65
   As are the glorious scenes at the great sights,
And that there be no fevery heats, nor colds,
   Oily expansions, or shrunk dirty folds,
But all so clear and led by reason's flame,
   As but to stumble in her sight were shame;   70
These I will honour, love, embrace, and serve,
   And free it from all question to preserve.
So short you read my character, and theirs
   I would call mine, to which not many stairs
Are asked to climb. First give me faith, who know   75
   Myself a little. I will take you so,
As you have writ yourself. Now stand, and then,
   Sir, you are sealed of the tribe of Ben.

48

The Dedication of the King's New Cellar
to Bacchus

Since, Bacchus, thou art father
Of wines, to thee the rather
We dedicate this cellar,
Where now, thou art made dweller,
And seal thee thy commission;   5
But 'tis with a condition
That thou remain here taster
Of all to the great master.

---

60 *dwell as in my centre*: cf. *Und*. 14. 30–3.   64 *well-tagged*: well-fastened.   69 *reason's flame*: cf. the lamp carried by the figure of Reason in *Hym*. 134.

48. *The King's New Cellar*: at Whitehall; built under Inigo Jones's direction.

And look unto their faces,
Their qualities, and races,   10
That both their odour take him
And relish merry make him.
    For, Bacchus, thou art freer
Of cares, and overseer
Of feast and merry meeting,   15
And still begin'st the greeting;
See then thou dost attend him,
Lyaeus, and defend him
By all the arts of gladness
From any thought like sadness.   20
    So mayst thou still be younger
Than Phoebus, and much stronger
To give mankind their eases,
And cure the world's diseases;
    So may the muses follow   25
Thee still, and leave Apollo,
And think thy stream more quicker
Than Hippocrene's liquor:
And thou make many a poet
Before his brain do know it;   30
So may there never quarrel
Have issue from the barrel;
But Venus and the graces
Pursue thee in all places,
And not a song be other   35
Than Cupid and his mother.
    That when King James, above here,
Shall feast it, thou mayst love there
The causes and the guests too,
And have thy tales and jests too,   40
Thy circuits and thy rounds free
As shall the feast's fair grounds be.
    Be it he hold communion
In great Saint George's union,

---

10 *races*: varieties.   18 *Lyaeus*: Bacchus.   28 *Hippocrene's liquor*: from the fountain of
the muses, on Mount Helicon.   29 *make many a poet*: including, no doubt, Jonson
himself: 'drink . . . is one of the elements in which he liveth', *Conv. Dr.* 683–4; cf. *Und.* 57.
24–5.   44 *Saint George's union*: the Feast of the Garter on St. George's Day, 23
April.

Or gratulates the passage   45
Of some well-wrought embassage,
Whereby he may knit sure up
The wished peace of Europe;
Or else a health advances,
To put his court in dances,   50
And set us all on skipping,
When with his royal shipping
The narrow seas are shady,
And Charles brings home the lady.

Accessit fervor capiti, numerusque lucernis.

49

*An Epigram
on the Court Pucelle*

Does the court pucelle then so censure me,
    And thinks I dare not her? Let the world see.
What though her chamber be the very pit
    Where fight the prime cocks of the game, for wit?
And that as any are struck, her breath creates   5
    New in their stead, out of the candidates?
What though with tribade lust she force a muse,
    And in an epicoene fury can write news

54 *Charles brings home the lady*: the Infanta; their arrival from Spain was expected in July 1623. Charles finally arrived without her on 15 Oct. 1623, and was greeted with jubilation.   55 Horace, *Sat.* II. i. 25: 'The heat has mounted to his head, and the lamps are double'; cf. *Poet.* III. v. 43–6.

49. *Court Pucelle*: i.e. whore. Two passages in *Conv. Dr.* (103–4, 646–8) identify her as Lady Bedford's friend Cecilia Bulstrode (1584–1609), of Buckinghamshire; Jonson's epitaph on this lady gives quite another picture of her qualities: see *U.V.* 9, and n.; and cf. Donne, 'Death, I recant', *Poetical Works*, ed. H. J. C. Grierson (Oxford, 1912), i. 282–3. For biographical details, see Percy Simpson, *TLS*, 6 March 1930, p. 187. James E. Savage in his introduction to a facsimile reprint of Overbury's *Conceited Newes* (1616; Gainesville, Florida, 1968) suggests that the 'news' of ll. 8–9 is a courtly game played in Cecilia Bulstrode's chamber; and that 'cocks' in l. 4 alludes to the name of one of the players, John Cocke or Cooke.   7 *tribade*: lesbian.

Equal with that which for the best news goes,
    As airy light, and as like wit as those?   10
What though she talk, and can at once with them
    Make state, religion, bawdry, all a theme?
And as lip-thirsty, in each word's expense,
    Doth labour with the phrase more than the sense?
What though she ride two mile on holidays   15
    To church, as others do to feasts and plays,
To show their 'tires, to view and to be viewed?
    What though she be with velvet gowns endued,
And spangled petticoats brought forth to eye,
    As new rewards of her old secrecy?   20
What though she hath won on trust, as many do,
    And that her truster fears her: must I too?
I never stood for any place: my wit
    Thinks itself nought, though she should value it.
I am no statesman, and much less divine;   25
    For bawdry, 'tis her language, and not mine.
Farthest I am from the idolatry
    To stuffs and laces: those my man can buy.
And trust her I would least, that hath foreswore
    In contract twice; what can she perjure more?   30
Indeed, her dressing some man might delight,
    Her face there's none can like by candle-light.
Not he that should the body have, for case
    To his poor instrument, now out of grace.
Shall I advise thee, pucelle? Steal away   35
    From court, while yet thy fame hath some small day;
The wits will leave you, if they once perceive
    You cling to lords, and lords, if them you leave
For sermoneers: of which now one, now other
    They say you weekly invite with fits of the mother,   40
And practise for a miracle; take heed
    This age would lend no faith to Darrel's deed:

---

11 *can at once*: Whalley's emendation; Folio (and H & S), 'cannot once'.   18 *endued*: clothed.   32 *by candle-light*: by which light all women are said to be attractive: cf. Tilley, W682, and *S.W.* V. ii. 37–8.   36 *day*: power.   39 *sermoneers*: a coinage.   40 *mother*: hysteria; a pun.   42 *Darrel's deed*: John Darrel (*fl.* 1562–1602), Puritan preacher; practised exorcism until imprisoned for imposture in 1599.

Or if it would, the court is the worst place,
    Both for the mothers and the babes of grace;
For there the wicked in the chair of scorn   45
    Will call it a bastard, when a prophet's born.

50

An Epigram
to the Honoured Elizabeth, Countess of Rutland

The wisdom, madam, of your private life
    Wherewith this while you live a widowed wife,
And the right ways you take unto the right,
    To conquer rumour and triumph on spite;
Not only shunning by your act to do   5
    Aught that is ill, but the suspicion too,
Is of so brave example, as he were
    No friend to virtue could be silent here.
The rather when the vices of the time
    Are grown so fruitful, and false pleasures climb   10
By all oblique degrees that killing height
    From whence they fall, cast down with their own weight.
And though all praise bring nothing to your name,
    Who, herein studying conscience and not fame,
Are in yourself rewarded; yet 'twill be   15
    A cheerful work to all good eyes, to see
Among the daily ruins that fall foul,
    Of state, of fame, of body, and of soul,
So great a virtue stand upright to view,
    As makes Penelope's old fable true: 20

---

44 *babes of grace*: probably children allegedly born by parthenogenesis. Newdigate compares Jonson's story of the lady who lay with a Puritan preacher hoping to conceive an angel or saint: 'which having obtained, it was but an ordinary birth' (*Conv. Dr.* 513–16).

50. *Elizabeth, Countess of Rutland*: the name is not found in the poem's title in Folio; the identification is Cunningham's. *Epig.* 79 and *For.* 12 are also addressed to her: see notes to those poems.   2 *widowed wife*: the Countess's husband was impotent: see *For.* 12. 93–100 n., and *Conv. Dr.* 220–2. It is a sign of Jonson's tact that the phrase might be taken as referring simply to the Earl's absence: see ll. 20 ff. (Homer likewise speaks of Penelope's 'wisdom' during Ulysses's absence.)   6 *the suspicion too*: *Disc.* 1323–5. 14 *conscience and not fame*: *Cat.* II. 378.

Whilst your Ulysses hath ta'en leave to go,
  Countries and climes, manners and men to know.
Only your time you better entertain,
  Than the great Homer's wit for her could feign;
For you admit no company but good,   25
  And when you want those friends, or near in blood,
Or your allies, you make your books your friends,
  And study them unto the noblest ends,
Searching for knowledge, and to keep your mind
  The same it was inspired, rich, and refined.   30
These graces, when the rest of ladies view,
  Not boasted in your life, but practised true,
As they are hard for them to make their own,
  So are they profitable to be known:
For when they find so many meet in one,   35
  It will be shame for them, if they have none.

51

Lord Bacon's Birthday

Hail, happy genius of this ancient pile!
How comes it all things so about thee smile?
The fire, the wine, the men! and in the midst
Thou stand'st as if some mystery thou didst!
Pardon, I read it in thy face, the day   5
For whose returns, and many, all these pray:
And so do I. This is the sixtieth year
Since Bacon, and thy lord was born, and here;
Son to the grave wise Keeper of the Seal,
Fame and foundation of the English weal.   10

22 *manners and men*: cf. *Epig.* 128. 2, *Und.* 14. 33. Rutland travelled extensively in
Europe. In 1603 he was engaged upon a mission to Christian IV of Denmark.   27–8
Cf. *Epig.* 102. 20 n.

51. *Lord Bacon's Birthday*: Francis Bacon (1561–1626), the essayist: Attorney-General,
1613; Privy Councillor, 1616; Lord Keeper, 1617; Lord Chancellor and first Baron
Verulam, 1618; Viscount St. Albans, 1621. Bacon's sixtieth birthday fell on 26 Jan.
1621. Later in the same year he was found guilty of corruption, and deprived of the
great seal. Jonson praises him highly in *Disc.* 884–98.   1 *ancient pile*: York House,
Bacon's residence on the Embankment.   2–3 Cf. *For.* 14. 1, 60.   9 *grave wise Keeper
of the Seal*: Sir Nicholas Bacon (1509–79); he held this office from 1558.

What then his father was, that since is he,
Now with a title more to the degree;
England's high Chancellor: the destined heir
In his soft cradle to his father's chair;
Whose even thread the fates spin round and full, 15
Out of their choicest and their whitest wool.
  'Tis a brave cause of joy, let it be known,
For 'twere a narrow gladness, kept thine own.
Give me a deep-crowned bowl, that I may sing,
In raising him, the wisdom of my king. 20

52

A Poem Sent Me by Sir William Burlase
The Painter to the Poet

To paint thy worth, if rightly I did know it,
And were but painter half like thee, a poet,
   Ben, I would show it;
But in this skill my unskilled pen will tire,
Thou, and thy worth, will still be found far higher, 5
   And I a liar.
Then what a painter's here! or what an eater
Of great attempts! when as his skill's no greater,
   And he a cheater!
Then what a poet's here! whom, by confession 10
Of all with me, to paint without digression,
   There's no expression.

19 *deep-crowned*: deep and brimming.

52. *Sir William Burlase*: of Little Marlow; sheriff of Buckinghamshire, 1601, d. 1629.

My Answer
The Poet to the Painter

Why? though I seem of a prodigious waist,
I am not so voluminous and vast
But there are lines wherewith I might be embraced.

'Tis true, as my womb swells, so my back stoops,
And the whole lump grows round, deformed, and droops, 5
But yet the tun at Heidelberg had hoops.

You were not tied by any painter's law
To square my circle, I confess, but draw
My superficies: that was all you saw.

Which if in compass of no art it came   10
To be described but by a monogram,
With one great blot, you had formed me as I am.

But whilst you curious were to have it be
An archetype for all the world to see,
You made it a brave piece, but not like me.   15

Oh, had I now your manner, mastery, might,
Your power of handling shadow, air, and sprite,
How I would draw, and take hold and delight.

But you are he can paint; I can but write:
A poet hath no more but black and white,   20
Ne knows he flattering colours, or false light.

Yet when of friendship I would draw the face,
A lettered mind and a large heart would place
To all posterity: I will write *Burlase*.

My Answer. 1 *prodigious waist*: Jonson weighed nearly twenty stone: see *Und.* 54. 12,
56. 11.   3 *lines*: playing on the sense of poetic lines, as in *Und.* 84. iv. 22.   6 *tun at
Heidelberg*: of great size, it was bound by 26 hoops; see Thomas Coryate's description
in *Crudities* (London, 1611), pp. 486–92, and the illustration facing p. 486.   8 *square
my circle*: attempt the impossible; with a play on words.   11 *monogram*: a picture
without shading or colour, a sketch; 'here the letter O, in allusion to Ben's girth'
(Newdigate).   21 *colours*: figures of rhetoric.   24 *posterity*: *Epig*. Ded. 14–17.

53

*An Epigram
to William, Earl of Newcastle*

When first, my lord, I saw you back your horse,
  Provoke his mettle, and command his force
To all the uses of the field and race,
  Methought I read the ancient art of Thrace,
And saw a centaur, past those tales of Greece;   5
  So seemed your horse and you both of a piece!
You showed like Perseus upon Pegasus,
  Or Castor mounted on his Cyllarus,
Or what we hear our home-born legend tell
  Of bold Sir Bevis and his Arundel;   10
Nay, so your seat his beauties did endorse
  As I began to wish myself a horse.
And surely had I but your stable seen
  Before, I think my wish absolved had been.
For never saw I yet the muses dwell,   15
  Nor any of their household, half so well.

53. *William, Earl of Newcastle*: William Cavendish (1592–1676), Viscount Mansfield, 1620; Earl (1628) and later Duke (1665) of Newcastle. He published two works on horsemanship (1657, 1667) and instructed the future Charles II in the art. The poem was written between 1620 and 1628 ('Viscount Mansfield' in one MS.). Newcastle is addressed again in *Und.* 59. 1 *back*: ride, break in. 5–6 Cf. Sidney, *Arcadia* (1590), II. v: ed. A. Feuillerat (Cambridge, 1912), p. 178; *Hamlet*, IV. iii. 85–8. 7 *Perseus upon Pegasus*: Pegasus is actually Bellerophon's horse, but the misattribution is traditional: see T. W. Baldwin, *PQ*, xx (1941), 361–70, G. B. Johnston, *RES*, vi (1955), 65–7, J. D. Reeves, ibid., 397–9. 8 *Castor . . . Cyllarus*: Castor was especially skilled in horsemanship; Cyllarus was his (or his brother Pollux's) horse. 10 *Sir Bevis . . . Arundel*: Bevis of Hampton, romance hero; his horse's name means 'swift as a swallow' (French *hirondelle*). 11 *endorse*: confirm; with a pun on the literal sense: to put on the back of, to ride. 12 *wish myself a horse*: cf. Sidney's remark about his Italian riding-master: 'I think he would have persuaded me to have wished myself a horse', *An Apology For Poetry*, ed. G. Shepherd (London, 1965), p. 95; Shepherd compares a similar wish expressed by one of Crato's pupils in *De disciplina scholarium*, once attributed to Boethius: J.-P. Migne, *Patrologiae Cursus Completus . . . Series Latina* (Paris, 1860), lxiv. 1230. A traditional reversal (the horse being customarily instanced in the logic-books as a non-rational creature): cf. Gulliver amongst the Houyhnhnms, and Rochester's 'Tunbridge Wells', 166–75, *Complete Poems*, ed. D. M.

So well, as when I saw the floor and room,
  I looked for Hercules to be the groom,
And cried, Away with the Caesarian bread!
  At these immortal mangers Virgil fed.   20

54

Epistle
to Mr. Arthur Squib

I am to dine, friend, where I must be weighed
  For a just wager, and that wager paid
If I do lose it: and, without a tale,
  A merchant's wife is regent of the scale.
Who, when she heard the match, concluded straight,   5
  An ill commodity! It must make good weight.
So that upon the point my corporal fear
  Is she will play Dame Justice too severe,
And hold me to it close; to stand upright
  Within the balance, and not want a mite,   10
Bur rather with advantage to be found
  Full twenty stone, of which I lack two pound:
That's six in silver; now within the socket
  Stinketh my credit, if into the pocket
It do not come. One piece I have in store;   15
  Lend me, dear Arthur, for a week five more

Vieth (New Haven & London, 1968), p. 80.   18 *Hercules*: who cleansed the Augean stables.   19–20 Alluding to a story that the young Virgil worked in the imperial stables, displaying such skill that he was rewarded with a double ration of bread: see A. C. Taylor, *TLS*, 28 Aug. 1937, p. 624.

54. *Arthur Squib*: see *Und.* 45. 'The wager, it seems, was that the poet weighed full twenty stone, but he found that he wanted two pounds of that weight. This he artfully turns to a reason for borrowing five pounds in money of his friend Mr. Squib, which added to the pound he had of his own, would make up the deficiency in his weight. Six pounds in silver, he says, will weigh two pounds in weight: it may be so; we will take his word' (Whalley).   13 *socket*: the image is of a candle burning low.

And you shall make me good, in weight and fashion,
    And then to be returned; or protestation
To go out after—till when, take this letter
    For your security. I can no better.   20

55

To Mr. John Burgess

Would God, my Burgess, I could think
Thoughts worthy of thy gift, this ink;
Then would I promise here to give
Verse that should thee and me outlive.
But since the wine hath steeped my brain   5
I only can the paper stain;
Yet with a dye that fears no moth,
But scarlet-like outlasts the cloth.

56

Epistle
to My Lady Covell

You won not verses, madam, you won me,
    When you would play so nobly, and so free.
A book to a few lines; but it was fit
    You won them too, your odds did merit it.
So have you gained a servant and a muse: 5
    The first of which, I fear, you will refuse;
And you may justly, being a tardy, cold,
    Unprofitable chattle, fat and old,
Laden with belly, and doth hardly approach
    His friends, but to break chairs or crack a coach.   10

---

18 *or protestation*: thus Folio; H & S 'on protestation'. A protestation was a formal
written declaration of non-acceptance or non-payment of a bill.

55. *John Burgess*: a clerk in the Exchequer; addressed again in *Und.* 57.

56. *Lady Covell*: see *Und.* 22. 23–4 and n.; she is otherwise unknown.   3 *a few lines*:
perhaps *Und.* 22.

His weight is twenty stone, within two pound,
   And that's made up as doth the purse abound.
Marry, the muse is one can tread the air,
   And stroke the water: nimble, chaste and fair;
Sleep in a virgin's bosom without fear,   15
   Run all the rounds in a soft lady's ear,
Widow or wife, without the jealousy
   Of either suitor or a servant by.
Such, if her manners like you, I do send:
   And can for other graces her commend,   20
To make you merry on the dressing stool
   A-mornings, and at afternoons to fool
Away ill company, and help in rhyme
   Your Joan to pass her melancholy time.
By this, although you fancy not the man,   25
   Accept his muse; and tell (I know you can)
How many verses, madam, are your due!
   I can lose none in tendering these to you.
I gain in having leave to keep my day,
   And should grow rich, had I much more to pay.   30

57

To Master John Burgess

Father John Burgess
Necessity urges
My woeful cry,
To Sir Robert Pye:
And that he will venture   5
To send my debenture.

---

11 *twenty stone, within two pound*: see *Und.* 54 n.   16 *rounds*: possibly playing on a secondary sense of the word, 'whispers'.

57. *John Burgess*: see *Und.* 55 and n. Jonson had been granted an annual pension by James in 1616; in 1630 Charles agreed to an increase: see *Und.* 68 and 76 and notes. Payment was evidently tardy. Jonson uses the Skeltonic verse-form again in the speeches of the Patrico in *Gyp. Met.* and of Skelton himself in *Fort. Is.*   4 *Sir Robert Pye*: 1585–1662, Remembrancer of the Exchequer from 1618; ironically, an ancestor of the poet laureate Henry James Pye, himself destined to run into acute financial difficulties.   6 *debenture*: voucher for payment from the Exchequer.

Tell him his Ben
Knew the time, when
He loved the muses;
Though now he refuses   10
To take apprehension
Of a year's pension,
And more is behind:
Put him in mind
Christmas is near;   15
And neither good cheer,
Mirth, fooling, nor wit,
Nor any least fit
Of gambol or sport
Will come at the court,   20
If there be no money;
No plover, or coney
Will come to the table,
Or wine to enable
The muse or the poet,   25
The parish will know it;
Nor any quick warming-pan help him to bed,
If the 'chequer be empty, so will be his head.

58

*Epigram
to My Bookseller*

Thou, friend, wilt hear all censures; unto thee
    All mouths are open, and all stomachs free:
Be thou my book's intelligencer, note
    What each man says of it, and of what coat

15–21 'Jonson has to prepare sport for the Court, viz. *Callipolis* and *Chloridia*': Fleay, who dates the poem Dec. 1630 (i. 331). 24–5 Cf. *Und.* 48. 29. 27 *warming-pan*: perhaps in the slang sense (cf. *N.I.* I. iii. 13), a bed-companion.

58. *My Bookseller*: probably Thomas Alchorne, who published *N.I.* in 1631 (Fleay, i. 331); possibly Robert Allot, who published *B.F.*, *D. is A.*, and *S. of N.* earlier the same year. Cf. *Epig.* 3, to John Stepneth. 1 *censures*: judgements (good or bad). 2 *stomachs*: dispositions. 4 *coat*: sort.

His judgement is; if he be wise and praise, 5
Thank him; if other, he can give no bays.
If his wit reach no higher but to spring
Thy wife a fit of laughter, a cramp-ring
Will be reward enough, to wear like those
That hang their richest jewels in their nose, 10
Like a rung bear or swine: grunting out wit
As if that part lay for a —— most fit!
If they go on, and that thou lov'st alife
Their perfumed judgements, let them kiss thy wife.

59

*An Epigram
to William, Earl of Newcastle*

They talk of fencing and the use of arms,
The art of urging and avoiding harms,
The noble science and the mastering skill
Of making just approaches, how to kill,
To hit in angles, and to clash with time: 5
As all defence or offence were a chime!
I hate such measured, give me mettled! fire,
That trembles in the blaze, but then mounts higher,
A quick and dazzling motion! When a pair
Of bodies meet like rarefied air! 10
Their weapons shot out with that flame and force
As they outdid the lightning in their course;
This were a spectacle! A sight to draw
Wonder to valour! No; it is the law

8 *cramp-ring*: a ring formerly hallowed by royalty on Good Friday, and thought to be efficacious against cramp, falling sickness, etc. Jonson implies that thoughtless laughter is a similar disease, which might be most appropriately warded off if the ring were worn not on the finger but through the nose. Cf. *Epig.* 2. 12, and n.   12 ——: a word, undoubtedly obscene, has been omitted from the Folio.   13 *alife*: dearly.

59. *William, Earl of Newcastle*: see *Und.* 53. Written after 1628, when the earldom was conferred. Newcastle wrote, but did not publish, a treatise on fencing.   1–6 Jonson may be glancing at a work such as *Vincentio Saviolo his Practice* (London, 1595), which makes much of the importance of 'time' and 'measure' in fencing.   10 *rarefied air*: *N.I.* IV. iii. 19–21.   14 ff. Cf. Lovel on true valour, *N.I.* IV. iv; and Seneca, *De Ben.*, II. xxxiv. 3: 'Bravery is the virtue that scorns legitimate dangers, knowing how to ward off, to meet, and to court dangers.'

Of daring not to do a wrong is true   15
    Valour: to slight it, being done to you;
To know the heads of danger, where 'tis fit
    To bend, to break, provoke, or suffer it!
All this, my lord, is valour. This is yours,
    And was your father's, all your ancestors'!   20
Who durst live great 'mongst all the colds and heats
    Of human life, as all the frosts and sweats
Of fortune, when or death appeared, or bands;
    And valiant were with, or without, their hands.

60

An Epitaph on Henry, Lord La Warr.
To the Passer-by

If, passenger, thou canst but read,
Stay, drop a tear for him that's dead:
Henry, the brave young Lord La Warr,
Minerva's and the muses' care!
What could their care do 'gainst the spite   5
Of a disease that loved no light
Of honour, nor no air of good,
But crept like darkness through his blood,
Offended with the dazzling flame
Of virtue, got above his name?   10
No noble furniture of parts,
No love of action and high arts,
No aim at glory, or, in war,
Ambition to become a star

60. *Henry, Lord La Warr*: Henry West, 13th Baron De La Warr (1603–1 June 1628); M.P., 1621; son of Francis West (1586–?1633), the American colonist, after whom the state of Delaware is named.   5 *What could their care do . . .*: cf. *Lycidas*, 57–8: 'Had ye been there—for what could that have done? / What could the muse herself, that Orpheus bore . . .'   13–14 *war . . . star*: cf. *Und.* 15. 195–6.

THE UNDERWOOD 225

Could stop the malice of this ill, 15
That spread his body o'er, to kill:
And only his great soul envied
Because it durst have noblier died.

61

An Epigram

That you have seen the pride, beheld the sport,
    And all the games of fortune played at court;
Viewed there the market, read the wretched rate
    At which there are would sell the prince and state;
That scarce you hear a public voice alive, 5
    But whispered councils, and those only, thrive:
Yet are got off thence with clear mind and hands
    To lift to heaven: who is't not understands
Your happiness, and doth not speak you blest,
    To see you set apart, thus, from the rest, 10
To obtain of God what all the land should ask?
    A nation's sin got pardoned—'twere a task
Fit for a bishop's knees! O bow them oft,
    My lord, till felt grief make our stone hearts soft,
And we do weep to water for our sin. 15
    He that in such a flood as we are in
Of riot and consumption, knows the way
    To teach the people how to fast and pray,
And do their penance, to avert God's rod:
    He is the man, and favourite of God. 20

61. Gifford deduced that this poem is addressed to John Williams (1582–1650), Dean of Westminster, 1620, Bishop of Lincoln, 1621; Lord Keeper of the Privy Seal (in succession to Bacon), 1621, and removed from that office by Charles, 25 Oct. 1628: Jonson's poem may relate to this last event. A Star Chamber prosecution was brought against Williams in the same year; he was imprisoned 1637–40.

62

An Epigram to King Charles,
for a Hundred Pounds He Sent Me in My Sickness. 1629

Great Charles, among the holy gifts of grace
    Annexed to thy person and thy place,
'Tis not enough (thy piety is such)
    To cure the called king's evil with thy touch;
But thou wilt yet a kinglier mastery try,   5
    To cure the poet's evil, poverty;
And in these cures dost so thyself enlarge
    As thou dost cure our evil at thy charge.
Nay, and in this thou show'st to value more
    One poet, than of other folk ten score.   10
O piety, so to weigh the poor's estates!
    O bounty, so to difference the rates!
What can the poet wish his king may do,
    But that he cure the people's evil too?

63

To King Charles and Queen Mary.
For the Loss of Their First-Born; an Epigram Consolatory

Who dares deny that all first fruits are due
    To God, denies the Godhead to be true;
Who doubts those fruits God can with gain restore,
    Doth by his doubt distrust his promise more.

---

62. *My Sickness*: Jonson suffered a paralytic stroke in 1626 and another in 1628. He spoke of his sickness and poverty in the epilogue to *N.I.* in Jan. 1629, obliquely soliciting royal assistance. Charles responded with the gift referred to here.   4 *king's evil*: scrofula, thought to be curable by royal touch; the parliamentarians soon attempted to discredit this practice.   10 *ten score*: those touched for the king's evil were given an angel; Jonson reckons Charles's gift to be worth 200 angels.   12 *difference*: differentiate.   14 *people's evil*: referring to the differences between king and parliament, which led to Charles's dissolution of parliament in March 1629.

63. *First-Born*: a son, Charles James, who was born prematurely on 13 May 1629, and lived only a few hours.   1–4 See Exod. 22 : 29 on the need to offer God first-fruits and first-born sons; and Exod. 34 : 20 on the redemption of the first-born. Cf. Dame Purecraft, *B.F.* I. vi. 64–6.

He can, he will, and with large interest, pay   5
    What, at his liking, he will take away.
Then, royal Charles and Mary, do not grutch
    That the Almighty's will to you is such;
But thank his greatness, and his goodness too,
    And think all still the best that he will do.   10
That thought shall make he will this loss supply
    With a long, large, and blessed posterity!
For God, whose essence is so infinite,
    Cannot but heap that grace he will requite.

64

An Epigram
to Our Great and Good King Charles,
on His Anniversary Day, 1629

How happy were the subject, if he knew
    Most pious king, but his own good in you!
How many times, Live long, Charles! would he say,
    If he but weighed the blessings of this day,
And as it turns our joyful year about,   5
    For safety of such majesty, cry out?
Indeed, when had Great Britain greater cause
    Than now, to love the sovereign and the laws?
When you that reign are her example grown,
    And what are bounds to her, you make your own?   10
When your assiduous practice doth secure
    That faith which she professeth to be pure?
When all your life's a precedent of days,
    And murmur cannot quarrel at your ways?
How is she barren grown of love, or broke,   15
    That nothing can her gratitude provoke!

7 *grutch*: grudge, complain.   9 *his greatness, and his goodness*: cf. *For*. 15. 1: 'Good and great God'.

64. 1–2 Possibly a memory of Virgil, *Georgics*, ii. 458–9: 'O happy husbandmen! too happy, should they come to know their blessings!'; but Jonson also alludes to the rising trouble in the land; cf. *Und*. 82. 5.   5 *turns our joyful year about*: Charles's Anniversary Day fell on 27 March; the calendar year began on 25 March.

O times! O manners! surfeit bred of ease,
The truly epidemical disease!
'Tis not alone the merchant, but the clown
Is bankrupt turned; the cassock, cloak, and gown   20
Are lost upon account! and none will know
How much to heaven for thee, great Charles, they owe!

65

An Epigram on the Prince's Birth, 1630

And art thou born, brave babe? Blest be thy birth,
That hath so crowned our hopes, our spring, and earth,
The bed of the chaste lily and the rose!
What month than May was fitter to disclose
This prince of flowers? Soon shoot thou up, and grow   5
The same that thou art promised; but be slow
And long in changing. Let our nephews see
Thee quickly come the garden's eye to be,
And there to stand so. Haste now, envious moon,
And interpose thyself ('care not how soon)   10
And threat' the great eclipse. Two hours but run,
Sol will re-shine. If not, Charles hath a son.

———*Non displicuisse meretur,*
Festinat, Caesar, qui placuisse tibi.

17 *O times! O manners!*: Cicero, *In Catilinam*, I. i. 2; cf. *Und.* 15. 161.

65. The future Charles II was born on 29 May 1630. Poems of celebration were also written by Corbett, Herrick, Hoskyns, King, Randolph, Shirley, and others. See also *Dubia* 5, and 'A Parallel of the Prince to the King', H & S, viii. 429. 3 *lily . . . rose*: the flowers of France and of England; Henrietta Maria and Charles. Cf. *Fort. Is.* 543, *Love's Tr.* 210–14, *Love's Welc. Bols.* 82–3, *U.V.* 33. 8. 7 *nephews*: i.e. grandchildren, posterity; cf. Latin *nepotes*. 8 *eye*: brightest spot. 9–12 An eclipse of the sun occurred two days after Charles's birth. 12 *son*: a pun. 13–14 'He deserves not to displease you, Caesar, who hastes to please you' (Martial, *On the Spectacles*, xxxi). Cf. *King's Ent.* 644.

66

An Epigram to the Queen, then Lying In, 1630

Hail Mary, full of grace! it once was said,
   And by an angel, to the blessed'st maid,
The mother of our Lord: why may not I
   Without prophaneness, yet a poet, cry
Hail Mary, full of honours! to my queen,   5
   The mother of our prince? When was there seen,
Except the joy that the first Mary brought,
   Whereby the safety of mankind was wrought,
So general a gladness to an isle,
   To make the hearts of a whole nation smile,   10
As in this prince? Let it be lawful so
   To compare small with great, as still we owe
Glory to God. Then, Hail to Mary! spring
   Of so much safety to the realm and king!

67

An Ode, or Song, by All the Muses, in Celebration of Her Majesty's Birthday, 1630

1. *Clio*   Up public joy, remember
              This sixteenth of November,
                 Some brave uncommon way;
              And though the parish steeple
              Be silent, to the people   5
                 Ring thou it holiday.

---

66. The lying-in at St. James's was elaborate, and attended with some apprehension; the queen wore a trinket around her neck, to avert miscarriage. 1 *Hail Mary*: Luke 1 : 28.

67. 4–12 The silence of the bells and of the Tower guns probably reflects Henrietta Maria's unpopularity; Davenant likewise refers to the Londoners' reluctance to ring church bells and light bonfires on her behalf: 'The Queen, returning to London after a long absence', *Shorter Poems*, ed. A. M. Gibbs (Oxford, 1972), p. 47.

2. *Melpomene*

What though the thrifty Tower
And guns there spare to pour
   Their noises forth in thunder;
As fearful to awake   10
This city, or to shake
   Their guarded gates asunder?

3. *Thalia*

Yet let our trumpets sound,
And cleave both air and ground
   With beating of our drums;   15
Let every lyre be strung,
Harp, lute, theorbo sprung
   With touch of dainty thumbs!

4. *Euterpe*

That when the choir is full
The harmony may pull   20
   The angels from their spheres;
And each intelligence
May wish itself a sense
   Whilst it the ditty hears.

5. *Terpsichore*

Behold the royal Mary,   25
The daughter of great Harry
   And sister to just Louis,
Comes in the pomp and glory
Of all her brother's story
   And of her father's prowess!   30

6. *Erato*

She shows so far above
The feigned queen of love,
   This sea-girt isle upon,
As here no Venus were;
But that she reigning here   35
   Had got the ceston on!

20–1 Cf. *Und.* 3. 21–4, and n.   22–3 An 'intelligence', in the old cosmology, was thought to control each of the heavenly spheres; 'a sense' = one equipped with human faculties of perception.   26 *Harry*: Henri IV.   27 *Louis*: Louis XIII.   36 *ceston*: Venus's girdle, which made the wearer irresistible: cf. *Und.* 2. v. 41 and n., *Und.* 28. 14.

7. *Calliope*

See, see our active king
Hath taken twice the ring
    Upon his pointed lance;
Whilst all the ravished rout   40
Do mingle in a shout,
    Hey! for the flower of France!

8. *Urania*

This day the court doth measure
Her joy in state and pleasure,
    And with a reverend fear   45
The revels and the play
Sum up this crowned day,
    Her two-and-twentieth year!

9. *Polyhymnia*

Sweet, happy Mary! All
The people her do call.   50
    And this the womb divine,
So fruitful and so fair,
Hath brought the land an heir,
    And Charles a Caroline!

68

An Epigram
to the Household, 1630

What can the cause be, when the king hath given
    His poet sack, the household will not pay?
Are they so scanted in their store, or driven,
    For want of knowing the poet, to say him nay?
Well, they should know him, would the king but grant   5
    His poet leave to sing his household true;

38 *taken twice the ring*: i.e. twice speared a suspended ring while riding at full gallop: a courtly pastime. Cf. Webster, *The Duchess of Malfi*, I. i. 88.   46 *the play*: Randolph's *Amyntas* (H & S).   54 *a Caroline*: (i) a little Charles; (ii) one loyal to Charles: predating the *OED*'s first recorded usage of 1652.

68. Increasing Jonson's pension from 100 marks (£66.13.4) to £100 p.a. in 1630, Charles added the grant of an annual tierce (42 gallons) of Canary wine from the royal cellars at Whitehall.

He'd frame such ditties of their store and want
  Would make the very Greencloth to look blue;
And rather wish, in their expense of sack,
  So the allowance from the king to use,   10
As the old bard should no Canary lack.
  'Twere better spare a butt than spill his muse.
For in the genius of a poet's verse
  The king's fame lives. Go now, deny his tierce!

69

Epigram
to a Friend, and Son

Son, and my friend, I had not called you so
  To me, or been the same to you, if show,
Profit, or chance had made us; but I know
  What by that name we each to other owe—
Freedom and truth, with love from those begot: 5
  Wise crafts, on which the flatterer ventures not.
His is more safe commodity, or none;
  Nor dares he come in the comparison.
But as the wretched painter, who so ill
  Painted a dog that now his subtler skill   10
Was to have a boy stand with a club, and fright
  All live dogs from the lane and his shop's sight,
Till he had sold his piece, drawn so unlike;
  So doth the flatterer with fair cunning strike
At a friend's freedom, proves all circling means   15
  To keep him off, and howsoe'er he gleans
Some of his forms, he lets him not come near
  Where he would fix, for the distinction's fear;
For as at distance few have faculty
  To judge, so all men coming near can spy.   20

---

8 *Greencloth*: the Board which controlled expenditure of the Royal household.   12
spill: spoil; playing on words.

69. Fleay (i. 332) suggests this may be addressed to Sir Lucius Cary (see *Und.* 70 and
n.).   3 *Profit, or chance*: *Und.* 70. 99–101.   9–13 Plutarch, *How to Tell a Flatterer*,
xxiv*.   15 *circling*: devious.

Though now of flattery, as of picture, are
More subtle works and finer pieces far
Than knew the former ages: yet to life
All is but web and painting; be the strife
Never so great to get them; and the ends 25
Rather to boast rich hangings than rare friends.

70

To the Immortal Memory and Friendship of That Noble Pair,
Sir Lucius Cary and Sir H. Morison

The Turn
Brave infant of Saguntum, clear
Thy coming forth in that great year
When the prodigious Hannibal did crown
His rage with razing your immortal town.
Thou, looking then about, 5
Ere thou wert half got out,
Wise child, didst hastily return,
And mad'st thy mother's womb thine urn.
How summed a circle didst thou leave mankind
Of deepest lore, could we the centre find! 10

70. Sir Lucius Cary (?1610–43), second Viscount Falkland, of Great Tew, Oxfordshire; son of Sir Henry Cary of *Epig.* 66; Secretary of State, 1642. He behaved generously towards Jonson in the latter's last years, and wrote an elegy on Jonson's death (printed in H & S, xi. 430–7). For biographical details, see Kurt Weber, *Lucius Cary, Second Viscount Falkland* (New York, 1940). Sir Henry Morison, son of Sir Richard Morison and nephew of Fynes Morison, the traveller, died at Carmarthen, probably of small-pox, in late July or Aug. 1629, on or near his 21st birthday: see Cary's elegy on Morison, ll. 181, 317–20: printed by K. B. Murdock, *HSNPL*, xx (1938), 29–42. Jonson's poem is the first sustained attempt in English to imitate the Pindaric ode; see Robert Shafer, *The English Ode to 1660* (New York, 1966), pp. 106 ff. 'Turn', 'counter-turn', and 'stand' are Jonson's renderings of Scaliger's *volta*, *rivolta*, and *stanza*, which in turn represent Greek 'strophe', 'antistrophe', and 'epode', the three sections of the choral ode. For criticism, see Ian Donaldson, *SLI*, vi (1973), 139–52. 1 *Saguntum*: Hannibal captured and destroyed this town in Spain in 219 B.C., thus breaking the truce between the Romans and the Carthaginians and beginning the second Punic war. The story of the infant is from Pliny, *Nat. Hist.* VII. iii. 40–2: 'an infant at Saguntum . . . at once went back into the womb in the year in which that city was destroyed by Hannibal'. 1 *clear*: noble, shining (Latin *clarus*). 3 *prodigious Hannibal*: Horace's *Hannibal dirus*, *Odes*, III. vi. 36. 8 *womb thine urn*: for the collocation, see *For.* 12. 45–7, and n. 9 *summed*: complete. 9 *circle*: the emblem of perfection: 'the perfectest figure is the round', *Hym.* 404; cf. *Epig.* 128. 8.

The Counter-Turn

Did wiser nature draw thee back
From out the horror of that sack?
Where shame, faith, honour, and regard of right
Lay trampled on; the deeds of death and night
Urged, hurried forth, and hurled   15
Upon the affrighted world:
Sword, fire, and famine with fell fury met,
And all on utmost ruin set;
As, could they but life's miseries foresee,
No doubt all infants would return like thee.   20

The Stand

For what is life, if measured by the space,
Not by the act?
Or masked man, if valued by his face
Above his fact?
Here's one outlived his peers   25
And told forth four-score years;
He vexed time, and busied the whole state;
Troubled both foes and friends,
But ever to no ends;
What did this stirrer, but die late?   30
How well at twenty had he fallen or stood!
For three of his four-score he did no good.

The Turn

He entered well by virtuous parts,
Got up and thrived with honest arts;
He purchased friends and fame and honours then,   35
And had his noble name advanced with men;

---

12 *sack*: devastation, plunder.   19 *life's miseries*: cf. *Epig*. 45. 5–8.   21–2 Seneca,
Epist. xciii. 4: 'Let us measure [our lives] by their performance, not by their dur-
ation'.   23 *face*: cf. the character of that name in *Alch*.; and Busy in *B.F*. I. iii. 136–7,
who 'stands more upon his face than upon his faith, at all times'.   24 *fact*:
deeds.   25–32 Seneca, *Epist*. xciii. 3*. 'The only well-known statesman who was
approaching four-score years old in 1629 was Sir Edward Coke' (Newdigate). Is his
name glanced at in l. 42? *Und*. 46 was written before Coke's fall.   30 *stirrer*:
agitator.   31 *twenty*: probably the age at which Morison died.

But weary of that flight,
He stooped in all men's sight
To sordid flatteries, acts of strife,
And sunk in that dead sea of life   40
So deep, as he did then death's waters sup,
But that the cork of title buoyed him up.

The Counter-Turn

Alas, but Morison fell young!
He never fell: thou fall'st, my tongue.
He stood, a soldier to the last right end,   45
A perfect patriot, and a noble friend;
But most, a virtuous son.
All offices were done
By him so ample, full, and round,
In weight, in measure, number, sound,   50
As, though his age imperfect might appear,
His life was of humanity the sphere.

The Stand

Go now, and tell out days summed up with fears;
And make them years;
Produce thy mass of miseries on the stage,   55
To swell thine age;
Repeat of things a throng,
To show thou hast been long,
Not lived; for life does her great actions spell
By what was done and wrought   60

38 *stooped*: predating *OED*'s first recorded usage in this (moral) sense, 1743; evidently by analogy with the stooping (plummeting) of a bird. 40 *dead sea*: Seneca's *mare mortuum*, *Epist*. lxvii. 4 (of 'an easy existence, untroubled by the attacks of fortune'). 43–52 Seneca, *Epist*. xciii. 4*. 43–4 *fell . . . fell . . . fall'st*: playing on the physical and moral senses of the verb. Cf. Seneca's *in nulla parte cessavit*: 'in no respect had he fallen short'. 45 *stood*: remained steadfast: cf. *Epig*. 98. 1. 49 *round*: complete. 50 Cf. Wisd. 11 : 21: 'But by measure and number and weight thou didst order all things'; *U.V*. 31. 23–4. 52 *humanity*: *Epig*. Ded. 31, *Epig*. 113. 8. 52 *the sphere*: 'the perfectest form', *Haddington* 276–80; cf. l. 9 above. 53–9 These lines are probably self-referring; cf. the self-pitying epilogue to *N.I*. The present passage is in the spirit of 'Come, leave the loathed stage', *Songs*, 14. 53 *summed up*: filled out; cf. l. 9 above. 57 *Repeat of*: reiterate; celebrate (*OED*, 1, 2. † d). 58–9 *been . . . lived*: Seneca, *Epist*. xciii. 4*; and cf. *Und*. 75. 153–4. 59 *spell*: discover, or denote (*OED*, *v*.², 2. a, 3).

In season, and so brought
To light: her measures are, how well
Each syllabe answered, and was formed, how fair;
These make the lines of life, and that's her air.

The Turn
It is not growing like a tree 65
In bulk, doth make man better be;
Or standing long an oak, three hundred year,
To fall a log at last, dry, bald, and sere:
A lily of a day
Is fairer far, in May, 70
Although it fall and die that night;
It was the plant and flower of light.
In small proportions we just beauty see,
And in short measures life may perfect be.

The Counter-Turn
Call, noble Lucius, then for wine, 75
And let thy looks with gladness shine;
Accept this garland, plant it on thy head;
And think, nay know, thy Morison's not dead.
He leaped the present age,
Possessed with holy rage 80
To see that bright eternal day,
Of which we priests and poets say
Such truths as we expect for happy men;
And there he lives with memory, and Ben

61–2 *brought | To light*: Jonson uses the phrase elsewhere with reference both to childbirth (*Und.* 75. 144) and to publication (*Epig.* 131. 1–2). 62 *measures*: criteria; but the musical and poetic senses are also hinted at. 63 *syllabe*: Jonson's usual form of this word, e.g. in *Eng. Gram.* 64 *lines*: lineaments; playing on the sense of lines of verse. 'Lines of life' also suggests the threads spun by the fates: *OED* 'line', *sb.²*, I. 1. g; for a similar wordplay, cf. Hugh Holland's verses on Shakespeare's 1st Folio: 'Though his line of life went soon about / The life yet of his lines shall never out'; and Shakespeare, Sonnet 16. 9. 64 *air*: manner, melody. 65 *a tree*: the hint is from Seneca, *Epist.* xciii. 4*. 72 *flower of light*: cf. *U.V.* 26. 29 and n. 73–4 Seneca, *Epist.* xciii.: 'Just as one of small stature can be a perfect man, so in a life of small compass can be a perfect life'. Cary himself was a small man: see *The Life of Edward Earl of Clarendon, written by himself* (Oxford, 1857), i. 36. 77 *garland*: the poem itself; perhaps suggested by the 'plant' of l. 72. 78–84: Seneca, *Epist.* xciii. 5*; cf. Milton, *Lycidas*, 165–7. 81 *bright eternal day*: cf. *Und.* 84. ix. 121.

The Stand

Jonson, who sung this of him, ere he went   85
Himself to rest,
Or taste a part of that full joy he meant
To have expressed
In this bright asterism;
Where it were friendship's schism   90
(Were not his Lucius long with us to tarry)
To separate these twi-
Lights, the Dioscuri;
And keep the one half from his Harry.
But fate doth so alternate the design,   95
Whilst that in heaven, this light on earth must shine.

The Turn

And shine as you exalted are;
Two names of friendship, but one star:
Of hearts the union. And those not by chance
Made, or indentured, or leased out to advance   100
The profits for a time.
No pleasures vain did chime,
Of rhymes, or riots, at your feasts,
Orgies of drink, or feigned protests:
But simple love of greatness, and of good;   105
That knits brave minds and manners, more than blood.

The Counter-Turn

This made you first to know the why
You liked; then after to apply
That liking; and approach so one the tother,
Till either grew a portion of the other: 110

92–3 *twi-* / *Lights, the Dioscuri*: twin-lights; Castor and Pollux. After Castor's death, Jupiter allowed him to share his brother Pollux's immortality: while one brother was on earth, the other was in the infernal regions; they changed places at regular intervals. They became the constellation Gemini, the twin stars which never appear simultaneously in the heavens. On the enjambment, see *Und.* 29. 10 n., and John Hollander, ed., *Ben Jonson* (New York, 1961), Introduction, pp. 19–20.   95 *alternate*: reverse. 98 *star*: 'Good men are the stars and planets of the ages wherein they live, and illustrate the times', *Disc.* 1100–1; cf. *Epig.* 94, *Und.* 15. 196, *U.V.* 26. 77.   99–101 Cf. *Und.* 69. 2–3.   103 *rhymes*: *Und.* 47. 26.   105 *of greatness, and of good*: *Epig.* Ded. 15–16 n.   107–10 Cf. *Epig.* 86. 7–8.

Each styled, by his end,
The copy of his friend.
You lived to be the great surnames
And titles by which all made claims
Unto the virtue. Nothing perfect done   115
But as a Cary, or a Morison.

The Stand
And such a force the fair example had,
As they that saw
The good and durst not practise it, were glad
That such a law   120
Was left yet to mankind;
Where they might read and find
Friendship in deed was written, not in words;
And with the heart, not pen,
Of two so early men   125
Whose lines her rolls were, and records.
Who, ere the first down bloomed on the chin,
Had sowed these fruits, and got the harvest in.

71

*To the Right Honourable,
the Lord High Treasurer of England
an Epistle Mendicant, 1631*

MY LORD,
Poor wretched states, pressed by extremities,
Are fain to seek for succours and supplies
Or princes' aids, or good men's charities.

125 *so early men*: Clarendon speaks of Cary's remarkable intellectual attainments at an early age: *Life*, i. 35, 40–1. Cary testifies to Morison's similar attainments in 'An Anniversary': see W. D. Briggs, *Anglia*, xxxvii (1938), 476. Cf. the phrase 'thy most early wit' (of Donne), *Epig.* 23. 3; and *U.V.* 18. 7.   126 *rolls . . . records*: cf. *Und.* 35. 3–6, 75. 155–6.

71. *Lord High Treasurer*: Richard Weston, 1st Earl of Portland (1577–1635); he held this office 1628–33. See *Und.* 73 and n., 75. 97 ff., 77, 78. 27–8.

Disease, the enemy, and his engineers,
Want, with the rest of his concealed compeers,   5
Have cast a trench about me, now, five years;

And made those strong approaches, by faussebraies,
Redoubts, half-moons, horn-works, and such close ways,
The muse not peeps out one of hundred days;

But lies blocked up and straitened, narrowed in,   10
Fixed to the bed and boards, unlike to win
Health, or scarce breath, as she had never been.

Unless some saving honour of the crown
Dare think it, to relieve, no less renown
A bed-rid wit than a besieged town.   15

72

To the King, on His Birthday
19 November 1632
an Epigram Anniversary

This is King Charles's day. Speak it, thou Tower,
   Unto the ships, and they from tier to tier,
Discharge it 'bout the island in an hour,
   As loud as thunder, and as swift as fire.
Let Ireland meet it out at sea, half-way,   5
   Repeating all Great Britain's joy, and more,
Adding her own glad accents to this day,
   Like echo playing from the other shore.

5 *Want*: G. B. Johnston (*N & Q*, cxcix (1954), 471) suggests a play on 'want' (*OED*, *sb.*¹) = 'mole', comparing Jonson's letter to Newcastle (H & S, i. 213–14). His suggested emendation, 'wants', seems unnecessary.   6 *five years*: an approximate figure: Jonson's first paralytic stroke occurred in 1626, his second in 1628.   7 *faussebraies*: artificial mounds or walls thrown up in front of a main rampart.

72. 1–4 Cf. *Hamlet*, V. ii. 267–9.

What drums or trumpets, or great ordnance can,
   The poetry of steeples, with the bells,   10
Three kingdoms' mirth, in light and airy man,
   Made lighter with the wind. All noises else,
At bonfires, rockets, fireworks, with the shouts
   That cry that gladness, which their hearts would pray,
Had they but grace of thinking at these routs   15
     On the often coming of this holiday:
       And ever close the burden of the song,
       Still to have such a Charles, but this Charles long.

       The wish is great, but where the prince is such,
       What prayers, people, can you think too much?   20

73

On the Right Honourable and Virtuous Lord Weston,
Lord High Treasurer of England,
upon the Day He Was Made Earl of Portland,
17 February 1632.
To the Envious

   Look up, thou seed of envy, and still bring
     Thy faint and narrow eyes to read the king
   In his great actions: view whom his large hand
     Hath raised to be the port unto his land!

9 *can*: i.e. can speak (see l. 1); 'speak' is again understood after 'bells' in l. 10, and 'All noises else', in l. 12.   11 *man*: possibly the obsolete northern form of 'moan' (see *OED*); possibly a reference to the Isle of Man; more probably an (unsolved) textual corruption.   12 *wind*: thus in two MSS.; H & S follow the Folio reading 'wine'.   18 Cf. *Panegyre*, 162 (of James): 'Still to have such a king, and this king long' (from Martial, *Epig.* XII. vi. 5–6, on the accession of Nerva); varied here by way of allusion to the young Prince Charles, now two years old.

73. *To the Envious*: When Weston first took office as Lord Treasurer in 1628, 'the extreme visible poverty of the exchequer sheltered that province from the envy it had frequently created' (Clarendon, *History of the Rebellion*, i. § 39). But not for long: Weston quickly became unpopular through his ambition, his timorousness, his apparent Catholic sympathies, his meanness with public funds and ostentatious spending with his own. See *Und*. 71 and n.   2 *faint and narrow eyes*: Envy is traditionally pictured as blind or narrow-eyed: see Ovid, *Metam*. ii. 760–805, Dante, *Purgatorio*, xiii.   4 *port . . . land*: punning on Weston's new title; 'port' = tribute, income, payment (*OED*, † *sb.*⁵).

Weston! That waking man, that eye of state,   5
    Who seldom sleeps, whom bad men only hate!
Why do I irritate or stir up thee,
    Thou sluggish spawn, that canst, but wilt not, see?
Feed on thyself for spite, and show thy kind:
    To virtue and true worth be ever blind.   10
Dream thou couldst hurt it; but before thou wake
    To effect it, feel thou hast made thine own heart ache.

74

To the Right Honourable Jerome, Lord Weston
an Ode Gratulatory,
for His Return from His Embassy, 1632

Such pleasure as the teeming earth
Doth take in easy nature's birth,
    When she puts forth the life of every thing,
And in a dew of sweetest rain
She lies delivered, without pain,   5
    Of the prime beauty of the year, the spring;
The rivers in their shores do run,
The clouds rack clear before the sun,
    The rudest winds obey the calmest air;
Rare plants from every bank do rise,   10
And every plant the sense surprise,
    Because the order of the whole is fair!
The very verdure of her nest,
Wherein she sits so richly dressed,
    As all the wealth of season there was spread,   15

---

5 *That waking man, that eye of state*: cf. *Und.* 75. 109–11. Weston's coat of arms
displayed an eagle, traditionally famed for its powers of sight (*Gentleman's Magazine*
(1823), i. 413; Pliny, *Nat. Hist.* X. iii. 10); the contrast is with the imperfect sight of
the envious.

74. *Jerome, Lord Weston*: 1605–63, 2nd Earl of Portland, son of Richard Weston of
Und. 71 & 73; styled Lord Weston from 1633; succeeded to the earldom, 1635. He
returned in March 1633 from a mission to Paris (1632) and Turin. See also *Und.* 75.

Doth show the graces and the hours
Have multiplied their arts and powers
    In making soft her aromatic bed.
Such joys, such sweets doth your return
Bring all your friends, fair lord, that burn   20
    With love, to hear your modesty relate
The business of your blooming wit,
With all the fruit shall follow it,
    Both to the honour of the king and state.
O how will then our court be pleased,   25
To see great Charles of travail eased,
    When he beholds a graft of his own hand
Shoot up an olive fruitful, fair,
To be a shadow to his heir,
    And both a strength and beauty to his land!   30

·75

Epithalamion;
or, a Song, Celebrating the Nuptials
of that Noble Gentleman, Mr. Jerome Weston,
Son and Heir of the Lord Weston, Lord High Treasurer of
England,
with the Lady Frances Stuart,
Daughter of Esmé, Duke of Lennox, Deceased,
and Sister of the Surviving Duke of the Same Name

Though thou hast passed thy summer standing, stay
    Awhile with us, bright sun, and help our light;
Thou canst not meet more glory on the way
    Between thy tropics, to arrest thy sight,

28 *an olive*: perhaps referring to Weston's role as a peace-negotiator.   29 *shadow*:
companion, Latin *umbra* (Newdigate).

75. For Jerome Weston, see *Und.* 74 and n.; for his father, *Und.* 71 & 73 and n.
Frances Stuart (1617–94) was the daughter of Jonson's former patron Esmé Stuart (see
Epig. 127 and n.); he died shortly after inheriting the dukedom in 1624, and was
succeeded by his son James (1612–55). The marriage took place at Roehampton
Chapel on 25 June 1632. Davenant also wrote an epithalamion for the occasion:
Shorter Poems, ed. Gibbs, pp. 45–6. Shirley's 'Epithalamion: to his noble friend I.W.'
perhaps celebrates the same event: *Works*, ed. W. Gifford and A. Dyce (London,
1833), vi. 438–9.   1 *summer standing*: solstice. Spenser's *Epithalamion*, written for his

Than thou shalt see today: 5
    We woo thee, stay
    And see what can be seen,
The bounty of a king, and beauty of his queen!

See the procession! What a holy day
    (Bearing the promise of some better fate)   10
Hath filled with caroches all the way
    From Greenwich hither to Roehampton gate!
        When looked the year, at best,
            So like a feast?
        Or were affairs in tune,   15
By all the spheres' consent, so in the heart of June?

What bevy of beauties, and bright youths at charge
    Of summer's liveries and gladding green,
Do boast their loves and braveries so at large
    As they came all to see, and to be seen!   20
        When looked the earth so fine,
            Or so did shine,
        In all her bloom and flower
To welcome home a pair, and deck the nuptial bower?

It is the kindly season of the time,   25
    The month of youth, which calls all creatures forth
To do their offices in nature's chime,
    And celebrate (perfection at the worth)
        Marriage, the end of life,
            That holy strife,   30
        And the allowed war:
Through which not only we, but all our species are.

Hark how the bells upon the waters play
    Their sister-tunes from Thames's either side!
As they had learned new changes for the day,   35
    And all did ring the approaches of the bride;

<hr />

own wedding on the year's longest day (St. Barnabas: 11 June), also has an invocation to the sun.   11 *caroches*: stately coaches.   17–18 *at charge | Of*: probably, expensively decked out in.   28 *perfection*: cf. Donne, 'Epithalamion Made at Lincoln's Inn': 'Today put on perfection, and a woman's name'; Marlowe, *Hero and Leander*, i. 268; *Hym.* 197, 295, 404, 559; Tilley, W718.   30–1 *Hym.* 455–8.

                     The Lady Frances, dressed
                          Above the rest
                     Of all the maidens fair,
          In graceful ornament of garland, gems, and hair.   40

          See how she paceth forth in virgin white,
               Like what she is, the daughter of a duke,
          And sister; darting forth a dazzling light
               On all that come her simplesse to rebuke!
                          Her tresses trim her back,   45
                               As she did lack
                          Nought of a maiden queen,
          With modesty so crowned, and adoration seen.

          Stay, thou wilt see what rites the virgins do!
               The choicest virgin-troop of all the land;   50
          Porting the ensigns of united two,
               Both crowns and kingdoms in their either hand;
                          Whose majesties appear,
                               To make more clear
                          This feast than can the day,   55
          Although that thou, O sun, at our entreaty stay!

          See, how with roses and with lilies shine
               (Lilies and roses, flowers of either sex)
          The bright bride's path, embellished more than thine
               With light of love, this pair doth intertex!   60
                          Stay, see the virgins sow
                               Where she shall go
                          The emblems of their way:
          Oh, now thou smil'st, fair sun, and shin'st, as thou wouldst stay!

---

40 *hair*: cf. Spenser, *Epithalamion*, 154–5.   41 *virgin white*: ibid., 151: 'Clad all in
white, that seems a virgin best'.   44 *simplesse*: i.e. simplicity of dress.   45 *tresses*: cf.
Und. 2. vi. 10.   51 *Porting the ensigns*: carrying the heraldic arms or badges.   54 *clear*:
illustrious.   57–8 The roses and lilies, as the flowers of England and France, represent
the union of Charles and Henrietta Maria, who were present at the marriage; cf. *Und*.
65. 3 and n. But they are also flowers traditionally associated with marriage: see *Hym*.
219–20 and Jonson's n., and cf. Spenser, *Epith*. 43, Statius, *Silvae*, I. ii. 22, etc.   59
thine: i.e. the sun's.   60 *intertex*: weave together.

With what full hands, and in how plenteous showers,   65
    Have they bedewed the earth where she doth tread,
As if her airy steps did spring the flowers,
    And all the ground were garden, where she led!
            See, at another door,
                On the same floor,   70
            The bridegroom meets the bride
With all the pomp of youth, and all our court beside.

Our court, and all the grandees; now, sun, look,
    And looking with thy best inquiry, tell,
In all the age of journals thou hast took   75
    Saw'st thou that pair became these rites so well,
            Save the preceding two?
                Who, in all they do,
            Search, sun, and thou wilt find,
They are the exampled pair, and mirror of their kind.   80

Force from the phoenix, then, no rarity
    Of sex, to rob the creature; but from man,
The king of creatures, take his parity
    With angels, muse, to speak these: nothing can
            Illustrate these, but they   85
                Themselves today,
            Who the whole act express;
All else we see beside are shadows, and go less.

It is their grace and favour that makes seen
    And wondered at the bounties of this day: 90
All is a story of the king and queen!
    And what of dignity and honour may

<hr />

73 *grandees*: James Stuart, the bride's brother, had been made a grandee of Spain in Jan. 1632.   75 *journals*: days' travels.   80 *exampled*: exemplary.   81–2 The phoenix is traditionally sexless; ever since Petrarch had compared Laura to that bird, however, it had been common for poets to refer to it as though it were feminine: see W. H. Matchett, *The Phoenix and Turtle* (The Hague, 1965), Introduction. Jonson's specific allusion may be to King James's *Phoenix*, in which Esmé Stuart (James's cousin, and the bride's grandfather) is represented as a female phoenix: *Poems*, ed. James Craigie (Scottish Text Society, Edinburgh & London, 1955), i. 39–59.   88 *go less*: are abated or diminished.

Be duly done to those
Whom they have chose,
And set the mark upon   95
To give a greater name and title to: their own!

Weston, their treasure, as their treasurer,
    That mine of wisdom and of counsels deep,
Great say-master of state, who cannot err,
    But doth his carat and just standard keep   100
            In all the proved assays,
                And legal ways
            Of trials, to work down
Men's love unto the laws, and laws to love the crown.

And this well moved the judgement of the king   105
    To pay with honours to his noble son,
Today, the father's service, who could bring
    Him up to do the same himself had done.
            That far all-seeing eye
                Could soon espy   110
            What kind of waking man
He had so highly set, and in what Barbican.

Stand there; for when a noble nature's raised
    It brings friends joy, foes grief, posterity fame;
In him the times, no less than prince, are praised,   115
    And by his rise, in active men his name
            Doth emulation stir;
                To the dull, a spur
            It is: to the envious meant
A mere upbraiding grief, and torturing punishment.   120

See, now the chapel opens, where the king
    And bishop stay to consummate the rites:
The holy prelate prays, then takes the ring,
    Asks first, Who gives her? (I, Charles): then he plights

99 *say-master*: assay-master.   109–11 Cf. *Und*. 73. 5.   113–20 Repeated in prose in
Disc. 1292–97.   119 *envious*: *Und*. 73 and n.   121 ff. Cf. Spenser, *Epith*. 204–41.   124
Charles: after Esmé Stuart's death in 1624, his children had passed into royal guar-
dianship. The wedding had been arranged by Charles.

One in the other's hand,   125
    Whilst they both stand
    Hearing their charge, and then
The solemn choir cries, Joy, and they return, Amen.

O happy bands! and thou more happy place,
    Which to this use wert built and consecrate!   130
To have thy God to bless, thy king to grace,
    And this their chosen bishop celebrate
            And knit the nuptial knot,
                Which time shall not,
            Or cankered jealousy,   135
With all corroding arts, be able to untie!

The chapel empties, and thou mayst be gone
    Now, sun, and post away the rest of day:
These two, now holy church hath made them one,
    Do long to make themselves so, another way: 140
            There is a feast behind
                To them of kind,
            Which their glad parents taught
One to the other, long ere these to light werè brought.

Haste, haste, officious sun, and send them night   145
    Some hours before it should, that these may know
All that their fathers and their mothers might
    Of nuptial sweets, at such a season, owe,
            To propagate their names,
                And keep their fames   150
            Alive, which else would die,
For fame keeps virtue up, and it posterity.

The ignoble never lived, they were awhile
    Like swine, or other cattle here on earth:
Their names are not recorded on the file   155
    Of life, that fall so; Christians know their birth

132 *bishop*: Laud, then Bishop of London.   141 *behind*: i.e. ahead.   145 Cf. Spenser,
Epith. 282 ff.: 'Haste thee, O fairest planet, to thy home', etc.   153–4 *Und.* 70. 58–9,
Pleas. Rec. 102–3, *Love's Tr.* 92–6.   155–6 *recorded on the file* | *Of life*: cf. *Und.* 35. 3–
6; 70. 126.

    Alone, and such a race
     We pray may grace
    Your fruitful spreading vine,
But dare not ask our wish in language fescennine.   160

Yet, as we may, we will with chaste desires
 (The holy perfumes of the marriage bed)
Be kept alive those sweet and sacred fires
 Of love between you and your lovelihead:
    That when you both are old,   165
     You find no cold
    There; but, renewed, say
(After the last child born), This is our wedding day.

Till you behold a race to fill your hall,
 A Richard, and a Jerome, by their names   170
Upon a Thomas, or a Francis call;
 A Kate, a Frank, to honour their granddames,
    And 'tween their grandsires' thighs,
     Like pretty spies,
    Peep forth a gem; to see   175
How each one plays his part of the large pedigree.

And never may there want one of the stem
 To be a watchful servant for this state;
But like an arm of eminence, 'mongst them
 Extend a reaching virtue, early and late: 180
    Whilst the main tree, still found
     Upright and sound,
    By this sun's noonstead's made
So great; his body now alone projects the shade.

---

160 *fescennine*: bawdy, obscene: as in many Roman weddings. 161 *will*: i.e. will ask. 165–8 Cf. *U.V.* 18. 23–4, and n. 170–2 Richard and Jerome, after the groom's father and the groom himself; Thomas, after the groom's brother, and Francis, perhaps, after his great-grandfather; Kate, after the bride's mother, formerly Katherine Clifton; Frank, after the bride herself. One son, Charles, was born in 1639. 178 *watchful servant*: recalling the terms in which the groom's father had been praised: ll. 109–10 above, and *Und.* 73. 5. 180 *reaching*: far-reaching. 184 Recalling Lucan's comparison of the aged Pompey to an old oak, *De Bello Civili*, i. 135–43; prompted, no doubt, by the idea of the genealogical tree.

They both are slipped to bed; shut fast the door,   185
    And let him freely gather love's first fruits;
He's master of the office; yet no more
    Exacts than she is pleased to pay: no suits,
        Strifes, murmurs, or delay,
            Will last till day;   190
        Night and the sheets will show
The longing couple all that elder lovers know.

76

The Humble Petition of Poor Ben.
To the Best of Monarchs, Masters, Men,
King Charles

—*Doth most humbly show it,*
To your Majesty your poet:

That whereas your royal father,   5
James the blessed, pleased the rather,
Of his special grace to letters,
To make all the muses debtors
To his bounty; by extension
Of a free poetic pension,   10
A large hundred marks' annuity,
To be given me in gratuity
For done service, and to come:
    And that this so accepted sum
Or dispensed in books, or bread,   15
(For with both the muse was fed)

187–8 The notion is a common one (cf. e.g. Donne's 'Epithalamion . . . on the Lady Elizabeth and Count Palatine', 87–96), but is sharpened by the fact that Weston's father's 'office' was that of Lord Treasurer. 192 *elder lovers*: cf. Marlowe, *Hero and Leander*, ii. 69: 'Which taught him all that elder lovers know'.

76. Jonson's petition met with success: see *Und.* 57 & 68 and notes. The warrant for the new pension (printed in H & S, i. 245–8) is dated 26 March 1630. Jonson's poem is based on Martial, *Epig.* IV. xxvii*. 2 *Monarchs, Masters, Men*: cf. *Und.* 79. 7.

Hath drawn on me from the times
All the envy of the rhymes
And the rattling pit-pat noise
Of the less poetic boys;   20
When their pot-guns aim to hit,
With their pellets of small wit,
Parts of me they judged decayed,
But we last out, still unlaid.
    Please your majesty to make   25
Of your grace, for goodness' sake,
Those your father's marks, your pounds;
Let their spite, which now abounds,
Then go on and do its worst;
This would all their envy burst,   30
And so warm the poet's tongue
You'd read a snake in his next song.

77

*To the Right Honourable, the Lord Treasurer of England
an Epigram*

If to my mind, great lord, I had a state,
    I would present you now with curious plate
Of Nuremburg, or Turkey; hang your rooms
    Not with the Arras, but the Persian looms.
I would, if price or prayer could them get,   5
    Send in what or Romano, Tintoret,

23 *decayed*: Owen Felltham had referred to Jonson's 'declining wit' in a poem replying to 'Come leave the loathed stage' (printed in H & S, xi. 339–40). In the same year (1629) Jonson had admitted in the epilogue to *N.I.* that his strength was waning.   24 *unlaid*: '?laid out (as a corpse); laid in the grave' (*OED*, 1. b, citing this example only).   31–2 Referring to Aesop's story of the compassionate farm-hand who puts a frozen snake to his bosom, and is eventually bitten for his pains. Charles is evidently being warned that it is dangerous to tolerate one of Jonson's rivals, though the point is clumsily made: Jonson forgets that he has already rendered Martial's *invidus* (= jealous man) by a plural (l. 20, etc.). In the margin of his own copy of Martial Jonson wrote 'Inigo'.

77. *Lord Treasurer*: Weston: see *Und.* 71 n. The poem is an adaptation of Horace's address to Censorinus, *Odes*, IV. viii.   1 i.e. if my circumstances were as fine as my wishes.   3 *Nuremburg*: a famous centre for silversmiths.   4 *Arras . . . Persian*: 'Arras' had by now become a generic term for any hanging; Persian carpets were naturally esteemed.   6 *Romano*: Giulio Romano, 1492–1546.

Titian, or Raphael, Michelangelo,
   Have left in fame to equal, or outgo
The old Greek hands in picture, or in stone.
   This I would do, could I think Weston one   10
Catched with these arts, wherein the judge is wise
   As far as sense, and only by the eyes.
But you I know, my lord, and know you can
   Discern between a statue and a man,
Can do the things that statues do deserve,   15
   And act the business which they paint, or carve.
What you have studied are the arts of life:
   To compose men and manners, stint the strife
Of murmuring subjects, make the nations know
   What worlds of blessings to good kings they owe,   20
And mightiest monarchs feel what large increase
   Of sweets and safeties they possess by peace.
These I look up at with a reverent eye,
   And strike religion in the standers-by;
Which, though I cannot as an architect   25
   In glorious piles or pyramids erect
Unto your honour: I can tune in song
   Aloud; and (haply) it may last as long.

78

An Epigram
to My Muse, the Lady Digby
on Her Husband, Sir Kenelm Digby

Though, happy muse, thou know my Digby well,
   Yet read him in these lines: he doth excel
In honour, courtesy, and all the parts
   Court can call hers, or man could call his arts.

12 Cf. *Hym.* 1–6.   18 *men and manners*: *Epig.* 128. 2 and n.   25 *Which*: referring to 'the arts of life' (l. 17), just listed.   25 *as an architect*: 'another expression of Jonson's perennial antagonism to Inigo Jones' (Newdigate).   26 *piles or pyramids*: cf. *Epig.* 60. 1–5.

78. Sir Kenelm Digby (1603–65), naval commander, adventurer, diplomatist, scientist, scholar; he married Venetia Stanley (see *Und.* 84 and n.) in 1625 after a period of enforced separation during which she believed him to be dead; he was later to act as Jonson's literary executor. For biographical details, R. T. Petersson, *Sir Kenelm Digby* (London, 1956).   1 *muse*: see *Und.* 84. iv. 21, *Und.* 84. ix, title.

He's prudent, valiant, just, and temperate;   5
  In him all virtue is beheld in state:
And he is built like some imperial room
  For that to dwell in, and be still at home.
His breast is a brave palace, a broad street,
  Where all heroic ample thoughts do meet;   10
Where nature such a large survey hath ta'en,
  As other souls, to his, dwell in a lane:
Witness his action done at Scanderoon,
  Upon my birthday, the eleventh of June;
When the apostle Barnaby the bright   15
  Unto our year doth give the longest light;
In sign the subject and the song will live,
  Which I have vowed posterity to give.
Go, muse, in, and salute him. Say he be
  Busy, or frown at first, when he sees thee   20
He will clear up his forehead, think thou bring'st
  Good omen to him in the note thou sing'st,
For he doth love my verses, and will look
  Upon them (next to Spenser's noble book)
And praise them too. Oh, what a fame 't will be!   25
  What reputation to my lines and me,
When he shall read them at the Treasurer's board
  (The knowing Weston) and that learned lord
Allows them! Then, what copies shall be had,
  What transcripts begged; how cried up, and how glad   30
Wilt thou be, muse, when this shall them befall!
  Being sent to one, they will be read of all.

---

8 *at home*: a favourite thought: *For.* 4. 68, *Und.* 14. 30.   13 *Scanderoon*: in Turkey;
defeating the French and Venetian fleet, 1628.   14 *my birthday*: sometimes emended
to 'his' (the reading of Quarto and Duodecimo editions), on the strength of Richard
Ferrar's *Epitaph* on Digby (1665); in defence of 'my', see W. D. Briggs, *MLN*, xxxiii
(1918), 137–45.   15 *Barnaby the bright*: St. Barnabas's Day, the longest day under the
old calendar. The phrase occurs in Spenser's *Epithalamion*, but is also traditional.   18
posterity: cf. *Epig.* Ded. 17.   19 ff. Cf. Martial, *Epig.* VII. xcvii*.   24 *Spenser's noble
book*: Digby wrote some *Observations* upon *The Faerie Queene*, a work for which he had
a particular affection.   28 *Weston*: for further evidence of Weston's interest in
Jonson's verse, see Cary's poem in *Jonsonus Virbius* (H & S, xi. 434), l. 168.

79

New Years expect new gifts: sister, your harp,
  Lute, lyre, theorbo, all are called today;
Your change of notes, the flat, the mean, the sharp,
  To show the rites and to usher forth the way
Of the New Year, in a new silken warp,   5
  To fit the softness of our Year's-gift, when
    We sing the best of monarchs, masters, men;
For, had we here said less, we had sung nothing then.

A New Year's Gift Sung to King Charles, 1635

Rector chori
Today old Janus opens the New Year
  And shuts the old. Haste, haste, all loyal swains,   10
That know the times and seasons when to appear,
  And offer your just service on these plains;
Best kings expect first-fruits of your glad gains.
1. Pan is the great preserver of our bounds,
2. To him we owe all profits of our grounds: 15
3. Our milk, 4. Our fells, 5. Our fleeces, 6. And first lambs;
7. Our teeming ewes, 8. And lusty-mounting rams.
9. See where he walks, 10. With Mira by his side;
Chorus
Sound, sound his praises loud, and with his, hers divide.

79. Gifts were customarily given at New Year rather than at Christmas; on the practice
of New Year poems, see *For.* 12 n. The present poem is a reworking of material from
Pan's Anniversary (1620): there James, in the figure of Pan, was praised for his policy
of peace; here Charles, in the figure of Pan, is praised in the same terms: but from l. 54
the identification of Charles with Pan breaks down. See E. Simpson, *RES*, xiv (1938),
175–8. 2 *called*: called for. 3 *mean*: 'applied to the tenor and alto parts and the tenor
clef, as intermediate between the bass and treble' (*OED*, adj.², 1. † b). 5 *warp*: woven
garment. 7 *monarchs, masters, men*: cf. *Und.* 76. 2. 9 Rector chori: leader of the
chorus. 18 *Mira*: Henrietta Maria; here figured as Pan's sister (l. 25) rather than wife
for decorous reasons: Pan is traditionally ruttish. Newdigate, taking 'sister' literally,
argues that Pan must represent Henrietta Maria's brother, Louis XIII, but it is
difficult to imagine that Jonson would address Louis in such extravagant terms (see
e.g. ll. 42–3).

Shepherds
Of Pan we sing, the best of hunters, Pan,   20
      That drives the hart to seek unused ways,
And in the chase more than Silvanus can;
Chorus
      Hear, O you groves, and hills resound his praise.
Nymphs
Of brightest Mira do we raise our song,
      Sister of Pan, and glory of the spring;   25
Who walks on earth as May still went along;
Chorus
      Rivers and valleys, echo what we sing.
Shepherds
Of Pan we sing, the chief of leaders, Pan,
      That leads our flocks and us, and calls both forth
To better pastures than great Pales can;   30
Chorus
      Hear, O you groves, and hills, resound his worth.
Nymphs
Of brightest Mira is our song, the grace
      Of all that nature yet to life did bring;
And were she lost, could best supply her place;
Chorus
      Rivers and valleys, echo what we sing.   35
            1. Where'er they tread the enamoured ground
                  The fairest flowers are always found;
            2. As if the beauties of the year
                  Still waited on them where they were.
            1. He is the father of our peace;   40
            2. She to the crown hath brought increase.
            1. We know no other power than his,
                  Pan only our great shepherd is,
Chorus
Our great, our good. Where one's so dressed
In truth of colours, both are best.   45

22 *Silvanus*: a rural god, half-goat, half-man.   30 *great Pales*: the Italian country
goddess (Virgil's *magna Pales, Georgics*, iii. 1).   34 *were she lost*: *N.I.* III. ii. 65–72;
Epig. 105.

Haste, haste you hither, all you gentler swains,
That have a flock, or herd, upon these plains;
This is the great preserver of our bounds,
To whom you owe all duties of your grounds;
Your milks, your fells, your fleeces, and first lambs,   50
Your teeming ewes, as well as mounting rams.
Whose praises let's report unto the woods,
That they may take it echoed by the floods.
   'Tis he, 'tis he; in singing, he,
    And hunting, Pan, exceedeth thee.   55
   He gives all plenty, and increase,
   He is the author of our peace.

Where'er he goes upon the ground,
The better grass and flowers are found.
To sweeter pastures lead he can,   60
Than ever Pales could, or Pan;
He drives diseases from our folds,
The thief from spoil his presence holds.
Pan knows no other power than his,
This only the great shepherd is.   65
   'Tis he, 'tis he, &c.

[For *Und.* 80 and 81, see *Dubia*, 2 and 3]

82

To My Lord the King,
on the Christening His Second Son, James

That thou art loved of God, this work is done,
   Great king, thy having of a second son;
And by thy blessing may thy people see
   How much they are beloved of God, in thee;
Would they would understand it! Princes are   5
   Great aids to empire, as they are great care

---

82. The future James II was born 14 Oct. 1633, and christened by Laud 24 Nov. 1633.   5 *Would they would understand it!*: alluding to the rising discontent in the land; cf. *Und.* 64. 1–2.

To pious parents, who would have their blood
    Should take first seisin of the public good,
As hath thy James; cleansed from original dross
    This day by baptism, and his Saviour's cross: 10
Grow up, sweet babe, as blessed in thy name,
    As in renewing thy good grandsire's fame;
Methought Great Britain in her sea before
    Sat safe enough, but now secured more.
At land she triumphs in the triple shade   15
    Her rose and lily, intertwined, have made.

Oceano secura meo, securior umbris.

83

An Elegy
on the Lady Jane Paulet,
Marchioness of Winton

What gentle ghost, besprent with April dew,
    Hails me so solemnly to yonder yew,
And beckoning woos me, from the fatal tree
    To pluck a garland for herself, or me?
I do obey you, beauty! for in death   5
    You seem a fair one. Oh, that you had breath
To give your shade a name! Stay, stay, I feel
    A horror in me, all my blood is steel!

---

8 *first seisin*: *primer seisin* was an ancient feudal right by which the Crown exacted from the heir of a deceased tenant his first year's profits.   12 *grandsire's*: James I's.   15 *triple shade*: the three (living) children born to Charles and Henrietta Maria at this stage: Charles, Mary, and James.   16 *rose and lily*: see *Und.* 65. 3 n.   17 'Safe in my ocean, safer in my shades'; source unidentified.

83. *Lady Jane Paulet*: b. 1607, daughter of Thomas, Viscount Savage, of Rock Savage in Cheshire; married in 1622 John Paulet, 5th Marquis of Winchester, loyal friend of Henrietta Maria and Charles. She died 'big with child' on 15 April 1631. Milton and other Cambridge poets also wrote tributes to her. Jonson's poem is discussed by O. B. Hardison, *The Enduring Monument* (Chapel Hill, 1962), pp. 142–5, 149, 166.   1–3 Echoed by Pope, *Elegy to the Memory of an Unfortunate Lady*, 1–2: 'What beck'ning ghost, along the moonlight shade / Invites my step, and points to yonder glade?' Cf. the beckoning ghost of *Hamlet*, I. iv. 58.   3 *fatal tree*: various superstitions concerning death are associated with the yew, primarily because it is commonly found in church-yards, and its leaves are poisonous.

Stiff, stark, my joints 'gainst one another knock!
  Whose daughter?—ha! Great Savage of the Rock?   10
He's good as great. I am almost a stone!
  And ere I can ask more of her, she's gone.
Alas, I am all marble! Write the rest
  Thou wouldst have written, fame, upon my breast:
It is a large fair table, and a true,   15
  And the disposure will be something new,
When I, who would her poet have become,
  At least may bear the inscription to her tomb.
She was the Lady Jane, and Marchioness
  Of Winchester—the heralds can tell this: 20
Earl Rivers' grandchild—serve not forms, good fame,
  Sound thou her virtues, give her soul a name.
Had I a thousand mouths, as many tongues,
  And voice to raise them from my brazen lungs,
I durst not aim at that: the dotes were such   25
  Thereof, no notion can express how much
Their carat was! I or my trump must break,
  But rather I, should I of that part speak!
It is too near of kin to heaven, the soul,
  To be described; fame's fingers are too foul   30
To touch these mysteries. We may admire
  The blaze and splendour, but not handle fire!
What she did here by great example well
  To enlive posterity, her fame may tell;
And, calling truth to witness, make that good   35
  From the inherent graces in her blood!
Else, who doth praise a person by a new
  But a feigned way, doth rob it of the true.
Her sweetness, softness, her fair courtesy,
  Her wary guards, her wise simplicity,   40

---

16 *disposure*: arrangement, disposition.   20 *heralds*: cf. *Epig.* 9. 4 and n.   21 *Earl Rivers'*: Thomas Darcy, her mother's father.   23–4 A common classical formula; cf. Homer, *Iliad*, ii. 489, Virgil, *Georgics*, ii. 42–4, *Aeneid*, vi. 625, etc.   25 *dotes*: endowments, natural qualities.   30–1 Cf. Milton on the limitations of fame, 'That last infirmity of noble mind', etc., *Lycidas*, 71.   34 *enlive*: impart spiritual life to.

Were like a ring of virtues 'bout her set,
    And piety the centre, where all met.
A reverend state she had, an awful eye,
    A dazzling, yet inviting majesty:
What nature, fortune, institution, fact   45
    Could sum to a perfection, was her act!
How did she leave the world, with what contempt!
    Just as she in it lived, and so exempt
From all affection. When they urged the cure
    Of her disease, how did her soul assure   50
Her sufferings, as the body had been away!
    And to the torturers, her doctors, say,
Stick on your cupping-glasses, fear not, put
    Your hottest caustics to, burn, lance, or cut:
'Tis but a body which you can torment,   55
    And I into the world all soul was sent!
Then comforted her lord, and blessed her son
    Cheered her fair sisters in her race to run,
With gladness tempered her sad parents' tears,
    Made her friends' joys to get above their fears,   60
And, in her last act, taught the standers-by
    With admiration and applause to die.
Let angels sing her glories, who did call
    Her spirit home to her original;
Who saw the way was made it, and were sent   65
    To carry and conduct the complement
'Twixt death and life; where her mortality
    Became her birthday to eternity.
And now, through circumfused light, she looks
    On nature's secrets, there, as her own books: 70

45 *institution*: education.   45 *fact*: noble deeds.   47 ff. Cf. Pliny, *Epist.* V. xvi*.   49 *affection*: passion.   50 *assure*: encourage, give confidence to.   53 *cupping-glasses*: for bleeding.   54 An abscess on the cheek had to be lanced.   57 *son*: Charles Paulet, b. *c.* 1625, 6th Marquis of Winchester, 1674; Duke of Bolton, 1689.   58 *sisters*: Elizabeth and Dorothy Savage, who had danced in *Chloridia*.   61–2 The Christian art of dying is implicitly compared with the art of the tragic actor, who merely feigns death: cf. Marvell, *An Horatian Ode*, 53–8.   66 *complement*: i.e. the body.   68 *birthday*: cf. Donne's *Second Anniversary*, 214.   69 ff. Cf. Seneca, *Ad Marciam de Consolatione*, xxv. 2.

Speaks heaven's language, and discourseth free
   To every order, every hierarchy;
Beholds her Maker, and in him doth see
   What the beginnings of all beauties be,
And all beatitudes that thence do flow: 75
   Which they that have the crown are sure to know.
Go now, her happy parents, and be sad,
   If you not understand what child you had;
If you dare grudge at heaven, and repent
   To have paid again a blessing was but lent   80
And trusted so, as it deposited lay
   At pleasure to be called for, every day;
If you can envy your own daughter's bliss
   And wish her state less happy than it is;
If you can cast about your either eye,   85
   And see all dead here, or about to die;
The stars, that are the jewels of the night,
   And day, deceasing with the prince of light,
The sun; great kings and mightiest kingdoms fall;
   Whole nations, nay mankind, the world, with all   90
That ever had beginning there, to have end!
   With what injustice should one soul pretend
To escape this common known necessity;
   When we were all born, we began to die;
And, but for that contention and brave strife   95
   The Christian hath to enjoy the future life,
He were the wretched'st of the race of men:
   But as he soars at that, he bruiseth then
The serpent's head; gets above death and sin,
   And, sure of heaven, rides triumphing in.   100

---

76 *the crown*: Jas. 1 : 12, Rev. 2 : 10.   80 *but lent*: see *Epig.* 45. 3 n.   85 ff. Cf.
Donne's *Anniversaries*.   94 Proverbial: Tilley, M73.   95–7 1 Cor. 15 : 19: 'If in this
life only we have hope in Christ, we are of all men most miserable'.   98–9 Gen. 3 : 15.

84

Eupheme;
or, the Fair Fame Left to Posterity
of that Truly Noble Lady, the Lady Venetia Digby,
Late Wife of Sir Kenelm Digby, Knight:
    A Gentleman Absolute in All Numbers.
    Consisting of These Ten Pieces:
        The Dedication of Her Cradle;
        The Song of Her Descent;
        The Picture of Her Body;
        Her Mind;
        Her Being Chosen a Muse;
        Her Fair Offices;
        Her Happy Match;
        Her Hopeful Issue;
        Her 'Αποθέωσις, or Relation to the Saints;
        Her Inscription, or Crown.

Vivam amare voluptas, defunctam religio.

                           Statius.

84. i

The Dedication of Her Cradle

Fair fame, who art ordained to crown
With evergreen, and great renown,
Their heads that envy would hold down
    With her, in shade

---

84. *Eupheme*: the word is Jonson's coinage, and 'fair fame' his translation of it. For Sir Kenelm and Lady Venetia Digby, see *Und*. 78 and n. Jonson evidently sent this (now imperfect) sequence of poems to Sir Kenelm on his wife's death in 1633; some of the poems appear, however, to have been written during Venetia's lifetime. 'Absolute in all numbers' (= perfect in all ways) is a rendering of Pliny's *omnibus numeris absolutum* (*Epist*. IX. xxxviii: of Rufus's book). The motto is from *Silvae*, V, preface: 'to love [a wife] is a joy, while she is alive, and a religion, when she is departed'; Statius is writing to console Abascantus, Secretary of State to Domitian, on the loss of his wife, Priscilla.

84. i. 1–5 The contrast between 'fair fame' and 'envy' is reminiscent of that between *fama bona* and *fama mala* in Jonson's poem on Raleigh's frontispiece, *Und*. 24, and may owe something to C. Ripa, *Iconologia* (Padua, 1611), pp. 154–5.

Of death and darkness, and deprive   5
Their names of being kept alive
By thee and conscience, both who thrive
        By the just trade

Of goodness still: vouchsafe to take
This cradle, and for goodness' sake,   10
A dedicated ensign make
        Thereof, to time.

That all posterity, as we,
Who read what the crepundia be,
May something by that twilight see   15
        'Bove rattling rhyme.

For though that rattles, timbrels, toys
Take little infants with their noise,
As properest gifts to girls and boys
        Of light expense;   20

Their corals, whistles, and prime coats,
Their painted masks, their paper boats,
With sails of silk, as the first notes
        Surprise their sense:

Yet here are no such trifles brought,   25
No cobweb cauls, no surcoats wrought
With gold or clasps, which might be bought
        On every stall.

But here's a song of her descent,
And call to the high parliament   30
Of heaven, where seraphim take tent
        Of ordering all.

This uttered by an ancient bard,
Who claims (of reverence) to be heard,
As coming with his harp, prepared   35
        To chant her gree,

---

13 *posterity*: *Epig.* Ded. 14–17.   14 *crepundia*: childish toy, rattle.   16 *rattling rhyme*: playing on 'crepundia'. Cf. Sidney, *Astrophil and Stella*, xv. 6; for the animus against rhyme, see *Und.* 29 and n.   17–28 Cf. *Disc.* 1437 ff.   21 *prime coats*: i.e. children's first skirt-like garments.   31 *tent*: care.   36 *gree*: degree, descent.

Is sung; as also her getting up
By Jacob's ladder to the top
Of that eternal port kept ope'
     For such as she.   40

84. ii

The Song of Her Descent

I sing the just and uncontrolled descent
   Of Dame Venetia Digby, styled the fair;
For mind and body the most excellent
   That ever nature, or the later air,
Gave two such houses as Northumberland   5
   And Stanley, to the which she was co-heir.
Speak it, you bold Penates, you that stand
   At either stem, and know the veins of good
Run from your roots; tell, testify the grand
   Meeting of graces, that so swelled the flood   10
Of virtues in her as, in short, she grew
   The wonder of her sex and of your blood.
And tell thou, Alderley, none can tell more true
   Thy niece's line than thou that gav'st thy name
Into the kindred, whence thy Adam drew   15
   Meschin's honour with the Cestrian fame
Of the first Lupus, to the family
   By Ranulf. . . .
         The rest of this song is lost

---

38 *Jacob's ladder*: Gen. 28 : 12.
84. ii. Her Descent: Jonson has versified a MS. account of Venetia's family tree. 1
uncontrolled: unchallenged, unreproved. 5–6 *Northumberland | And Stanley*: on her
father's and her mother's sides, respectively; relating her to the Percy family, and the
Earls of Derby. 13 *Alderley*: Adam de Alderley (the name is variously spelt) was the
father of the first Stanley: Camden, *Remains Concerning Britain* (London, 1870), p.
116. 16–18 Hugh de Avranches (d. 1101), nicknamed *lupus* (the wolf), and his grand-
son Ranulf or Randulf (d. *c.* 1129), nicknamed *le meschin* (the mean), were Earls of
Chester. There were later two other famous Randulfs in the family.

84. iii

The Picture of the Body

Sitting, and ready to be drawn,
    What make these velvets, silks, and lawn,
    Embroideries, feathers, fringes, lace,
    Where every limb takes like a face?

Send these suspected helps to aid   5
    Some form defective, or decayed;
    This beauty without falsehood fair
    Needs nought to clothe it but the air.

Yet something, to the painter's view,
    Were fitly interposed; so new,   10
    He shall, if he can understand,
    Work with my fancy his own hand.

Draw first a cloud, all save her neck,
    And out of that make day to break;
    Till, like her face it do appear,   15
    And men may think all light rose there.

Then let the beams of that disperse
    The cloud, and show the universe;
    But at such distance as the eye
    May rather yet adore than spy.   20

---

84. iii. Jonson may be remembering the opening of Statius's *Silvae*, V. i, from which he twice draws mottoes for this sequence: Abascantus, trying to recall the memory of the dead Priscilla, commissions craftsmen to model her likeness in metal and wax; Statius at first laments his lack of skill in these crafts, then asserts that his way of celebrating her is more lasting. This poem and the next also echo Lucian's *Essays in Portraiture*, in which Lycinus models a statue of a girl's body, and Polystratus attempts a portrait of her soul. For other painting poems, see Carew, *Poems*, ed. Dunlap, pp. 106–7, Shirley, *Works*, ed. Gifford/Dyce, vi. 414–18, and Herrick, pp. 38, 47–8. Van Dyke painted Venetia on several occasions; a death-bed portrait of her is in the Dulwich Portrait Gallery.   13–14 Cf. *Beauty*, 181–2: 'She was drawn in a circle of clouds, her face and body breaking through'.   16 *all light rose there*: cf. *Und.* 2. iv. 17–18 and n.

The heaven designed, draw next a spring,
 With all that youth or it can bring:
 Four rivers branching forth like seas,
 And Paradise confining these.

Last, draw the circles of this globe,   25
 And let there be a starry robe
 Of constellations 'bout her hurled;
 And thou hast painted beauty's world.

But, painter, see thou do not sell
 A copy of this piece, nor tell   30
 Whose 'tis: but if it favour find,
 Next sitting we will draw her mind.

84. iv

The Mind

Painter, you're come, but may be gone;
 Now I have better thought thereon,
 This work I can perform alone;
 And give you reasons more than one.

Not that your art I do refuse;   5
 But here I may no colours use.
 Beside, your hand will never hit
 To draw a thing that cannot sit.

You could make shift to paint an eye,
 An eagle towering in the sky,   10
 The sun, a sea, or soundless pit,
 But these are like a mind, not it.

23 *Four rivers*: as in Eden, Gen. 2 : 10–14.   24 *confining*: bordering upon.   28 *beauty's world*: cf. *Und*. 2. iv. 12, 'Love's world'.

84. iv. 6 *colours*: playing on the sense of rhetorical colours.   9–12 Ripa associates the mind with the eye, *Iconologia* (Padua, 1611), p. 157 (under *Fatica*), and takes the

No, to express a mind to sense
   Would ask a heaven's intelligence;
   Since nothing can report that flame   15
   But what's of kin to whence it came.

Sweet mind, then speak yourself, and say
   As you go on, by what brave way
   Our sense you do with knowledge fill
   And yet remain our wonder still.   20

I call you muse, now make it true:
   Henceforth may every line be you;
   That all may say that see the frame,
   This is no picture, but the same.

A mind so pure, so perfect fine,   25
   As 'tis not radiant, but divine:
   And so disdaining any trier;
   'Tis got where it can try the fire.

There, high exalted in the sphere,
   As it another nature were,
   It moveth all, and makes a flight   30
   As circular as infinite.

Whose notions when it will express
   In speech, it is with that excess
   Of grace, and music to the ear,   35
   As what it spoke, it planted there.

The voice so sweet, the words so fair,
   As some soft chime had stroked the air;
   And though the sound were parted thence,
   Still left an echo in the sense.   40

towering eagle as an emblem of noble thought, p. 415 (*Pensiero*). The sun (associated with Apollo, and hence with intellection) is an emblem of wisdom: see Guy de Tervarent, *Attributs et symboles dans l'art profane 1450–1600* (Geneva, 2 vols., 1958, 1959), ii. 355. 21 *muse*: cf. the titles of *Und.* 78 & 84. ix. 22 *line*: playing on the poetic and artistic senses of the word, as in *Und.* 52. 3 ('My Answer'). 32 The circle was an emblem of perfection and of infinity: cf. *Hym.* 404, *N.I.* III. ii. 107, *Epig.* 128. 8, Vaughan, 'The World', ll. 1–2 (*Works*, ed. L. C. Martin (Oxford, 1957), p. 466).

But that a mind so rapt, so high,
   So swift, so pure, should yet apply
   Itself to us, and come so nigh
   Earth's grossness: there's the how and why.

Is it because it sees us dull,   45
   And stuck in clay here, it would pull
   Us forth by some celestial sleight
   Up to her own sublimed height?

Or hath she here, upon the ground
   Some paradise, or palace found   50
   In all the bounds of beauty fit,
   For her to inhabit? There is it.

Thrice happy house, that hast receipt
   For this so lofty form, so straight,
   So polished, perfect, round, and even,   55
   As it slid moulded off from heaven.

Not swelling like the ocean proud,
   But stooping gently as a cloud,
   As smooth as oil poured forth, and calm
   As showers, and sweet as drops of balm.   60

Smooth, soft, and sweet, in all a flood
   Where it may run to any good;
   And where it stays, it there becomes
   A nest of odorous spice and gums.

In action, winged as the wind;   65
   In rest, like spirits left behind
   Upon a bank, or field of flowers,
   Begotten by that wind, and showers.

In thee, fair mansion, let it rest,
   Yet know with what thou art possessed: 70
   Thou entertaining in thy breast
   But such a mind, mak'st God thy guest.

55 *perfect, round*: see l. 32, n.   72 *God thy guest*: Seneca, *Epist.* xxxi. 11: 'What else could you call such a soul than a god dwelling as a guest in a human body?'; cf. *Disc.* 1211–12.

84. viii

A whole quaternion in the midst of this poem is lost,
containing entirely the three next pieces of it, and all of the
fourth (which in the order of the whole, is the eighth)
excepting the very end: which at the top of the next
quaternion goeth on thus:

But for you, growing gentlemen, the happy branches of
two so illustrious houses as these, wherefrom your
honoured mother is in both lines descended, let me leave
you this last legacy of counsel; which, so soon as you
arrive at years of mature understanding, open you, sir,   5
that are the eldest, and read it to your brethren, for it will
concern you all alike. Vowed by a faithful servant and
client of your family, with his latest breath expiring it,

                                                    B.J.

To Kenelm, John, George

Boast not these titles of your ancestors,   10
    Brave youths, they're their possessions, none of yours;
When your own virtues equalled have their names,
    'Twill be but fair to lean upon their fames,
For they are strong supporters; but till then,
    The greatest are but growing gentlemen.   15
It is a wretched thing to trust to reeds,
    Which all men do that urge not their own deeds
Up to their ancestors; the river's side
    By which you're planted shows your fruit shall bide.
Hang all your rooms with one large pedigree: 20
    'Tis virtue alone is true nobility.
Which virtue from your father, ripe, will fall;
    Study illustrious him, and you have all.

---

84. viii. *quaternion*: a quire of four sheets folded in two.   9 Kenelm, 1625–48; John,
b. 1627; George, b. 1632, d. during the Civil War. Another son, Everard, was b. 1629
and d. in infancy. There was also a daughter.   11 *none of yours*: cf. Ulysses in Ovid's
Metam. xiii. 140–1: 'For as to race and ancestry and the deeds that others than
ourselves have done, I call those in no true sense our own', and the mottoes of Sir
Philip Sidney and the 2nd Earl of Essex.   16–18 Juvenal, *Sat.* viii. 76–7: 'It is a poor
thing to lean upon the fame of others, lest the pillars give way and the house fall down
in ruin'; cf. Tilley, R61.   20–1 Ibid., 19–20: 'Though you deck your hall from end to
end with ancient waxen images, virtue is the one and only true nobility'; cf. Tilley,
V85, *C.R.* V. i. 31.

84. ix

Elegy on My Muse,
the Truly Honoured Lady, the Lady Venetia Digby:
who, Living, Gave Me Leave to Call Her So.
Being Her ᾿Αποθέωσις, or Relation to the Saints.

Sera quidem tanto struitur medicina dolori.

'Twere time that I died too, now she is dead,
  Who was my muse, and life of all I said,
The spirit that I wrote with, and conceived;
  All that was good or great in me she weaved,
And set it forth; the rest were cobwebs fine,   5
  Spun out in name of some of the old nine,
To hang a window or make dark the room,
  Till, swept away, they were cancelled with a broom:
Nothing that could remain, or yet can stir
  A sorrow in me, fit to wait to her!   10
Oh, had I seen her laid out a fair corse
  By death on earth, I should have had remorse
On nature for her, who did let her lie,
  And saw that portion of herself to die.
Sleepy, or stupid nature, couldst thou part   15
  With such a rarity, and not rouse art
With all her aids to save her from the seize
  Of vulture death, and those relentless clees?
Thou wouldst have lost the phoenix, had the kind
  Been trusted to thee, not to itself assigned.   20
Look on thy sloth, and give thyself undone:
  For so thou art with me, now she is gone.
My wounded mind cannot sustain this stroke;
  It rages, runs, flies, stands, and would provoke
The world to ruin with it; in her fall,   25
  I sum up mine own breaking, and wish all.

84. ix. *Muse*: as in *Und.* 78, title, 84. iv. 21. ᾿Αποθέωσις: apotheosis. *Sera quidem . . . dolori*: 'Late indeed is the balm composed for so great a sorrow'; Statius, *Silvae*, V. i. 16; see note to motto at head of this sequence. 6 *the old nine*: the muses; cf. the dismissiveness of *For.* 10. 12–13 *remorse | On*: pity for. 18 *clees*: claws.

Thou hast no more blows, fate, to drive at one:
  What's left a poet, when his muse is gone?
Sure, I am dead, and know it not! I feel
  Nothing I do, but like a heavy wheel,   30
Am turned with another's powers. My passion
  Whirls me about, and to blaspheme in fashion,
I murmur against God, for having ta'en
  Her blessed soul hence, forth this valley vain
Of tears, and dungeon of calamity.   35
  I envy it the angels' amity!
The joy of saints, the crown for which it lives,
  The glory and gain of rest which the place gives!
Dare I prophane, so irreligious be
  To greet or grieve her soft euthanasy?   40
So sweetly taken to the court of bliss,
  As spirits had stolen her spirit, in a kiss,
From off her pillow and deluded bed,
  And left her lovely body unthought dead!
Indeed, she is not dead, but laid to sleep   45
  In earth, till the last trump awake the sheep
And goats together, whither they must come
  To hear their judge and his eternal doom;
To have that final retribution,
  Expected with the flesh's restitution.   50
For, as there are three natures, schoolmen call
  One corporal only, the other spiritual,
Like single; so there is a third, commixt
  Of body and spirit together, placed betwixt
Those other two; which must be judged or crowned: 55
  This, as it guilty is, or guiltless found,
Must come to take a sentence, by the sense
  Of that great evidence, the conscience,
Who will be there, against that day prepared,
  To accuse, or quit all parties to be heard.   60
O day of joy and surety to the just,
  Who in that feast of resurrection trust!

---

37 *crown*: Jas. 1 : 12, Rev. 2 : 10.   40 *greet*: weep for.   40 *euthanasy*: euthanasia,
gentle and easy death.   46–7 Matt. 25 : 33.   51–5 Cf. Aquinas's discussion of the
relationship of soul and body, *Summa Theologiae*, I. lxxv; and Donne's 'Letter to the
Countess of Bedford' ('T' have written then . . .'), ll. 57 ff., *Satires*, p. 97.

That great eternal holy-day of rest
   To body and soul, where love is all the guest,
And the whole banquet is full sight of God: 65
   Of joy the circle and sole period!
All other gladness with the thought is barred,
   Hope hath her end, and faith hath her reward.
This being thus, why should my tongue or pen
   Presume to interpel that fulness, when   70
Nothing can more adorn it than the seat
   That she is in, or make it more complete?
Better be dumb, than superstitious;
   Who violates the godhead is most vicious
Against the nature he would worship. He   75
   Will honoured be in all simplicity,
Have all his actions wondered at, and viewed
   With silence and amazement, not with rude,
Dull, and prophane, weak, and imperfect eyes,
   Have busy search made in his mysteries.   80
He knows what work he hath done to call this guest
   Out of her noble body to this feast,
And give her place, according to her blood,
   Amongst her peers, those princes of all good:
Saints, martyrs, prophets, with those hierarchies,   85
   Angels, archangels, principalities,
The dominations, virtues, and the powers,
   The thrones, the cherub, and seraphic bowers,
That planted round, there sing before the Lamb
   A new song to his praise, and great *I am*.   90
And she doth know, out of the shade of death,
   What 'tis to enjoy an everlasting breath!
To have her captived spirit freed from flesh,
   And on her innocence a garment fresh
And white as that, put on; and in her hand   95
   With boughs of palm, a crowned victrix stand!
And will you, worthy son, sir, knowing this,
   Put black and mourning on, and say you miss

---

65 *full sight*: Rev. 22 : 4; 1 Cor. 13 : 12.   70 *interpel*: break in on, disturb.   78–80 Cf.
Exod. 33 : 20–3; Luke 8 : 10.   90 *new song*: Rev. 5 : 9.   90 I am: Exod. 3 : 14.   95–6
Rev. 5 : 9.   96 *victrix*: female victor.

A wife, a friend, a lady, or a love,
  Whom her redeemer honoured hath above   100
Her fellows, with the oil of gladness, bright
  In heaven's empyrean, with a robe of light?
Thither you hope to come, and there to find
  That pure, that precious and exalted mind
You once enjoyed; a short space severs ye   105
  Compared unto that long eternity
That shall rejoin ye. Was she then so dear
  When she departed? You will meet her there
Much more desired, and dearer than before,
  By all the wealth of blessings, and the store   110
Accumulated on her by the Lord
  Of life and light, the Son of God, the Word!
There all the happy souls that ever were,
  Shall meet with gladness in one theatre;
And each shall know, there, one another's face,   115
  By beatific virtue of the place.
There shall the brother with the sister walk,
  And sons and daughters with their parents talk,
But all of God; they still shall have to say,
  But make him all in all, their theme that day,   120
That happy day, that never shall see night!
  Where he will be all beauty to the sight,
Wine or delicious fruits unto the taste,
  A music in the ears, will ever last,
Unto the scent a spicery or balm,   125
  And to the touch a flower like soft as palm.
He will all glory, all perfection be,
  God, in the Union, and the Trinity!
That holy, great, and glorious mystery
  Will there revealed be in majesty!   130
By light and comfort of spiritual grace,
  The vision of our Saviour, face to face
In his humanity! To hear him preach
  The price of our redemption, and to teach
Through his inherent righteousness, in death,   135
  The safety of our souls, and forfeit breath:

101 *oil of gladness*: Ps. 45 : 7.   120 *all in all*: I Cor. 15 : 28.   121 Cf. *Und.* 70 : 81,
'that bright eternal day'.

What fulness of beatitude is here!
   What love with mercy mixed doth appear!
To style us friends, who were by nature foes;
   Adopt us heirs by grace, who were of those   140
Had lost ourselves, and prodigally spent
   Our native portions and possessed rent;
Yet have all debts forgiven us, and advance
   By imputed right to an inheritance
In his eternal kingdom, where we sit   145
   Equal with angels, and co-heirs of it!
Nor dare we under blasphemy conceive
   He that shall be our supreme judge should leave
Himself so uninformed of his elect,
   Who knows the hearts of all, and can dissect   150
The smallest fibre of our flesh; he can
   Find all our atoms from a point to a span,
Our closest creeks and corners, and can trace
   Each line, as it were graphic, in the face.
And best he knew her noble character,   155
   For 'twas himself who formed and gave it her.
And to that form lent two such veins of blood,
   As nature could not more increase the flood
Of title in her. All nobility
   (But pride, that schism of incivility)   160
She had, and it became her; she was fit
   To have known no envy but by suffering it.
She had a mind as calm as she was fair,
   Not tossed or troubled with light lady-air,
But kept an even gait; as some straight tree   165
   Moved by the wind, so comely moved she.
And by the awful manage of her eye
   She swayed all business in the family!
To one she said, Do this; he did it. So
   To another, Move; he went. To a third, Go;   170
He run. And all did strive with diligence
   To obey and serve her sweet commandments.

---

140 *heirs*: Rom. 8 : 17.   141–2 Like the prodigal son, Luke 15; *possessed* = owing to God.   146 *Equal with angels*: Luke 20 : 36.   150 *knows the hearts*: Rev. 2 : 23.   152 *a span*: i.e. its full extent.   153 *creeks*: nooks, secret places.   157 *two such veins*: the Northumberlands and the Stanleys.   164 *lady-air*: airs thought befitting to a lady.   167 *manage*: command.

She was, in one, a many parts of life;
    A tender mother, a discreeter wife,
A solemn mistress, and so good a friend, 175
    So charitable to religious end,
In all her petty actions so devote,
    As her whole life was now become one note
Of piety and private holiness.
    She spent more time in tears herself to dress 180
For her devotions, and those sad essays
    Of sorrow, than all pomp of gaudy days:
And came forth ever cheered with the rod
    Of divine comfort, when she'd talked with God.
Her broken sighs did never miss whole sense, 185
    Nor can the bruised heart want eloquence:
For prayer is the incense most perfumes
    The holy altars, when it least presumes.
And hers were all humility; they beat
    The door of grace, and found the mercy-seat. 190
In frequent speaking by the pious psalms
    Her solemn hours she spent, or giving alms,
Or doing other deeds of charity,
    To clothe the naked, feed the hungry. She
Would sit in an infirmary whole days 195
    Poring, as on a map, to find the ways
To that eternal rest, where now she hath place
    By sure election, and predestined grace.
She saw her Saviour, by an early light,
    Incarnate in the manger, shining bright 200
On all the world. She saw him on the cross,
    Suffering and dying to redeem our loss.
She saw him rise, triumphing over death
    To justify, and quicken us in breath.
She saw him too in glory to ascend 205
    For his designed work, the perfect end
Of raising, judging, and rewarding all
    The kind of man, on whom his doom should fall.
All this by faith she saw, and framed a plea,
    In manner of a daily apostrophe, 210

173 i.e. she combined, in the one personality, many roles in life. 183 *rod*: Ps.
23 : 4. 187 *the incense*: Ps. 141 : 2; Rev. 8 : 3–4.

To him should be her judge, true God, true man,
   Jesus, the only-gotten Christ, who can
(As being redeemer, and repairer too
   Of lapsed nature) best know what to do,
In that great act of judgement: which the Father   215
   Hath given wholly to the Son (the rather
As being the Son of Man) to show his power,
   His wisdom, and his justice, in that hour,
The last of hours, and shutter-up of all;
   Where first his power will appear, by call   220
Of all are dead to life; his wisdom show
   In the discerning of each conscience, so;
And most his justice, in the fitting parts
   And giving dues to all mankind's deserts.
In this sweet ecstasy she was rapt hence.   225
   Who reads, will pardon my intelligence,
That thus have ventured these true strains upon,
   To publish her a saint. My muse is gone.

> *In pietatis memoriam*
> *quam praestas*
> Venetiae *tuae illustrissim.*
> *marit. dign.* Digbeie
> *Hanc* Ἀποθέωσιν, *tibi, tuisque sacro.*

The tenth, being her Inscription, or Crown, is lost.

85

Horace, Epode ii
The Praises of a Country Life

Happy is he, that from all business clear
   As the old race of mankind were,
With his own oxen tills his sire's left lands,
   And is not in the usurer's bands;

---

226 *intelligence*: conveyance of news; information.   229–33 'In memory of the devotion which you show to your most illustrious Venetia, I offer to you and to yours, worthy husband Digby, this apotheosis.'

85. Robert Shafer points out that in this translation and in *Und.* 86 & 87 Jonson is trying to represent the form as well as the content of his originals, by using couplets

Nor, soldier-like, started with rough alarms,   5
    Nor dreads the sea's enraged harms;
But flees the Bar and courts, with the proud boards
    And waiting-chambers of great lords.
The poplar tall he then doth marrying twine
    With the grown issue of the vine;   10
And with his hook lops off the fruitless race,
    And sets more happy in the place;
Or in the bending vale beholds afar
    The lowing herds there grazing are;
Or the pressed honey in pure pots doth keep   15
    Of earth, and shears the tender sheep;
Or when that autumn through the fields lifts round
    His head, with mellow apples crowned,
How, plucking pears his own hand grafted had,
    And purple-matching grapes, he's glad!   20
With which, Priapus, he may thank thy hands,
    And, Silvane, thine, that kept'st his lands.
Then, now beneath some ancient oak he may,
    Now in the rooted grass, him lay,
Whilst from the higher banks do slide the floods;   25
    The soft birds quarrel in the woods,
The fountains murmur as the streams do creep,
    And all invite to easy sleep.
Then when the thundering Jove his snow and showers
    Are gathering by the wintry hours,   30
Or hence, or thence, he drives with many a hound
    Wild boars into his toils pitched round;
Or strains on his small fork his subtle nets
    For the eating thrush, or pitfalls sets;
And snares the fearful hare and new-come crane,   35
    And 'counts them sweet rewards so ta'en.

---

and the same number of lines as Horace: *The English Ode to 1660* (New York, 1966), p. 100. Drummond says that Jonson read the translation to him, 'and admired it' (*Conv. Dr.* 75–7). 9–10 Vines were trained to grow on black poplars and elms; the 'marriage' analogy was traditional (cf. *Paradise Lost*, v. 215–17). 13–14 In the original, these lines precede ll. 11–12, but were sometimes editorially rearranged. *bending vale* is Jonson's rendering of *reducta valle*, 'in a remote valley'. 21 *Priapus*: god of generation, and of gardens and orchards. 22 *Silvane*: Silvanus, who presided over gardens and boundaries. 24 *rooted*: Latin *tenaci*, firmly rooted. 26 *quarrel*: evidently aiming at an oxymoronic effect; translating Latin *queruntur*. 35 *crane*: which visited Italy in the summer.

Who, amongst these delights, would not forget
    Love's cares so evil, and so great?
But if, to boot with these, a chaste wife, meet
    For household aid and children sweet,   40
Such as the Sabines'; or a sun-burnt blowze,
    Some lusty, quick Apulian's spouse,
To deck the hallowed hearth with old wood fired
    Against the husband comes home tired;
That penning the glad flock in hurdles by,   45
    Their swelling udders doth draw dry;
And from the sweet tub wine of this year takes,
    And unbought viands ready makes:
Not Locrine oysters I could then more prize,
    Nor turbot, nor bright golden-eyes;   50
If with bright floods, the winter troubled much,
    Into our seas send any such,
The Ionian godwit, nor the Guinea-hen
    Could not go down my belly then
More sweet than olives that new-gathered be   55
    From fattest branches of the tree;
Or the herb sorrel that loves meadows still,
    Or mallows, loosing body's ill;
Or at the feast of bounds, the lamb then slain,
    Or kid forced from the wolf again.   60
Among these cates how glad the sight doth come
    Of the fed flocks approaching home!
To view the weary oxen draw, with bare
    And fainting necks, the turned share!
The wealthy household swarm of bondmen met,   65
    And 'bout the steaming chimney set!

---

41 *Sabines'*: from the central Apennines; the women had the reputation of being excellent farmers' wives. 41 *blowze*: red-faced woman. 42 *Apulian's*: from Apulia (now Puglia), between Daunia and Calabria; the inhabitants were known for their industry. 51 *bright floods*: 'east floods' in MSS., translating Horace's *Eois fluctibus*. The scar fish ('golden-eyes') was common in the eastern Mediterranean; it was believed that storms in the east would drive the fish westward. 53 *godwit*: a marsh bird like the curlew; for Horace's *attagen*, a kind of grouse, which came to Italy from the east. 53 *Guinea-hen*: a new delicacy in Italy. 59 *feast of bounds*: in honour of Terminus, Roman god of bounds and limits: a chance to eat fresh meat. 60 *the wolf*: customarily said to kill the best young animal of the flock.

These thoughts when usurer Alfius, now about
　　To turn mere farmer, had spoke out,
'Gainst the ides his moneys he gets in with pain
　　At the calends, puts all out again.　　　　　　70

86

Horace, Ode the First, the Fourth Book:
To Venus

Venus, again thou mov'st a war
Long intermitted: pray thee, pray thee, spare;
　　I am not such as in the reign
Of the good Cinara I was; refrain,
　　　Sour mother of sweet loves, forbear　　　　5
To bend a man, now at his fiftieth year
　　Too stubborn for commands so slack;
Go where youth's soft entreaties call thee back.
　　More timely hie thee to the house,
With thy bright swans, of Paulus Maximus:　　　10
　　There jest and feast, make him thine host,
If a fit liver thou dost seek to toast;
　　For he's both noble, lovely, young,
And for the troubled client files his tongue,
　　　Child of a hundred arts, and far　　　　15
Will he display the ensigns of thy war.
　　And when he smiling finds his grace
With thee 'bove all his rivals' gifts take place,
　　He'll thee a marble statue make
Beneath a sweet-wood roof, near Alba lake:　　　20
　　There shall thy dainty nostril take
In many a gum, and for thy soft ear's sake
　　Shall verse be set to harp and lute,
And Phrygian hautboy, not without the flute.

69–70 *ides . . . calends*: Roman days for settling debts and making new financial arrange-
ments. Despite his praise of the country life, Alfius will stay where he is.

86. 6 *fiftieth year*: Cf. *Und.* 2. i. 3.　10 *bright swans*: sacred to Venus; cf. *Und.* 2. iv.
3.　*Paulus Maximus*: probably Ovid's patron, and friend of Augustus.　12 *liver*: the
seat of love.　14 *files his tongue*: makes smooth his speech.

There twice a day in sacred lays　　　　　　25
The youths and tender maids shall sing thy praise;
　　And in the Salian manner meet
Thrice 'bout thy altar with their ivory feet.
　　Me now nor wench, nor wanton boy
Delights, nor credulous hope of mutual joy,　　30
　　Nor care I now healths to propound,
Or with fresh flowers to girt my temple round.
　　But why, O why, my Ligurine,
Flow my thin tears down these pale cheeks of mine?
　　Or why, my well-graced words among,　　35
With an uncomely silence fails my tongue?
　　Hard-hearted, I dream every night
I hold thee fast! but fled hence, with the light,
　　Whether in Mars's field thou be,
Or Tiber's winding streams, I follow thee.　　40

87

Horace, Ode ix, 3rd Book:
To Lydia
Dialogue of Horace and Lydia

Horace　　Whilst, Lydia, I was loved of thee,
　　　　　And, 'bout thy ivory neck, no youth did fling
　　　　　　His arms more acceptable free,
　　　　　I thought me richer than the Persian king.

Lydia　　Whilst Horace loved no mistress more,　　5
　　　　　Nor after Chloe did his Lydia sound,
　　　　　　In name I went all names before,
　　　　　The Roman Ilia was not more renowned.

27 *Salian manner*: the Salii, priests of Mars, danced and postured through the streets beating their shields.　31 *propound*: propose.　39 *Mars's field*: Campus Martius, the large plain outside Rome.

87. Also translated by Herrick (p. 70), and others.　6 *sound*: celebrate.　8 *Ilia*: mother of Romulus and Remus.

Horace   'Tis true, I'm Thracian Chloe's, I,
　　　　　Who sings so sweet, and with such cunning plays,　10
　　　　　　As, for her, I'd not fear to die,
　　　　　So fate would give her life, and longer days.

Lydia   And I am mutually on fire
　　　　　With gentle Calais, Thurine Ornith's son;
　　　　　　For whom I doubly would expire,　15
　　　　　So fates would let the boy a long thread run.

Horace   But say old love return should make,
　　　　　And us disjoined, force to her brazen yoke,
　　　　　　That I bright Chloe off should shake,
　　　　　And to left Lydia now the gate stood ope?　20

Lydia   Though he be fairer than a star,
　　　　　Thou lighter than the bark of any tree,
　　　　　　And than rough Adria angrier far,
　　　　　Yet would I wish to love, live, die with thee.

88

A Fragment of Petronius Arbiter

Doing a filthy pleasure is, and short;
And done, we straight repent us of the sport;
Let us not then rush blindly on unto it,
Like lustful beasts that only know to do it:
For lust will languish, and that heat decay.　5
But thus, thus, keeping endless holiday,
Let us together closely lie, and kiss,
There is no labour, nor no shame in this;
This hath pleased, doth please, and long will please; never
Can this decay, but is beginning ever.　10

14 *Thurine*: from Thuriae (the name is variously spelt), a town in southern Italy.　23
Adria: the Adriatic.

88. Not in fact by Petronius, but printed in the edition of his works published in Paris
in 1585. Read to Drummond: *Conv. Dr.* 78–9.

89

Martial, Epigram lxxvii, Book VIII

Liber, of all thy friends, thou sweetest care,
    Thou worthy in eternal flower to fare,
If thou be'st wise, with Syrian oil let shine
    Thy locks, and rosy garlands crown thy head;
Dark thy clear glass with old Falernian wine,   5
    And heat with softest love thy softer bed.
He that but living half his days dies such,
    Makes his life longer than 'twas given him, much.

90

Martial, Epigram xlvii, Book X

The things that make the happier life are these,
Most pleasant Martial: substance got with ease,
Not laboured for, but left thee by thy sire;
A soil not barren; a continual fire;
Never at law; seldom in office gowned;   5
A quiet mind; free powers; and body sound;
A wise simplicity; friends alike-stated;
Thy table without art, and easy rated;
Thy night not drunken, but from cares laid waste,
No sour or sullen bed-mate, yet a chaste;   10
Sleep that will make the darkest hours swift-paced;
Will to be what thou art, and nothing more;
Nor fear thy latest day, nor wish therefore.

---

89. 7–8 Cf. *Epig.* 119. 15–16.

90. Also translated by Surrey, *Poems*, ed. Emrys Jones (Oxford, 1964), pp. 34–5, and by Randolph, *Poems*, ed. G. Thorn-Drury (London, 1929), p. 88. See also *Conv. Dr.* 15–16.   5 *in office gowned*: a mistranslation; Jonson mistakes the kind of toga to which Martial refers.   6 *free powers*: Martial's *vires ingenuae*, 'i.e. the natural strength of a gentleman, not the coarse strength of a labourer', W. C. A. Ker, *Martial's Epigrams* (London & Cambridge, Mass., 1968), ii. 189.   7 *alike-stated*: similarly placed.   9 *laid waste*: Jonson evidently means 'set free' (Martial: *soluta curis*, freed from cares).

UNGATHERED VERSE

UNGATHERED VERSE

I

From Thomas Palmer's The Sprite of Trees and Herbs, *1598–9*

When late, grave Palmer, these thy grafts and flowers,
So well disposed by thy auspicious hand,
Were made the objects to my weaker powers,
I could not but in admiration stand.
First, thy success did strike my sense with wonder,   5
That 'mongst so many plants transplanted hither
Not one but thrives, in spite of storms and thunder,
Unseasoned frosts, or the most envious weather.
Then I admired the rare and precious use
Thy skill hath made of rank despised weeds,   10
Whilst other souls convert to base abuse
The sweetest simples, and most sovereign seeds.
Next, that which rapt me was, I might behold
How, like the carbuncle in Aaron's breast,
The seven-fold flower of art, more rich than gold,   15
Did sparkle forth in centre of the rest;
Thus, as a ponderous thing in water cast
Extendeth circles into infinites,
Still making that the greatest that is last,
Till the one hath drowned the other in our sights: 20
So in my brain the strong impression
Of thy rich labours worlds of thoughts created,
Which thoughts being circumvolved in gyre-like motion
Were spent with wonder as they were dilated,
Till giddy with amazement I fell down   25
In a deep trance; * * * * *

1. Thomas Palmer was Principal of Gloucester Hall, Oxford, 1563–4, and Fellow and Lecturer in Rhetoric at St. John's College, Oxford, until 1566; deprived of these posts because of his Catholicism, he retired to Exeter, where persecution continued. *The Sprite of Trees and Herbs*, a botanical emblem book, was never published (BM Additional MS. 18040). 7 *storms and thunder*: alluding to the persecution which Palmer suffered. 8 *Unseasoned*: unseasonable. 8 *envious weather*: cf. *Love's Labour's Lost*, I. i. 100–1: 'envious-sneaping frost'. 14 *carbuncle*: Exod. 28 : 17; 39 : 10. 15 *seven-fold flower of art*: the seven liberal arts. 23 *circumvolved*: turned around.

* * * * * when, lo! to crown thy worth
I struggled with this passion that did drown
My abler faculties; and thus brake forth:
    Palmer, thy travails well become thy name,   30
    And thou in them shalt live as long as fame.

Dignum laude virum musa vetat mori.

2

From Nicholas Breton's Melancholic Humours, *1600*
In Authorem

Thou that wouldst find the habit of true passion,
    And see a mind attired in perfect strains;
Not wearing modes, as gallants do a fashion
    In these pied times, only to show their brains:

Look here on Breton's work, the master print,   5
    Where such perfections to the life do rise.
If they seem wry to such as look asquint
    The fault's not in the object, but their eyes.

For as one coming with a lateral view
    Unto a cunning piece wrought perspective   10
Wants faculty to make a censure true;
    So with this author's readers will it thrive:

---

30 travails: with a pun on 'travels': a palmer being a religious traveller.   32 ''Tis the
muse forbids the hero worthy of renown to perish', Horace, *Odes*, IV. viii. 28.

2. Nicholas Breton, ?1545–?1626, is dismissively referred to in *Und.* 43. 77.   1 *habit*: a
pun.   3 *modes*: 'moodes' in the original spelling: playing again on words.   5 *master
print*: from which a print (of a book, a picture, or a fabric) is taken: the phrase unites
the poem's various metaphors.   7 *look asquint*: cf. Ovid, *Metam.* ii. 787, where envy
looks *obliquo*, askance; and see *Und.* 73. 2 n.   10 *perspective*: H & S cf. Chapman and
Shirley, *The Tragedy of Chabot*, I. i. For a partial explanation of the phenomenon
Jonson describes, see E. H. Gombrich, *Art and Illusion* (London & New York, 1960),
pp. 254–5. The perspective effects in the court masques were perfect only from the
direct sight-line of the king's chair: see Allardyce Nicoll, *Stuart Masques and the
Renaissance Stage* (London, 1937), p. 34.   11 *censure*: judgement.

Which, being eyed directly, I divine,
His proof their praise will meet, as in this line.

3

Fragments from England's Parnassus, *1600*

Murder

Those that in blood such violent pleasure have,
Seldom descend but bleeding to their grave.

Peace

War's greatest woes, and misery's increase,
Flows from the surfeits which we take in peace.

Riches

Gold is a suitor never took repulse;   5
It carries palm with it where'er it goes,
Respect, and observation; it uncovers
The knotty heads of the most surly grooms,
Enforcing iron doors to yield it way,
Were they as strong rammed-up as Aetna gates.   10
It bends the hams of Gossip Vigilance,
And makes her supple feet as swift as wind.
It thaws the frostiest and most stiff disdain,
Muffles the clearness of election,
Strains fancy unto foul apostacy,   15
And strikes the quickest-sighted judgement blind.
Then why should we despair? Despair, away!
Where gold's the motive, women have no nay.

---

14 *this line*: the line of the poem, and the line of vision.

3. Three of the fourteen quotations from Jonson's work in *England's Parnassus*, ed.
Robert Allot: the others are from *E.M.O.*, *E.M.I.*, *For.* 11, and *Und.* 25. 1–2
Juvenal, *Sat.* x. 112–13: 'Few indeed are the kings who go down to Ceres' son-in-law
[Pluto] save by sword and slaughter—few the tyrants that perish by a bloodless
death'. 3–4 Cf. *Und.* 15. 121. 5 ff. Cf. *Volp.* I. i. 22–7. 6 *palm*: honour, glory (a
Latinism). 10 *Aetna gates*: by which Jupiter secured the giants imprisoned under this
mountain.

4

From Love's Martyr, *1601*
The Phoenix Analysed

Now, after all, let no man
    Receive it for a fable
    If a bird so amiable
Do turn into a woman.

Or, by our turtle's augur,   5
    That nature's fairest creature
    Prove of his mistress' feature
But a bare type and figure.

5

From Love's Martyr, *1601*
Ode ἐνθουσιαστικὴ

Splendour, O more than mortal!
For other forms come short all
Of her illustrate brightness,
As far as sin's from lightness.

Her wit as quick and sprightful   5
As fire, and more delightful
Than the stol'n sports of lovers,
When night their meeting covers.

Judgement adorned with learning
Doth shine in her discerning,   10
Clear as a naked vestal
Closed in an orb of crystal.

---

4. See *For.* 10 and n.   4 *a woman*: see *Und.* 75. 81–2 and n.

5. *ἐνθουσιαστικὴ*: inspired. Newdigate discovered a MS. copy of this poem addressed to Lucy, Countess of Bedford: see notes to *For.* 10. W. H. Matchett considers that the poem was originally written for the Countess of Bedford, and then used again in the quite different context of *Love's Martyr* as a final compliment to the 'Phoenix', whom he believes to be Queen Elizabeth: see his *The Phoenix and Turtle* (The Hague, 1965), pp. 100–1.   3 *illustrate*: resplendent, illustrious.

Her breath for sweet exceeding
The phoenix' place of breeding,
But, mixed with sound, transcending   15
All nature of commending.

Alas! Then whither wade I
In thought to praise this lady,
When seeking her renowning,
Myself am so near drowning?   20

Retire and say: Her graces
Are deeper than their faces;
Yet she's nor nice to show them,
Nor takes she pride to know them.

6

From Pancharis, *1603*
Ode ἀλληγορική

Who saith our times nor have, nor can
    Produce us a black swan?
  Behold, where one doth swim:
      Whose note and hue,
Besides the other swans' admiring him,   5
      Betray it true;
    A gentler bird than this
Did never dint the breast of Tamesis.

6. Like Jonson, Hugh Holland, the author of *Pancharis*, had been at Westminster
School, and had been subsequently converted to Catholicism. He died in 1633. Cam-
den, whose life Holland wrote, ranked him among the 'most pregnant wits of these our
times, whom succeeding ages may justly admire', *Remains Concerning Britain*
(London, 1870), p. 344. For biographical details, see L. I. Guiney, *Recusant Poets*
(London & New York, 1938), pp. 361 ff. In his prefatory verses to *Pancharis* and in
the poem itself, Holland represents himself as a swarthy and unattractive lover, and an
incompetent and inexperienced poet. Jonson neatly turns these disclaimers by figuring
Holland as a swan, a bird associated with Venus (and hence with love) and also with
Apollo (and hence with poetry: see esp. *Emblemata D.A. Alciati* (Lugd. 1551), p. 197,
Insignia Poetarum). (According to one Greek legend, Apollo had been turned into a
swan.) The swan's, and Holland's, blackness is taken as a sign that he is more than
ordinarily dear to Apollo, god not only of poetry but of the sun (see ll. 16 ff.). The
existence of real black swans was not known in Europe at this time; 'a black swan' (l. 2)
meant a person of mythical rarity (for Juvenal, a chaste wife: *Sat.* vi. 165); cf. ll. 32–3,
below. ἀλληγορική = allegorical. 5 *other swans*': Andrew Downes, Nicholas Hill,
and E.B. (probably Edmund Bolton), who had also written commendatory verses.

Mark, mark, but when his wing he takes
  How fair a flight he makes,   10
  How upward and direct!
    Whilst pleased Apollo
Smiles in his sphere to see the rest affect
    In vain to follow;
  This swan is only his   15
And Phoebus' love cause of his blackness is.

He showed him first the hoof-cleft spring
  Near which the Thespiades sing;
  The clear Dircaean fount
    Where Pindar swam,   20
The pale Pyrene, and the forked mount;
    And when they came
  To brooks and broader streams
From Zephyr's rape would close him with his beams.

This changed his down, till this, as white   25
  As the whole herd in sight,
  And still is in the breast:
    That part nor wind
Nor sun could make to vary from the rest,
  Or alter kind.   30
  So much doth virtue hate,
For style of rareness, to degenerate.

---

17 *hoof-cleft spring*: Hippocrene, the fountain of the muses, produced by a blow from Pegasus's hoof. Contrast Holland himself: 'My lips I never yet have soused / In Hippocrene. . . . / The climate where I was begotten / Of father Phoebus is forgotten', etc. (sig. A11). 18 *Thespiades*: the nine muses. 19 *Dircaean fount*: at Thebes, the birthplace of Pindar, whom Horace calls 'the Dircaean swan', *Odes*, IV. ii. 25. 21 *pale Pyrene*: an inspiring spring near Corinth; 'pale', because poets were supposed to look wan after their exertions: cf. Persius, *Sat*. Prologue, 4. 21 *the forked mount*: Parnassus; cf. Persius, *Sat*. Prologue, 2. 24 *close*: cover, protect. 25 *his down*: in *Pancharis* (sig. A10ᵛ) Holland describes the 'black down' of his beard. 31–2 These lines and ll. 47–8 were originally printed within inverted commas, to highlight their sententious character: on this typographical practice, see G. K. Hunter, *The Library*, 5th series, vi (1951), 171–88.

Be then both rare and good, and long
　　Continue thy sweet song.
　　Nor let one river boast   35
　　　Thy tunes alone;
But prove the air, and sail from coast to coast:
　　Salute old Mon;
　　But first to Clwyd stoop low,
The vale that bred thee pure as her hills' snow.   40

From thence display thy wing again,
　　Over Iërna main
　　To the Eugenian dale;
　　　There charm the rout
With thy soft notes, and hold them within pale   45
　　That late were out.
　　Music hath power to draw,
Where neither force can bend, nor fear can awe.

Be proof, the glory of his hand,
　　Charles Mountjoy, whose command   50
　　Hath all been harmony;
　　　And more hath won
Upon the kerne, and wildest Irishry
　　　Than time hath done,
　　Whose strength is above strength,   55
And conquers all things (yea, itself) at length.

---

35 *one river*: Holland briefly celebrates the Thames in *Pancharis*, sig. B7ᵛ–B8.　38
Mon: the Welsh name for Anglesey.　39 *Clwyd*: the valley in Denbighshire, Holland's
home.　42 *Iërna main*: the Irish sea.　43 *Eugenian dale*: Munster, Ireland.　44–54
Referring to the rebellion of Hugh O'Neill, 2nd Earl of Tyrone. Charles, Lord
Mountjoy, defeated Irish and Spanish forces at Kinsale in Munster in 1601, and
thereafter negotiated with Tyrone, who submitted formally in 1603.　56 Proverbial:
cf. Tilley, T326.

Whoever sipped at Baphyre river,
   That heard but spite deliver
   His far-admired acts,
     And is not rapt   60
With entheate rage to publish their bright tracts?
     (But this more apt
   When him alone we sing);
Now must we ply our aim: our swan's on wing.

Who (see!) already hath o'er-flown   65
   The Hebrid isles, and known
   The scattered Orcades;
     From thence is gone
To utmost Thule; whence he backs the seas
     To Caledon,   70
   And over Grampius mountain
To Lomond lake, and Tweed's black-springing fountain.

Haste, haste, sweet singer, nor to Tyne,
   Humber, or Ouse decline,
   But overland to Trent;   75
     There cool thy plumes,
And up again in skies and air to vent
   Their reeking fumes;
     Till thou at Thames alight,
From whose proud bosom thou began'st thy flight.   80

---

57 *sipped at Baphyre river*: tried to write poetry (the Baphyrus in Macedonia was thought to be the same river as Helicon). Mountjoy was praised by John Davies of Hereford in *Microcosmus* (1603); by Sylvester in three sonnets prefixed to the second *Week* of du Bartas (*c.* 1598; published 1641); by Daniel in his *Funeral Poem on the . . . Earl of Devonshire* (1606); and by Ford in *Fame's Memorial* (1606). 58 *spite*: which might have been aroused by Mountjoy's association with Essex and the rebellion of 1600–1; by his methods of government in Ireland (from which he was recalled on 26 May 1603); or by his liaison with Penelope, Lady Rich, whom he eventually married on 26 Dec. 1605. 61 *entheate*: inspired. 61 *tracts*: course, career. 67 *Orcades*: Orkneys. 69 *Thule*: traditionally, an island in the extreme north, of uncertain location; here perhaps the Shetland Islands. 69–70 *backs the seas | To Caledon*: goes back across the seas to Scotland. 71 *Grampius*: Grampians. 72 *Tweed's black-springing fountain*: Tweeds Well, near Moffat in Peebleshire. 78 *reeking fumes*: probably from the pottery kilns, already in operation in this area by the end of the sixteenth century.

Thames, proud of thee, and of his fate
  In entertaining late
The choice of Europe's pride:
  The nimble French;
The Dutch whom wealth, not hatred, doth divide;   85
  The Danes that drench
Their cares in wine; with sure
Though slower Spain, and Italy mature.

All which, when they but hear a strain
  Of thine, shall think the main   90
Hath sent her mermaids in
  To hold them here;
Yet looking in thy face, they shall begin
  To lose that fear;
And, in the place, envy   95
So black a bird, so bright a quality.

But should they know (as I) that this,
  Who warbleth *Pancharis*,
Were Cygnus, once high flying
  With Cupid's wing,   100
Though now by love transformed, and daily dying
  (Which makes him sing
With more delight and grace);
Or thought they Leda's white adulterer's place

84 *nimble French*: perhaps a reference to the visit of the Duke of Biron and a large party of French noblemen on 5 Sept. 1601; they arrived at Tower Wharf: see J. Nichols, *Progresses . . . Of Queen Elizabeth* (London, 1823), iii. 565. 85 *The Dutch*: see *Epig.* 32. 3 n. The division in fact reflected religious hostility. 86–7 Cf. Hamlet on the Danes' heavy drinking, which 'makes us traduced and taxed of other nations': *Hamlet*, I. iii. 18. 88 *slower Spain*: 'The Spartans and Spaniards have been noted to be of small dispatch', Bacon, 'Of Dispatch'. 99 *Cygnus*: son of Neptune; after his death, Apollo changed him into a swan, and placed him in the heavens as a constellation. 100–1 At the beginning of *Pancharis*, Holland asks Cupid for a pinion from his wing to assist him in writing; throughout the poem he mentions the fact that he is in love. 104 *Leda's white adulterer's place*: Jupiter, who changed into a swan in order to rape Leda.

Among the stars should be resigned   105
    To him, and he there shrined;
    Or Thames be rapt from us
        To dim and drown
In heaven the sign of old Eridanus:
        How they would frown!   110
    But these are mysteries
Concealed from all but clear prophetic eyes.

It is enough, their grief shall know
    At their return, nor Po,
    Iberus, Tagus, Rhine,   115
        Scheldt, nor the Maas,
Slow Arar nor swift Rhone, the Loire nor Seine,
        With all the race
        Of Europe's waters can
Set out a like, or second, to our swan.   120

7

From Thomas Wright's
The Passions of the Mind in General, *1604*
To the Author

In picture, they which truly understand
Require (besides the likeness of the thing)
Light, posture, heightening, shadow, colouring,
All which are parts commend the cunning hand;

109 *Eridanus*: a constellation; also the river Po.   114–17 European rivers: the Iberus is in Spain; the Tagus, in Spain and Portugal; the Scheldt, in France, Belgium, and Holland; the Maas, in Holland; the Arar is the Saône, in France.

7. Thomas Wright was a Yorkshireman and Jesuit who had taught in several Jesuit colleges in Europe; he was perhaps the priest who converted Jonson to Catholicism in 1598: see L. I. Guiney, *Recusant Poets* (New York and London, 1938), p. 335. An account of his career is given by Theodore A. Stroud, *ELH*, xiv (1947), 274–82. *The Passions of the Mind in General* was first published in an unauthorized edition in 1601; Jonson's verses are in the second edition of 1604, along with others by H.H. (probably Hugh Holland). Wright's work may have influenced Jonson's notions about dramatic character: see Robert E. Knoll, *Ben Jonson's Plays: An Introduction* (Nebraska, 1964), *passim*. For the analogy with painting, cf. *Disc.* 1541, 1549 ff., and ll. 15–23 of Chapman's address to Royden prefixed to *The Banquet of Sense*, in *Poems*, ed. P. B. Bartlett (New York & London, 1941), p. 19.   1 *truly understand*: see *Epig.* I. 2 and n.

And all your book, when it is throughly scanned,   5
Will well confess: presenting, limiting
Each subtlest passion, with her source and spring,
So bold, as shows your art you can command.
But now your work is done, if they that view
The several figures languish in suspense   10
To judge which passion's false and which is true,
Between the doubtful sway of reason and sense;
'Tis not your fault if they shall sense prefer,
Being told there, reason cannot, sense may, err.

8

From The Faithful Shepherdess
To the Worthy Author, Mr. John Fletcher

The wise and many-headed bench that sits
    Upon the life and death of plays and wits
(Composed of gamester, captain, knight, knight's man,
    Lady or pucelle, that wears mask or fan,
Velvet or taffeta cap, ranked in the dark   5
    With the shop's foreman, or some such brave spark
That may judge for his sixpence) had, before
    They saw it half, damned thy whole play, and more;
Their motives were, since it had not to do
    With vices, which they looked for, and came to.   10
I, that am glad thy innocence was thy guilt,
    And wish that all the muses' blood were spilt
In such a martyrdom, to vex their eyes
    Do crown thy murdered poem: which shall rise
A glorified work to time, when fire   15
    Or moths shall eat what all these fools admire.

5 *throughly*: thoroughly.   6 *limiting*: outlining.   12 *sense*: sensory perception. Wright's view of the relationship between reason, passion, and sense is set out in Pt. I, ch. ii, of his work.

8. *The Faithful Shepherdess* was in print by 1610. Jonson in *Conv. Dr.* 226–7 referred to it as 'a tragicomedy well done'.   1 *many-headed*: Horace, *Epist.* I. i. 76.   1 *bench*: cf. *B.F.* Ind. 104.   4 *pucelle*: whore.   5 *cap*: see *Und.* 42. 28.   7 *his sixpence*: the lowest price of admission mentioned in *B.F.* Ind. 88.   7–8 *before | They saw it half*: Jonson speaks with feeling: *Sejanus* had been hissed off the stage in 1603.   14 *crown*: punning on a now obsolete sense of the verb, to hold a coroner's inquest on (*OED*, *v.*²); cf. *Epig.* 17. 4.   15–16 *fire | Or moths*: Matt. 6 : 19–20, Luke 12 : 33.

The title-page of Coryate's *Crudities*, 1611, engraved by William Hole.

9

Epitaph on Cecilia Bulstrode

Stay, view this stone; and if thou beest not such,
Read here a little, that thou mayst know much.
It covers, first, a virgin; and then one
That durst be that in court: a virtue alone
To fill an epitaph. But she had more: 5
She might have claimed to have made the graces four,
Taught Pallas language, Cynthia modesty,
As fit to have increased the harmony
Of spheres, as light of stars; she was earth's eye;
The sole religious house and votary,   10
With rites not bound, but conscience. Wouldst thou all?
She was Cil Bulstrode. In which name, I call
Up so much truth as, could I it pursue,
Might make the fable of good women true.

10

From Coryate's Crudities, *1611*
Certain Opening and Drawing Distichs

To be applied as mollifying cataplasms to the
tumours, carnosities, or difficult pimples full of
matter appearing in the author's front, con-
flated of styptic and glutinous vapours arising

9. For Cecilia Bulstrode, see *Und.* 49 and n. She died at Lady Bedford's house in Twickenham on 4 Aug. 1609, evidently in physical and spiritual distress: see Lord Herbert of Cherbury's epitaph, *Poems*, ed. G. C. Moore Smith (Oxford, 1923), pp. 20–1. Jonson's epitaph is quite at odds with his attack on Miss Bulstrode in *Und.* 49. The epitaph was sent to his friend George Garrard with a covering note explaining that he wrote at speed and in sorrow. 2 *a little . . . much*: cf. *Epig.* 124. 2. 9 *eye*: bright spot, centre of intelligence: cf. *Und.* 65. 8, *Conv. Dr.* 408. 14 *fable of good women*: alluding to Chaucer's work of that name.

10. Thomas Coryate (?1577–1617) of Odcombe in Somersetshire, traveller and buffoon, journeyed nearly 2,000 miles across Europe, mainly on foot, between 14 May and 3 Oct. 1608. Failing at first to find a publisher for his account of his travels, *Crudities*, he appealed to people of eminence to write commendatory verses upon himself and his undertaking. The response was enthusiastic, if largely facetious. There is no evidence to support the notion that Jonson edited this volume: see Michael Strachan, *The Life and Adventures of Thomas Coryate* (London, etc., 1962), p. 125. Jonson refers scornfully to Coyrate in *Epig.* 129. 17 and *Und.* 13. 128. 2 *carnosities*: morbid fleshy growths. The reference is to William Hole's title-page illustrations: see p. 294. 3 *front*: forehead; frontispiece.

out of the *Crudities*; the heads whereof are par-   5
ticularly pricked and pointed out by letters
for the reader's better understanding. . . .

Here follow certain other verses, as charms to
unlock the mystery of the *Crudities*.

A

Here, like Arion, our Coryate doth draw   10
All sorts of fish with music of his maw.

B

Here, not up Holborn, but down a steep hill,
He's carried 'twixt Montreuil and Abbeville.

C

A horse here is saddled, but no Tom him to back:
It should rather have been Tom that a horse did lack.   15

D

Here, up the Alps (not so plain as to Dunstable)
He's carried like a cripple, from constable to constable.

E

A punk here pelts him with eggs. How so?
For he did but kiss her, and so let her go.

F

Religiously here he bids, Row from the stews!   20
He will expiate this sin with converting the Jews.

5 Crudities: in the old physiology, the word meant imperfectly 'concocted' humours.   5–6 *heads . . . pricked . . . pointed*: playing on words.   10–11 Coryate explains that his sea-sickness on the crossing to France came from a wish 'to satiate the gormandizing paunches of the hungry haddocks', *Crudities* (1611), p. 1. Arion, about to be pushed overboard during a sea voyage to Italy, played melodiously to a dolphin, who carried him ashore on its back.   12–13 See *Crudities*, p. 9. Holborn Hill leads to Tyburn.   14–15 Probably referring to a misadventure outside Lyons, p. 55.   16–17 Coryate paid 18*d*. to be carried the last, steep half-mile over the Alps (pp. 69–70). 'As plain as Dunstable highway' is proverbial: Tilley, D646.   18–19 Coryate took a scholarly interest in the courtesans of Venice; he visited one in order to 'view her own amorous person, hear her talk, observe her fashion of life'; but no more (p. 271).   21–3 Coryate visited the Jewish ghetto in Venice, and was rescued by Sir Henry Wotton from an angry debate with a rabbi (pp. 230–7).

G

And there, while he gives the zealous bravado,
A rabbin confutes him with the bastinado.

H

Here, by a boor too, he's like to be beaten,
For grapes he had gathered before they were eaten. 25

I

Old hat here, torn hose, with shoes full of gravel,
And louse-dropping case, are the arms of his travel.

K

Here, finer than coming from his punk you him see,
F shows what he was, K what he will be.

L

Here France and Italy both to him shed 30
Their horns, and Germany pukes on his head.

M

And here he disdained not, in a foreign land,
To lie at livery, while the horses did stand.

N

But here, neither trusting his hands nor his legs,
Being in fear to be robbed, he most learnedly begs. 35

24–5 A 'German boor' seized Coryate's hat after Coryate had taken some grapes from a vineyard (pp. 524–6). 26–7 Coryate hung up in Odcombe church the clothes he had worn on his travels; his shoes remained there until the eighteenth century. 29 'Not meaning by F and K as the vulgar may peevishly and unwittingly mistake: but that he was then coming from his courtesan, a Freshman, and now, having seen their fashions and written a description of them, he will shortly be reputed a Knowing, proper, and well-travelled scholar, as by his starched beard and printed ruff may be as properly insinuated' (Jonson's note). 31 *horns*: of plenty, but also of the cuckold: the ambiguity is apparent in the illustration itself. Hence perhaps the gibe of l. 29, at Coryate's inadequacy with (past) whore and (future) wife. 31 *Germany*: Coryate discusses German drinking habits on pp. 438–9. 32–3 At Bergamo (p. 350). 34–5 Fearing he was to be robbed by two 'ragged boors' outside Baden, Coryate pretended he was a beggar; they gave him 4½d. (p. 465).

11

From Coryate's Crudities ... *To the Right Noble Tom, Tell-Troth of his Travails, the Coryate of Odcombe, and his Book, Now Going to Travel.*

T ry and trust *Roger* was the word, but now   [80]
H onest Tom Tell-Troth puts down Roger: how?
O f travel he discourseth so at large,
M arry, he sets it out at his own charge;
A nd therein (which is worth his valour too)
S hows he dares more than Paul's Churchyard durst do. [85]

C ome forth thou bonny bouncing book then, daughter
O f Tom of Odcombe, that odd jovial author;
R ather his son, I should have called thee: why?
Y es, thou wert born out of his travelling thigh
A s well as from his brains, and claim'st thereby   [90]
T o be his Bacchus as his Pallas: be
E ver his thighs male then, and his brains she.

12

From Coryate's Crambe, *1611*
Certain Verses Written upon Coryate's Crudities

Which should have been printed with the other panegyric lines, but then were upon some occasions omitted, and now communicated to the world.

Incipit Ben Jonson
To the London Reader, on the Odcombian Writer,
Polytopian Thomas the Traveller.

---

11. This 'Characterism Acrostic' (acrostic character-sketch) follows a prose 'Character' of Coryate, probably also by Jonson: it is printed by H & S, whose subsequent line-numbers are given here. *Travails ... Travel*: the usual play on words; cf. 1. 89 below. [80] Try and trust Roger: a proverbial saying: 'trust the plain man'. [83] *at his own charge*: Coryate seems to have published *Crudities* at his own expense. [85] *Paul's Churchyard*: home of publishers. [91] Bacchus was born from Jupiter's thigh, Pallas Athene from his head.

12. Coryate's *Crambe* was one of two appendices to *Crudities* published in 1611. *Polytopian*: a nonce word: one who visits many places.

Whoever he be, would write a story at
The height, let him learn of Mr. Tom Coryate;
Who, because his matter in all should be meet
To his strength, hath measured it out with his feet.
And that, say philosophers, is the best model.   5
Yet who could have hit on't but the wise noddle
Of our Odcombian, that literate elf,
To line out no stride, but paced by himself,
And allow you for each particular mile
By the scale of his book, a yard of his style?   10
Which unto all ages for his will be known,
Since he treads in no other man's steps but his own.
And that you may see he most luckily meant
To write it with the self-same spirit he went,
He says to the world, Let any man mend it!   15
In five months he went it, in five months he penned it.
But who will believe this, that chanceth to look
The map of his journey, and sees in his book
France, Savoy, Italy, and Helvetia,
The Low Countries, Germany and Rhoetia   20
There named to be travelled? For this our Tom saith:
Pize on't! You have his historical faith.
Each leaf of his journal and line doth unlock
The truth of his heart there, and tells what a-clock
He went out at each place, and at what he came in,   25
How long he did stay, at what sign he did inn.
Besides he tried ship, cart, waggon, and chair,
Horse, foot, and all but flying in the air;
And therefore, however the travelling nation
Or builders of story have oft imputation   30
Of lying, he fears so much the reproof
Of his foot or his pen, his brain or his hoof,
That he dares to inform you (but somewhat meticulous)
How scabbed, how ragged, and how pediculous

4–5 Horace, *Epist*. I. vii. 98.   14 *self-same spirit*: i.e. pedestrian.   15 ff. At the end of
Crudities, Coryate gives the statistics of his travels: 1975 miles and 45 cities (etc.) in 5
months.   22 *Pize*: pox.   29–31 Travellers were proverbially said to be liars: Tilley,
T476.

He was in his travail, how like to be beaten   35
For grapes he had gathered before they were eaten.
How fain for his venery he was to cry *tergum o!*
And lay in straw with the horses at Bergamo;
How well and how often his shoes too were mended,
That sacred to Odcombe are there now suspended: 40
I mean that one pair wherewith he so hobbled
From Venice to Flushing; were not they well cobbled?
Yes. And thanks God in his pistle or his book
How many learned men he have drawn with his hook
Of Latin and Greek, to his friendship. And seven   45
He there doth protest he saw of the eleven.
Nay, more in his wardrobe, if you will laugh at a
Jest, he says: *item*, one suit of black taffeta
(Except a doublet), and bought of the Jews.
So that not them, his scabs, lice, or the stews,   50
Or anything else that another should hide
Doth he once dissemble, but tells he did ride
In a cart 'twixt Montreuil and Abbeville.
And being at Flushing enforced to feel
Some want, they say in a sort he did crave: 55
I writ he only his tail there did wave;
Which he not denies. Now, being so free,
Poor Tom, have we cause to suspect just thee?
No! As I first said, who would write a story at
The height, let him learn of Mr. Tom Coryate.   60

[U.V. *13, 14, 15—three short poems in Latin—are omitted from this
edition: see H & S, viii. 381–2.*]

36 *grapes*: see *U.V.* 10. 24–5 and n.   37 *tergum o!*: O my back!   38 *Bergamo*: *U.V.*
10. 32–3.   40 *suspended*: see *U.V.* 10. 26–7 and n.   43–6 In his 'Epistle to the reader',
Coryate names twelve learned men he has met and conversed with during his
travels.   48 *black taffeta*: admired by Coryate in Venice, *Crudities*, pp. 248, 259.   53
cart: *U.V.* 10. 12–13 and n.   54–5 Perhaps referring to Coryate's visit to Sir William
Browne of the English garrison at Flushing, *Crudities*, p. 653.

16

A Speech Presented unto King James at a Tilting,
in the Behalf of the Two Noble Brothers,
Sir Robert and Sir Henry Rich, Now Earls of Warwick and
Holland

Two noble knights, whom true desire and zeal
Hath armed at all points, charge me humbly kneel
Unto thee, king of men, their noblest parts
To tender thus: their lives, their loves, their hearts!
The elder of these two, rich hope's increase,   5
Presents a royal altar of fair peace;
And as an everlasting sacrifice
His life, his love, his honour which ne'er dies
He freely brings; and on this altar lays
As true oblations. His brother's emblem says   10
Except your gracious eye, as through a glass
Made prospective, behold him, he must pass
Still that same little point he was; but when
Your royal eye, which still creates new men,
Shall look, and on him so, then art's a liar
If from a little spark he rise not fire.   15

17

From Cynthia's Revenge, *or* Menander's Ecstacy, *1613*
To His Much and Worthily Esteemed Friend, the Author

Who takes thy volume to his virtuous hand
Must be intended still to understand;
Who bluntly doth but look upon the same
May ask, What author would conceal his name?

16. Sir Robert Rich, 2nd Earl of Warwick (1587–1658), and Sir Henry Rich, 1st Baron
Kensington and 1st Earl of Holland (1590–1649) were sons of Robert Rich, 1st Earl of
Warwick, and Penelope, *née* Devereux, Sidney's Stella. The tilting took place on 24
March 1613.   12 *prospective*: magnifying.

17. *Cynthia's Revenge* was published anonymously, the Dedication being signed I.S.:
identifiable from certain title-pages as John Stephens of Lincoln's Inn.   2 *understand*:
cf. *Epig.* 1. 2 and n.

Who reads may rove, and call the passage dark,   5
Yet may, as blind men, sometimes hit the mark.
Who reads, who roves, who hopes to understand
May take thy volume to his virtuous hand.
Who cannot read, but only doth desire
To understand, he may at length admire.   10

18

To the Most Noble, and Above His Titles,
Robert, Earl of Somerset

They are not those, are present with their face
    And clothes, and gifts, that only do thee grace
At these thy nuptials; but whose heart and thought
    Do wait upon thee: and their love not bought.
Such wear true wedding robes, and are true friends   5
    That bid, God give thee joy! And have no ends.
Which I do, early, virtuous Somerset,
    And pray thy joys as lasting be as great.
Not only this, but every day of thine
    With the same look, or with a better, shine.   10
May she whom thou for spouse today dost take
    Out-be that wife in worth thy friend did make;
And thou to her, that husband, may exalt
    Hymen's amends, to make it worth his fault.
So be there never discontent or sorrow   15
    To rise with either of you on the morrow.

5 *rove*: a term from archery: to shoot arrows at random; hence guess, conjecture.

18. King James's favourite, Robert Carr, Earl of Somerset, married the divorced
Countess of Essex (formerly Frances Howard) on 26 Dec. 1613. Carr's friend Sir
Thomas Overbury, who opposed the marriage, had been sent to the Tower (by Carr's
contrivance) in April of that year, and poisoned (by the Countess's) in September.
Jonson also wrote *A Challenge at Tilt* and *The Irish Masque* for this occasion.   6 *ends*:
ulterior motives.   7 *early*: cf. *Epig.* 23. 3, *Und.* 70. 125.   12 *that wife*: Sir Thomas
Overbury's poem, *The Wife*, which was published in 1614; Jonson knew it in MS.
(*Conv. Dr.* 214–16).   13–14 i.e. do better than Essex: the bride's first marriage
had been annulled because of Essex's alleged physical incapacity (the 'fault' of l. 14).

So be your concord still as deep, as mute,
    And every joy in marriage turn a fruit.
So may those marriage-pledges comforts prove,
    And every birth increase the heat of love.   20
So in their number may you never see
    Mortality, till you immortal be.
And when your years rise more than would be told,
    Yet neither of you seem to th'other old.
That all that view you then and late may say,   25
    Sure, this glad pair were married but this day!

19

From The Ghost of Richard the Third, *1614*
To His Friend the Author,
Upon His Richard

When these, and such, their voices have employed,
    What place is for my testimony void?
Or to so many and so broad seals had
    What can one witness, and a weak one, add
To such a work, as could not need theirs? Yet   5
    If praises, when they're full, heaping admit,
My suffrage brings thee all increase, to crown
    Thy *Richard*, raised in song, past pulling down.

23-4 Cf. Martial, *Epig*. IV. xiii. 9–10; as H & S observe, it is disconcerting to notice that Jonson had expressed much the same wish at the bride's first wedding, *Hym.* 561–4. Cf. *Und.* 75. 165–8.

19. The author was Christopher Brooke (d. 1628), of Lincoln's Inn; Donne's friend, and witness at his wedding; associated with the pastoral revival of 1613–16.   1 *such*: Chapman, Browne, Wither, and others.   3 *broad seals*: warrants, authorities (after the Broad or Great Seal of England).

20

From The Husband, *1614*
To the Worthy Author, on The Husband

It fits not only him that makes a book
    To see his work be good, but that he look
Who are his test, and what their judgement is,
    Lest a false praise do make their dotage his.
I do not feel that ever yet I had   5
    The art of uttering wares if they were bad,
Or skill of making matches in my life;
    And therefore I commend unto *The Wife*
That went before, a *Husband*. She, I'll swear,
    Was worthy of a good one; and this, here,   10
I know for such, as (if my word will weigh)
    She need not blush upon the marriage-day.

21

From Britannia's Pastorals, *The Second Book, 1616*
To My Truly-Beloved Friend, Mr. Browne, On His Pastorals

Some men, of books or friends not speaking right,
    May hurt them more with praise, than foes with spite.
But I have seen thy work, and I know thee:
    And, if thou list thyself, what thou canst be.
For though but early in these paths thou tread,   5
    I find thee write most worthy to be read.
It must be thine own judgement, yet, that sends
    This thy worth forth: that judgement mine commends.
And where the most read books on authors' fames,
    Or, like our money-brokers, take up names   10

20. The authorship of *The Husband* is unknown. The work followed Sir Thomas Overbury's *The Wife* (*U.V*. 18. 12 n).

21. The first book of William Browne's *Britannia's Pastorals* was published in 1613; the second, dedicated to William, Earl of Pembroke, in 1616: Jonson's *Epigrams* were dedicated to Pembroke the same year. The volume carried other commendatory verses. Browne praises Jonson's talents in his second song.   2 Cf. *U.V*. 26. 14.

On credit, and are cozened, see that thou
    By offering not more sureties than enow
Hold thine own worth unbroke: which is so good
    Upon the exchange of letters, as I would
More of our writers would, like thee, not swell   15
    With the *how much* they set forth, but the *how well*.

22

Charles Cavendish to His Posterity

Sons, seek not me among these polished stones,
These only hide part of my flesh and bones,
Which, did they ne'er so neat or proudly dwell,
Will all turn dust, and may not make me swell.
Let such as justly have outlived all praise   5
Trust in the tombs their careful friends do raise;
I made my life my monument, and yours,
To which there's no material that endures,
Nor yet inscription like it. Write but that,
And teach your nephews it to emulate: 10
It will be matter loud enough to tell
Not when I died, but how I lived. Farewell.

23

From George Chapman's The Georgics of Hesiod, *1618*
To My Worthy and Honoured Friend, Mr. George Chapman,
On His Translation of Hesiod's Works and Days

Whose work could this be, Chapman, to refine
Old Hesiod's ore and give it us, but thine,
Who hadst before wrought in rich Homer's mine?

---

22. Sir Charles Cavendish, of Welbeck, Notts, was the husband of Lady Katherine Ogle of *U.V.* 31, and the father of Jonson's patron, William Cavendish, of *Und.* 53 & 59. He died 4 April 1619; Jonson's lines are on his monument in the Cavendish chantry at Bolsover. 8 *To which*: compared with which. 10 *nephews*: descendants (Latin *nepotes*).

23. 3 *rich Homer's mine*: Chapman had published a specimen of his *Iliad* in 1598, the whole *Iliad* in 1611, and his *Odyssey* in 1614.

What treasure hast thou brought us! And what store
Still, still, dost thou arrive with at our shore,   5
To make thy honour and our wealth the more!

If all the vulgar tongues that speak this day
Were asked of thy discoveries, they must say
To the Greek coast thine only knew the way.

Such passage hast thou found, such returns made,   10
As now, of all men, it is called thy trade:
And who make thither else, rob or invade.

24

From The Rogue, *1622*
On the Author, Work, and Translator

Who tracks this author's or translator's pen
Shall find that either hath read books and men:
To say but one, were single. Then it chimes
When the old words do strike on the new times,
As in this Spanish Proteus; who, though writ   5
But in one tongue, was formed with the world's wit;
And hath the noblest mark of a good book,
That an ill man dares not securely look
Upon it, but will loathe, or let it pass,
As a deformed face doth a true glass.   10
Such books deserve translators of like coat
As was the genius wherewith they were wrote;
And this hath met that one that may be styled
More than the foster-father of this child;
For though Spain gave him his first air and vogue   15
He would be called henceforth the English *Rogue*,
But that he's too well-suited, in a cloth
Finer than was his Spanish, if my oath
Will be received in court; if not, would I
Had clothed him so. Here's all I can supply   20

---

24. *The Rogue* was a translation (1622) of Matheo Aleman's *Guzman de Alfarache* (Pt.
I 1599; Pt. II 1604) by James Mabbe (1572–?1642), Spanish scholar and Fellow of
Magdalen College, Oxford.   10 *true glass*: cf. *For.* 13. 26 ff.

To your desert, who've done it, friend. And this
Fair emulation and no envy is;
When you behold me wish myself the man
That would have done that which you only can.

25

From Mr. William Shakespeare's Comedies, Histories,
and Tragedies, *1623*
To the Reader

This figure that thou here seest put,
　　It was for gentle Shakespeare cut;
Wherein the graver had a strife
　　With nature, to out-do the life.
Oh, could he but have drawn his wit   5
　　As well in brass as he hath hit
His face, the print would then surpass
　　All that was ever writ in brass.
But since he cannot, reader, look
　　Not on his picture but his book.   10

26

From Mr. William Shakespeare's Comedies, Histories,
and Tragedies, *1623*
To the Memory of My Beloved,
The Author, Mr. William Shakespeare,
And What He Hath Left Us

To draw no envy, Shakespeare, on thy name,
　　Am I thus ample to thy book and fame;
While I confess thy writings to be such
　　As neither man nor muse can praise too much:

25. Printed with the Droeshout portrait in the First Folio.   3 *graver*: engraver.
3 *strife*: cf. Herrick, 'To the Painter to Draw Him a Picture', 3–4, Herrick, p. 38; *Venus
and Adonis*, 291–2, etc.   5–8 Cf. Martial, *Epig.* X. xxxii.   5–6: 'Would that art could
limn his character and mind! More beautiful in all the world would no painting be' (on
Marcus Antonius Primus). Cf. *U.V.* 26. 22 n.

26. *My Beloved*: 'for I loved the man, and do honour his memory (on this side
idolatry) as much as any', *Disc.* 654–5.   1 *envy*: a constant preoccupation: cf. *Epig.*
102. 16, 111. 14, *Und.* 73, 76. 17 ff., 84. i. 3, etc.

'Tis true, and all men's suffrage. But these ways 5
Were not the paths I meant unto thy praise:
For silliest ignorance on these may light,
Which, when it sounds at best, but echoes right;
Or blind affection, which doth ne'er advance
The truth, but gropes, and urgeth all by chance; 10
Or crafty malice might pretend this praise,
And think to ruin where it seemed to raise.
These are as some infamous bawd or whore
Should praise a matron: what could hurt her more?
But thou art proof against them, and indeed 15
Above the ill fortune of them, or the need.
I therefore will begin. Soul of the age!
The applause, delight, the wonder of our stage!
My Shakespeare, rise: I will not lodge thee by
Chaucer or Spenser, or bid Beaumont lie 20
A little further, to make thee a room;
Thou art a monument without a tomb,
And art alive still while thy book doth live,
And we have wits to read, and praise to give.
That I not mix thee so, my brain excuses: 25
I mean with great, but disproportioned, muses;
For if I thought my judgement were of years
I should commit thee surely with thy peers:
And tell how far thou didst our Lyly outshine,
Or sporting Kyd, or Marlowe's mighty line. 30

14 *what could hurt her more?*: cf. *U.V.* 21. 2. 18 Cf. Martial, *Epig.* IX. xxviii. 1–2: 'The darling pride of the stage, the glory of the games ... the favourite of your applause'. 19–21 Tilting at the opening of William Basse's *Elegy on Shakespeare*: 'Renowned Spenser, lie a thought more nigh / To learned Chaucer; and rare Beaumont lie / A little nearer Spenser, to make room / For Shakespeare in your threefold, fourfold tomb'. 22 *a monument without a tomb*: cf. Horace, *Odes*, III. xxx. 1–2: 'I have finished a monument more lasting than bronze and loftier than the pyramids' royal pile': localized by a further allusion to Shakespeare's Stratford monument (cf. Leonard Digges's poem in the First Folio, declaring that Shakespeare's work will outlive his monument). The line is echoed by William Cavendish and varied by John Taylor in their poems on Jonson's death in 1637 (H & S, xi. 489, 426), and remembered in Milton's 'On Shakespeare', l. 8: 'hast built thyself a livelong monument'. 27 *of years*: mature. 28 *commit*: in the sense of Latin *committo*, match, bring together for comparison. 29 *our Lyly outshine*: cf. the lily as the 'flower of light' in *Und.* 70. 72, and Virgil, *Aeneid*, vi. 708–9.

And though thou hadst small Latin, and less Greek,
  From thence to honour thee I would not seek
For names, but call forth thundering Aeschylus,
  Euripides, and Sophocles to us,
Pacuvius, Accius, him of Cordova dead, 35
  To life again, to hear thy buskin tread
And shake a stage; or, when thy socks were on,
  Leave thee alone for the comparison
Of all that insolent Greece or haughty Rome
  Sent forth, or since did from their ashes come. 40
Triumph, my Britain, thou hast one to show
  To whom all scenes of Europe homage owe.
He was not of an age, but for all time!
  And all the muses still were in their prime
When like Apollo he came forth to warm 45
  Our ears, or like a Mercury to charm!
Nature herself was proud of his designs,
  And joyed to wear the dressing of his lines,
Which were so richly spun and woven so fit
  As, since, she will vouchsafe no other wit. 50
The merry Greek, tart Aristophanes,
  Neat Terence, witty Plautus, now not please,
But antiquated and deserted lie
  As they were not of nature's family.
Yet must I not give nature all: thy art, 55
  My gentle Shakespeare, must enjoy a part.

---

31 *small Latin, and less Greek*: J. E. Spingarn finds a source for this phrase in A. S. Minturno, *L'Arte Poetica* (Venice, 1564), p. 158: *A History of Literary Criticism in the Renaissance* (New York, 1908), p. 89, n. 35 *Pacuvius, Accius*: Roman tragedians. 35 *him of Cordova dead*: Seneca. 39 *insolent Greece or haughty Rome*: cf. *Disc.* 916–18, on Bacon's having 'performed that in our tongue which may be compared or preferred either to insolent Greece or haughty Rome'; recalling Seneca the elder, *Controversiarum*, i. Pref. § 6: 'all that Roman eloquence could put beside or above that of insolent Greece flourished about the time of Cicero'. 45–6 *Apollo . . . Mercury*: gods of poetry and eloquence, respectively. 49 *spun and woven*: perhaps remembering the literal and figurative senses of Greek ῥάπτω: to sew together, to plot or contrive. Cf. Pindar, *Nemean Odes*, ii. 2, and Jonson, *U.V.* 42. 14–18, *M.L.* Chor. before Act I, 133–41, before Act V, 1–9. 51 *merry Greek*: playing on words: a 'Greek' was a merry fellow (*OED*, 5: cf. the related phrase 'merry as a grig'). 52 *Neat*: elegant, pithy. 55 ff. Contrast *Conv. Dr.* 50, 'That Shakespeare wanted art'. For the background to this opposition of art and nature, see T. W. Baldwin, *William Shakspere's Small Latine and Lesse Greeke* (Urbana, 1944), i. 13 ff., 41; and for the place of these terms in the developing debate about the nature of Shakespeare's genius, see E. N. Hooker (ed.), *The Critical Works of John Dennis* (Baltimore, 1943), ii. 428–31.

For though the poet's matter nature be,
   His art doth give the fashion. And that he
Who casts to write a living line must sweat
   (Such as thine are) and strike the second heat   60
Upon the muses' anvil: turn the same
   (And himself with it) that he thinks to frame;
Or for the laurel he may gain a scorn:
   For a good poet's made, as well as born;
And such wert thou. Look how the father's face   65
   Lives in his issue: even so, the race
Of Shakespeare's mind and manners brightly shines
   In his well-turned and true-filed lines:
In each of which he seems to shake a lance,
   As brandished at the eyes of ignorance.   70
Sweet swan of Avon! What a sight it were
   To see thee in our waters yet appear,
And make those flights upon the banks of Thames
   That so did take Eliza, and our James!
But stay, I see thee in the hemisphere   75
   Advanced, and made a constellation there!
Shine forth, thou star of poets, and with rage
   Or influence chide or cheer the drooping stage;
Which, since thy flight from hence, hath mourned like night,
   And despairs day, but for thy volume's light.   80

[For *U.V.* 27, see *Dubia*, 4]

---

59 *casts*: intends (*OED*, † 44); and plans, shapes, disposes: of material things and artistic works (*OED*, † 45). 59 *living line*: as in ll. 48–9 and 67–8, Shakespeare's lines are seen as being at once artificial and organic. 62 *himself with it*: Briggs, *MP*, xv (1917), 279–80, compares Horace's assertion that the poet who wishes to write a 'legitimate poem' must turn and twist himself like a mime: *Epist.* II. ii. 124–5. 63 *for*: instead of. 64 Cf. *E.M.I.* V. v. 38–40. 66 *race*: combining several senses: (i) offspring, posterity (*OED*, *sb.*² 1), i.e. his writings; (ii) liveliness, piquancy, raciness (*OED*, *sb.*² 10. b: of writing); (iii) onward movement, rush (*OED*, *sb.*¹ II. 5). 69 *shake a lance*: playing on the poet's name, as in l. 37. 71 *Sweet swan of Avon*: as Homer had been known as the swan of Meander, Virgil as the swan of Mantua, and Pindar as the Dircaean swan. On the association of the swan with poetry, see *U.V.* 6 and n. 76 *a constellation*: the stellification of the dead person is a common theme in classical epitaphs: see Richmond Lattimore, *Themes in Greek and Latin Epitaphs* (Urbana, 1962), pp. 311–13; and cf. *Und.* 70. 89 ff., *U.V.* 6. 97–112. 80 Cf. *Cat.* IV. 761.

28

To the Memory of That Most Honoured Lady Jane,
Eldest Daughter to Cuthbert, Lord Ogle,
And Countess of Shrewsbury

I could begin with that grave form, *Here lies*,
And pray thee, reader, bring thy weeping eyes
To see who 'tis: a noble countess, great
In blood, in birth, by match, and by her seat;
Religious, wise, chaste, loving, gracious, good;   5
And number attributes unto a flood:
But every table in this church can say
A list of epithets, and praise this way.
No stone in any wall here but can tell
Such things of everybody, and as well.   10
Nay, they will venture one's descent to hit,
And Christian name, too, with a herald's wit.
But I would have thee to know something new,
Not usual in a lady, and yet true:
At least so great a lady. She was wife   15
But of one husband; and since he left life,
But sorrow, she desired no other friend:
And her she made her inmate to the end,
To call on sickness still to be her guest,
Whom she with sorrow first did lodge, then feast,   20
Then entertain, and as death's harbinger;
So wooed at last, that he was won to her
Importune wish, and by her loved lord's side
To lay her here, enclosed, his second bride.
Where, spite of death, next life, for her love's sake,   25
This second marriage will eternal make.

---

28. Lady Jane Ogle, Countess of Shrewsbury, was the eldest daughter of Cuthbert, Lord Ogle, and sister of Lady Katherine Ogle of *U.V.* 31; Jonson's patron William Cavendish, Earl (later Duke) of Newcastle was her nephew (see *Und.* 53 & 59). She died 7 Jan. 1625, and was buried in St. Edmund's Chapel, Westminster Abbey, next to her husband. Jonson's lines are not on the monument.   1 *grave*: a pun.   12 *a herald's wit*: cf. *Epig.* 9. 4 and n.   18 *inmate*: lodger; cf. *U.V.* 31. 26 and Donne, 'The Anniversary', l. 18.   23 *Importune*: importunate.

29

From Lucan's Pharsalia, *1627*
To My Chosen Friend,
The Learned Translator of Lucan,
Thomas May Esq.

When, Rome, I read thee in thy mighty pair,
And see both climbing up the slippery stair
Of fortune's wheel, by Lucan driven about,
And the world in it, I begin to doubt:
At every line some pin thereof should slack   5
At least, if not the general engine crack.
But when again I view the parts so peised,
And those in number so, and measure raised,
As neither Pompey's popularity,
Caesar's ambition, Cato's liberty,   10
Calm Brutus' tenor start, but all along
Keep due proportion in the ample song,
It makes me, ravished with just wonder, cry
What muse, or rather god of harmony
Taught Lucan these true modes? Replies my sense: 15
What gods but those of arts and eloquence,
Phoebus and Hermes? They whose tongue or pen
Are still the interpreters 'twixt gods and men!
But who hath them interpreted, and brought
Lucan's whole frame unto us, and so wrought   20
As not the smallest joint or gentlest word
In the great mass or machine there is stirred?
The self-same genius! so the work will say:
The sun translated, or the son of May.

29. Thomas May (1595–1650) also translated Martial and Virgil's *Georgics*. Marvell later satirized him in *Tom May's Death*. 1 *mighty pair*: Caesar and Pompey. 7 *peised*: balanced. 11 *start*: swerve. 18 Cf. *Disc.* 1883–4. 24 *sun*: Phoebus Apollo. 24 *son of May*: Mercury (Hermes) was the son of Maia; a strained pun on the translator's name.

30

From The Battle of Agincourt, *1627*
The Vision of Ben Jonson,
On the Muses of His Friend
M. Drayton

It hath been questioned, Michael, if I be
A friend at all; or, if at all, to thee:
Because who make the question have not seen
Those ambling visits pass in verse between
Thy muse and mine, as they expect. 'Tis true;   5
You have not writ to me, nor I to you;
And though I now begin, 'tis not to rub
Haunch against haunch, or raise a rhyming club
About the town; this reckoning I will pay
Without conferring symbols. This is my day.   10
    It was no dream: I was awake, and saw!
Lend me thy voice, O fame, that I may draw
Wonder to truth, and have my vision hurled,
Hot from thy trumpet, round about the world!
    I saw a beauty from the sea to rise   15
That all earth looked on: and that earth, all eyes!
It cast a beam as when the cheerful sun
Is fair got up, and day some hours begun,
And filled an orb as circular as heaven.
The orb was cut forth into regions seven,   20

30. Jonson told Drummond in 1618–19 that 'Drayton feared him, and he esteemed not of him' (*Conv. Dr.* 153). Jonson's poem, a late insertion in Drayton's volume, was perhaps written in response to Drayton's own praise of him later in the same volume (see note to l. 91 below). J. W. Hebel considers Jonson's poem 'sly satire rather than compliment' (*PMLA*, xxxix (1924), 830–2); B. H. Newdigate sees it as 'too fulsome to be taken seriously . . . a bit of good-tempered leg-pulling' (*Michael Drayton and His Circle* (Oxford, 1941, 1961), p. 136). The extravagance of the poem may perhaps merely reflect Jonson's anxiety to make lavish amends to an unexpected admirer.   **8** *rhyming club*: cf. the 'play club' of *Songs*, 14. 26.   **10** *conferring symbols*: contributing shares, as in an eating club: Plautus's *symbolarum collatores*, in *Curculio*, IV. i (474).   **11, 13** *dream . . . vision*: Jonson distinguishes between a fantasy and a revelation.   **20** *regions seven*: the seven poems (or groups of poems) set out on the title-page of Drayton's *Poems*, 1619.

And those so sweet and well-proportioned parts
As it had been the circle of the arts!
When, by thy bright *Ideas* standing by,
I found it pure and perfect poesy;
There read I straight thy learned *Legends* three,   25
Heard the soft airs between our swains and thee,
Which made me think the old Theocritus
Or rural Virgil come to pipe to us!
But then thy epistolar *Heroic Songs*,
Their loves, their quarrels, jealousies, and wrongs   30
Did all so strike me, as I cried, Who can
With us be called the Naso, but this man?
And looking up, I saw Minerva's fowl
Perched overhead, the wise Athenian *Owl*;
I thought thee then our Orpheus, that wouldst try   35
Like him, to make the air one volary;
And I had styled thee Orpheus: but before
My lips could form the voice, I heard that roar
And rouse, the marching of a mighty force,
Drums against drums, the neighing of the horse,   40
The fights, the cries; and, wondering at the jars,
I saw and read it was thy *Barons' Wars*!
Oh, how in those dost thou instruct these times
That rebels' actions are but valiant crimes;
And, carried, though with shout and noise, confess   45
A wild and an authorized wickedness!
Sayst thou so, Lucan? But thou scorn'st to stay
Under one title. Thou hast made thy way
And flight about the isle well-near by this
In thy admired periegesis.   50

23 Ideas: Drayton's *Idea, The Shepherd's Garland*, 1593.   25 Legends *three*: there
were actually four: *Pierce Gaveston*, 1593/4; *Matilda*, 1594; *Robert, Duke of Normandy*,
1596; *Great Cromwell*, 1607.   29 *epistolar* Heroic Songs: *England's Heroical Epistles*,
1597.   32 *Naso*: Ovid.   34 Owl: Drayton's *The Owl*, 1604.   42 Barons' Wars: 1603; a
rewriting of *Mortimeriados*, 1596.   43 *these times*: referring to the rising political
trouble between Charles and his parliament.   46 *authorized*: i.e. permitted when it
should not be permitted: cf. a similar sense in 'licensed'.   47 *Lucan*: who, in
Pharsalia, had written about the Roman civil wars.   50–1 *periegesis . . . circumduction*:
both words mean 'leading around'; the first recalls the title of a versified geography of
the world written *c.* 300 A.D. by Dionysius of Alexandria.

Or universal circumduction
Of all that read thy *Poly-Olbion.*
That read it? That are ravished! Such was I
With every song, I swear, and so would die;
But that I hear again thy drum to beat   55
A better cause, and strike the bravest heat
That ever yet did fire the English blood:
Our right in France! (if rightly understood).
There, thou art Homer! Pray thee, use the style
Thou hast deserved, and let me read the while   60
Thy catalogue of ships, exceeding his;
Thy list of aids and force, for so it is
The poet's act; and for his country's sake
Brave are the musters that the muse will make.
And when he ships them where to use their arms,   65
How do his trumpets breathe! What loud alarms!
Look how we read the Spartans were inflamed
With bold Tyrtaeus' verse: when thou art named,
So shall our English youth urge on, and cry
An Agincourt! An Agincourt! Or die.   70
This book, it is a catechism to fight,
And will be bought of every lord or knight
That can but read; who cannot, may in prose
Get broken pieces, and fight well by those.
The *Miseries of Margaret the Queen*   75
Of tender eyes will more be wept, than seen;
I feel it by mine own, that overflow
And stop my sight in every line I go.
But then refreshed with thy *Fairy Court*,
I look on *Cynthia*, and *Sirena's* sport,   80
As on two flowery carpets that did rise
And with their grassy green restored mine eyes.

---

52 Poly-Olbion: 1612 and 1622. Contrast *Conv. Dr.* 25–8: 'That Michael Drayton's *Poly-Olbion*, if he had performed what he promised to write (the deeds of all the Worthies) had been excellent; his long verses pleased him not'.   55 ff. Referring to the contents of Drayton's *Battle of Agincourt* volume: first, to the title poem; to *The Miseries of Margaret the Queen* (75); *Nymphidia, The Court of Fairy* (79); *The Quest of Cynthia* and *The Shepherd's Sirena* (80), and *The Mooncalf* (84); *Ends* (89) refers to the *Elegies Upon Sundry Occasions*.   61 *catalogue of ships*: in *Agincourt*, 361 ff.; in Homer's *Iliad*, ii. 484 ff.   68 *Tyrtaeus*: Greek lyric poet, who roused the Spartans just as they wished to raise the siege of Ithome, and helped them to defeat the Messenians.

Yet give me leave to wonder at the birth
Of thy strange *Mooncalf*, both thy strain of mirth
And gossip-got acquaintance, as to us   85
Thou hadst brought Lapland, or old Cobalus,
Empusa, Lamia, or some monster more
Than Afric knew, or the full Grecian store!
I gratulate it to thee, and thy *Ends*,
To all thy virtuous and well-chosen friends;   90
Only my loss is that I am not there:
And till I worthy am to wish I were,
I call the world that envies me to see
If I can be a friend, and friend to thee.

31

Epitaph on Katherine, Lady Ogle
'Ο Ζεὺς κατεῖδε χρόνιος εἰς τὰς διφθερὰς.

'Tis a record in heaven. You, that were
Her children and grandchildren, read it here!
Transmit it to your nephews, friends, allies,
Tenants and servants, have they hearts and eyes
To view the truth and own it. Do but look   5
With pause upon it: make this page your book;
Your book? Your volume! Nay, the state and story,
Code, digests, pandects of all female glory!

86 *Lapland*: traditionally the home of witches and wizards.   86 *Cobalus*: demon of the
mines.   87 *Empusa*: a spectre who had power to change her shape: see Aristophanes,
Frogs, 293.   87 *Lamia*: in classical mythology, a female demon who devoured chil-
dren; sometimes identified with Empusa.   91 *not there*: Jonson is in fact praised in
'To My Most Dearly-Loved Friend, Henry Reynolds, Esq', *Works*, ed. J. W. Hebel
(Oxford, 1961), iii. 229; what he regrets is that he is not a recipient of a verse letter.

31. Katherine, Lady Ogle (Baroness Ogle, 1628) was the widow of Sir Charles
Cavendish (*U.V.* 22); the mother of William Cavendish, Earl (and later Duke) of
Newcastle (*Und.* 53 & 59); and the younger sister of Lady Jane Ogle, Countess of
Shrewsbury (*U.V.* 28). She died 18 April 1629, and was buried at Bolsover. The
Greek motto is a proverb: 'In the fullness of time Zeus observes the records'.   3
nephews: descendants (Latin *nepotes*).   8 *digests, pandects*: collections of Roman civil

Diphthera Jovis

She was the light (without reflex
Upon herself) to all her sex;   10
The best of women: her whole life
Was the example of a wife,
Or of a parent, or a friend!
All circles had their spring and end
In her, and what could perfect be   15
Or without angles, it was she.
All that was solid in the name
Of virtue, precious in the frame,
Or else magnetic in the force,
Or sweet or various in the course;   20
What was proportion, or could be
By warrant called just symmetry
In number, measure, or degree
Of weight or fashion, it was she.
Her soul possessed her flesh's state   25
In fair freehold, not an inmate;
And when the flesh here shut up day,
Fame's heat upon the grave did stay;
And hourly brooding o'er the same
Keeps warm the spice of her good name,   30
Until the dust returned be
Into a phoenix, which is she.

For this did Katherine, Lady Ogle, die
To gain the crown of immortality,
Eternity's great charter; which became   35
Her right, by gift and purchase of the Lamb:
Sealed and delivered to her in the sight
Of angels and all witnesses of light,
Both saints and martyrs, by her loved Lord.
And this a copy is of the record.   40

laws made under Justinian. *Diphthera Jovis*: i.e. Jove's record; διφθέρα being prepared hide used for writing upon.   9 *reflex*: reflection.   14–16 The circle was regarded as the perfect figure: cf. *Und*. 84. iv. 32 n.   23–4 See *Und*. 70. 50 and n.   26 *inmate*: temporary lodger; cf. *U.V.* 28. 18.

32

From Bosworth Field, *1629*
On the Honoured Poems of His Honoured Friend,
Sir John Beaumont, Baronet

This book will live: it hath a genius; this
   Above his reader or his praiser is.
Hence, then, profane! Here needs no words' expense
   In bulwarks, ravelins, ramparts, for defence,
Such as the creeping common pioneers use   5
   When they do sweat to fortify a muse.
Though I confess a Beaumont's book to be
   The bound and frontier of our poetry;
And doth deserve all muniments of praise
   That art or engine on the strength can raise.   10
Yet who dares offer a redoubt to rear,
   To cut a dyke, or stick a stake up here
Before this work, where envy hath not cast
   A trench against it, nor a battery placed?
Stay till she make her vain approaches: then   15
   If, maimed, she come off, 'tis not of men
This fort of so impregnable access,
   But higher power, as spite could not make less
Nor flattery; but, secured by the author's name,
   Defies what's cross to piety or good fame;   20
And like a hallowed temple free from taint
   Of ethnicism, makes his muse a saint.

33

From French Court Airs, *1629*
To My Worthy Friend, Master Edward Filmer,
On His Work Published

   What charming peals are these,
That, while they bind the senses, do so please?
   They are the marriage-rites
Of two, the choicest pair of man's delights,

32. Sir John Beaumont (1583–1627), brother of the dramatist Francis Beaumont.
Jonson is perhaps punning on his name: *beau mont*. 1 *a genius*: 'A book, to live, must
have a genius', Martial, *Epig*. VI. lxi. 10. 10 *engine*: wit; with a play on military
engine. 22 *ethnicism*: paganism.

Music and poesy: 5
French air and English verse here wedded lie.
Who did this knot compose
Again hath brought the lily to the rose;
    And, with their chained dance,
Re-celebrates the joyful match with France.   10
    They are a school to win
The fair French daughter to learn English in;
    And, graced with her song,
To make the language sweet upon her tongue.

34

An Expostulation with Inigo Jones

Master Surveyor, you that first began
From thirty pound in pipkins, to the man
You are: from them leaped forth an architect
Able to talk of Euclid, and correct

---

33. (opposite) Filmer (d. 1669) translated the songs in this volume from the French of
Pierre Guedron and Antoine Boesset.   8 ff. The marriage of French music and
English verse is likened to the marriage of the French Henrietta Maria and the English
Charles; for the lily and the rose, see *Und.* 65. 3 and n.   14 *sweet upon her tongue*: cf.
Chaucer's Friar, Prologue to *The Canterbury Tales*, 264–5: 'Somewhat he lipsed, for his
wantownesse, / To make his Englissh sweete upon his tonge'; echoed again in *N.I.*
I. iii. 68–9.

34. The quarrel between Jonson and his collaborator in his court masques, Inigo
Jones (1573–1652), flared up again in 1631 when Jones took objection to his name
appearing second to Jonson's on the title-page of *Love's Triumph Through Callipolis*
('The Inventors, Ben Jonson, Inigo Jones'). Jonson responded by omitting Jones's
name altogether from the title-page of the Quarto of *Chloridia*, and by writing this
poem and *U.V.* 35 & 36. As presented in the *Expostulation*, the quarrel was a fun-
damental one concerning the relative places of visual and verbal elements in the court
masque: its intellectual background is analysed by D. J. Gordon, *JWCI*, xii (1949),
152–78. Yet Jones's and Jonson's theories about the masque were in fact surprisingly
similar, and their partnership more deeply harmonious than Jonson's satire allows: see
Stephen Orgel's Introduction to *The Complete Masques of Ben Jonson* (New Haven,
1969), and Stephen Orgel and Roy Strong, *Inigo Jones: the Theatre of the Stuart Court*
(London, Berkeley, and Los Angeles, 1973) *passim*. The *Expostulation* must have been
written in or after July 1631, not immediately after Shrovetide, as H & S suggest: see
note to l. 104 below.   1 *Master Surveyor*: Jones had been appointed Surveyor of the
King's Works in 1615.   2 *thirty pound in pipkins*: pipkins are pots and pans; H & S,
comparing *S.W.* II. v. 118–19, take this as a reference to 'commodity', a species of
usury: see *Epig.* 12. 13 and n. Little is known about Jones's early life: his father was a
cloth-worker of humble means. Jonson's taunts about social origins are reminiscent of
those in the comedies: cf. the quarrels of Face, Subtle, and the Otters.

Both him and Archimede; damn Archytas,   5
The noblest engineer that ever was;
Control Ctesibius, overbearing us
With mistook names out of Vitruvius;
Drawn Aristotle on us: and thence shown
How much architectonike is your own!   10
(Whether the building of the stage or scene,
Or making of the properties it mean,
Vizors or antics, or it comprehend
Something your sirship doth not yet intend!)
By all your titles and whole style at once   15
Of Tire-man, Mountebank, and Justice Jones
I do salute you! Are you fitted yet?
Will any of these express your place or wit?
Or are you so ambitious 'bove your peers,
You would be an asinigo, by your ears?   20
Why, much good do't you! Be what beast you will,
You'll be, as Langley said, an Inigo still.

---

5 *Archytas*: Pythagorean philosopher and mathematician, *c.* 430–390 B.C.   7 *Ctesibius*: Alexandrian engineer of the third century B.C.   8 *Vitruvius*: Jones probably could not read Vitruvius's *De architectura* in Latin; his copy of the Italian translation of 1567 survives. Jonson had read (and annotated) at least the first chapter of the Latin text.   10 *architectonike*: the term used by Aristotle (ἀρχιτεκτονικὴ τέχνη, *Nicomachean Ethics*, I. i) to denote the ultimate end to which all knowledge is directed and subordinated, viz., virtuous action. To illustrate his notion of a hierarchy of kinds of knowledge, Aristotle takes over an analogy used by Plato (*Politics*, 259E): 'The architect conceives the design, the labourers carry out the details'. Jonson implies that it is now not merely in an analogical sense that this proposition holds good: Jones has literally set up architecture as the highest kind of knowledge; yet his understanding of the science remains absurdly superficial. Cf. also Sidney's *Apology for Poetry*, ed. G. Shepherd (London, 1964), p. 104.   14 *your sirship*: i.e. your reverence; perhaps a coinage (*OED*'s first usage 1873).   16 *Tire-man*: man in charge of costumes.   16 *Mountebank*: cf. *Epig.* 129. 13 n.   16 *Justice*: Jones was a J.P. for Westminster.   18 *wit*: cf. the gibes at Jones's 'wit' in *Epig.* 115. 28, 129. 14.   20 *asinigo*: Spanish *asnico*, little ass; glancing at Inigo's name and his social ambitions (see *U.V.* 35).   22 *Langley*: Francis Langley, owner of the Paris Garden and builder of the Swan theatre.   22 *Inigo*: punning on Italian *iniquo*, 'impious, wicked, unrighteous' (Florio, *Queen Anna's New World of Words*, 1616); cf. *Conv. Dr.* 467–9: 'He said to Prince Charles of Inigo Jones, that when he wanted words to express the greatest villain in the world, he would call him an Inigo'.

What makes your wretchedness to bray so loud
In town and court; are you grown rich and proud?
Your trappings will not change you: change your mind.   25
No velvet sheath you wear will alter kind;
A wooden dagger is a dagger of wood,
Though gold or ivory hafts would make it good.
What is the cause you pomp it so? (I ask)
And all men echo, You have made a masque.   30
I chime that too; And I have met with those
That do cry up the machine, and the shows,
The majesty of Juno in the clouds,
And peering-forth of Iris in the shrouds!
The ascent of Lady Fame, which none could spy,   35
Not those that sided her, Dame Poetry,
Dame History, Dame Architecture, too,
And Goody Sculpture, brought with much ado
To hold her up. O shows! Shows! Mighty shows!
The eloquence of masques! What need of prose,   40
Or verse, or sense, to express immortal you?
You are the spectacles of state! 'Tis true
Court hieroglyphics, and all arts afford
In the mere perspective of an inch-board.
You ask no more than certain politic eyes,   45
Eyes that can pierce into the mysteries

---

26 *velvet sheath*: the fashionable velvet scabbard.   27 *wooden dagger*: traditionally
carried by the Vice in the interludes: cf. *Epig.* 115. 5, 115. 27.   33–4 Juno and Iris
appear in *Chlor.* 248 ff.   34 *shrouds*: fly-ropes.   35–9 Cf. *Chlor.* 306–15, where Fame
ascends to the heavens: '*Poesy*: We that sustain thee, learned Poesy, / *History*: And I,
her sister, severe History, / *Architecture*: With Architecture, who will raise thee high, /
Sculpture: And Sculpture, that can keep thee from to die, / *Chorus*: All help to lift
thee to eternity.'   36 *sided*: flanked.   41 *sense*: cf. *Disc.* 1886–9: 'The sense is as the
life and soul of language, without which all words are dead. Sense is wrought out of
experience, the knowledge of human life and actions, or of the liberal arts . . .'.   42
spectacles of state: (i) spectacular entertainments of the state; (ii) (ironically) models or
patterns of behaviour for the state (*OED*, II. 5. b).   43 *hieroglyphics*: the word was
used in relation to symbolic figures in the court masque: see *Blackness*, 269, and
Daniel's Address to Lucy, Countess of Bedford (l. 33), prefixed to *The Vision of
Twelve Goddesses*. But Jonson also implies that the shallowness of the masque illusion
is a hieroglyph or symbol of the shallowness of the court itself.   46 *mysteries*: secrets;
playing on the technical, theological, and political senses of the word (for the latter, see
OED 'mystery', *sb.*[1] 5. c, and cf. the French *mystère d'état*). The colours of masque
costumes frequently had symbolic significance: in ll. 54–5 Jonson suggests that Jones
had carried such symbolism to the point of obscurity.

Of many colours, read them, and reveal
Mythology there painted on slit deal.
Oh, to make boards to speak! There is a task!
Painting and carpentry are the soul of masque.   50
Pack with your peddling poetry to the stage:
This is the money-get, mechanic age!
To plant the music where no ear can reach,
Attire the persons as no thought can teach
Sense what they are: which, by a specious, fine   55
Term of the architects, is called *design*!
But in the practised truth destruction is
Of any art beside what he calls his.
Whither, O whither will this tire-man grow?
His name is ἐκενοποιός we all know,   60
The maker of the properties, in sum,
The scene, the engine! But he now is come
To be the music-master, fabler, too;
He is, or would be, main Dominus Do-
All in the work! And so shall still, for Ben: 65
Be Inigo the whistle, and his men.

50 *soul*: a bitter reversal of the distinction made in *Hym.* 1–10, where the spectacular side of the masque is compared to the human body, which must perish, while the words of the masque are compared to the soul, which will endure. Cf. *Blackness*, 8.   56 *design*: D. J. Gordon (pp. 169–70) gives instances of contemporary architectural usages of this term. Jonson also hints at the theological sense of the word, as in l. 96; hence the contrast with 'destruction', l. 57.   60 ἐκενοποιός: maker of masks and other stage properties. Perhaps recalling Aristotle, *Poetics*, vi. § 28.   63 *music-master*: according to Cicero, *De Oratore*, iii. 174, the same artist was originally responsible for both music and poetry. There is no record of Jones's attempting to write music for a masque.   63 *fabler*: on the importance of the fable as 'the form and soul of any poetical work or poem', see *Disc.* 2354–5. Jonson had in fact praised a scene which Jones had written for *Queens*, 680–709. In 1632 Jones wrote the fable for *Tempe Restored*.   64–5 *Dominus Do- | All*: Jones had to do almost everything in the way of preparing the masques for performance, including direction. In-and-In Medlay—a character modelled on Jones—also wants to 'do all' in *T. of T.*, V. ii. 35–40, V. vii. 13–14.   66 *whistle*: the one who gives the orders; but the word is contemptuous ('mouth-piece').

He's warm on his feet now, he says, and can
Swim without cork: why, thank the good Queen Anne.
I am too fat to envy him; he too lean
To be worth envy. Henceforth I do mean   70
To pity him, as smiling at his feat
Of lantern-lurry: with fuliginous heat
Whirling his whimsies, by a subtlety
Sucked from the veins of shop-philosophy.
What would he do now, giving his mind that way,   75
In presentation of some puppet-play
Should but the king his justice-hood employ
In setting-forth of such a solemn toy?
How would he firk, like Adam Overdo,
Up and about, dive into cellars, too,   80
Disguised, and thence drag forth enormity:
Discover vice, commit absurdity

---

67 *warm on his feet*: doing nicely: see *OED*, 'warm', 8, and *Volp.* II. ii. 40–1 for the 'Lombard proverb', 'cold on my feet' (= forced to sell one's goods cheaply because of poverty).   68 *Swim without cork*: get on without the assistance of others; cf. 'swim without bladders', G. L. Apperson, *English Proverbs and Proverbial Phrases* (London, etc. 1929).   68 *Queen Anne*: referring not merely to her central role in many of the court masques, but to her architectural commissions for Jones: e.g. the Queen's House at Greenwich, commissioned in 1617 (two years before her death) but not finally completed until 1635. Her brother, the King of Denmark, was possibly Jones's earliest patron.   69 *fat . . . lean*: the figure of Envy is traditionally represented as lean: cf. Maciente, 'a lean mongrel', in *E.M.O.* I. ii. 212, and Marlowe's Envy in *Dr. Faustus* (B), II. ii. 699.   72 *lantern-lurry*: an effect of moving lights ('lurry' = a confused mass), e.g. Jones's region of fire in *Hym.* 223–4, which 'with a continual motion was seen to whirl circularly', and for which Jonson once professed admiration (*Hym.* 668–9). W. A. Armstrong notices that Jonson's characteristic attack is 'against scenes, lights and machines which *moved* before the spectators' eyes; i.e. against those most likely to distract attention from the spoken word', *Jacobean Theatre*, Stratford-Upon-Avon Studies, 1, ed. J. R. Brown and B. Harris (London, 1960), p. 51.   72 *fuliginous*: sooty; but the word was also used of thick noxious vapours that trouble the head: hence it refers to Jones himself, as well as his invention.   73 *whimsies*: (i) whirligig devices (*OED*, II † 5): (ii) capricious notions.   76 *puppet-play*: cf. *Epig.* 97. 1–2 n., 129. 16; and the indirect satire on Jones in the puppet-plays of *B.F.* and *T. of T.*   78 *toy*: (i) amusement, entertainment (*OED*, 2); (ii) foolish or idle fancy, whim, caprice (*OED*, † 4). Cf. Bacon, 'Of Masques and Triumphs': 'These things are but toys, to come amongst such serious observations'.   79 *firk*: move briskly.   79 ff. In 1630 Jones and other J.P.s for Westminster carried out inspections of houses suspected of being infected by the plague: see J. A. Gotch, *Inigo Jones* (London, 1928), pp. 147–8. Jones likens this to the officiousness of Justice Overdo in *B.F.*: see esp. II. i, for the cellar-diving and quest for enormities (alluding in turn to the activities of a former Lord Mayor of London, Sir Thomas Hayes).

Under the moral? Show he had a pate
Moulded, or stroked up, to survey a state?
O wise surveyor! Wiser architect!   85
But wisest Inigo! Who can reflect
On the new priming of thy old sign-posts,
Reviving with fresh colours the pale ghosts
Of thy dead standards; or (with miracle) see
Thy twice-conceived, thrice-paid-for imagery   90
And not fall down before it, and confess
Almighty architecture: who no less
A goddess is, than painted cloth, deal-boards,
Vermillion, lake, or cinnabar affords
Expression for; with that unbounded line   95
Aimed at in thy omnipotent design.
What poesy e'er was painted on a wall
That might compare with thee? What story shall,
Of all the Worthies, hope to outlast thy one,
So the materials be of Purbeck stone?   100
Live long the Feasting Room! And ere thou burn
Again, thy architect to ashes turn!
Whom not ten fires nor a parliament can,
With all remonstrance, make an honest man.

---

83 *Under the moral*: i.e. as one commissioned by, or representing, the king; see *OED*, 'moral', † 3, and examples, and cf. *B.F.* II. i. 1.   84 *stroked up*: playing on a secondary sense, 'flattered' (*OED*, 1 † e).   84 *survey*: referring primarily to Jones's position as Surveyor of the King's Works, but implying also that he is setting himself up as a surveyor of state morality.   89 *dead standards*: literally, faded posts, properties in the masques; metaphorically, dead moral standards.   90 Jones had re-used material from *Nept. Tr.* in *Fort. Is.* in 1625, but so too had Jonson himself. 'Imagery' refers primarily to masquing properties but has religious overtones ('Thou shalt not make unto thyself any graven image', Exod. 20 : 4) which are developed in the lines which follow.   94 *cinnabar*: crimson.   97 *on a wall*: i.e. graffiti; cf. *Alch*. V. v. 41–2.   101 *the Feasting Room*: the Banqueting House at Whitehall, burnt down on 12 Jan. 1619, was rebuilt on grander lines by Jones (completed 31 March 1622). The bottom of the vault of the new building was paved with Purbeck stone, from Dorset.   104 *remonstrance*: in June 1631, Jones—acting on the authority of the Commissioners for Pious Uses—stopped the parishioners of St. Gregory's church from digging a vault which he judged to be a threat to the foundations of St. Paul's. The parishioners petitioned the Privy Council to overthrow this order; the Privy Council turned down the petition in July, but the affair was to drag on for many years. See Gotch, *Inigo Jones*, ch. xi.

35

To Inigo, Marquis Would-Be
A Corollary

But 'cause thou hear'st the mighty king of Spain
Hath made his Inigo marquis, wouldst thou fain
Our Charles should make thee such? 'Twill not become
All kings to do the self-same deeds with some;
Besides, his man may merit it, and be   5
A noble honest soul: what's this to thee?
He may have skill and judgement to design
Cities and temples: thou, a cave for wine
Or ale! He build a palace: thou a shop
With sliding windows, and false lights a-top;   10
He draw a forum with quadrivial streets:
Thou paint a lane, where Thumb the pygmy meets!
He, some Colossus to bestride the seas
From the famed pillars of old Hercules:
Thy canvas giant at some channel aims,   15
Or Dowgate torrent, falling into Thames,
And, straddling, shows the boys' brown-paper fleet,
Yearly set out there, to sail down the street.

---

35. For background, see *U.V.* 34, n.   1–2 Philip IV of Spain had rewarded his
architect Giovanni Baptista Crescenzio for his work on the Escurial by creating him
Marquis della Torre.   8 *a cave for wine*: see *Und.* 48 and n.   9–10 *a shop . . . a-top*: H
& S suspect the allusion is to Jones's plans for the new Whitehall; W. A. Armstrong
(*Jacobean Theatre*, ed. Brown and Harris, p. 51) suggests it may rather be to Jones's
machina ductilis in *Oberon* (esp. 139–40, 420–2). Both identifications have their difficul-
ties. But an extant drawing by Jones in the library of Worcester College, Oxford (i.
53 B, HT 80) shows a four-storey arcade of shops which appears closely to match
Jonson's description (drawn to my attention by John Harris and Stephen Orgel). Mr.
Harris dates the drawing pre-1630.   11 *quadrivial*: leading in four directions.   12
Thumb: Tom Thumb, who appeared in an antimasque in *Fort. Is.* 423.   12 *pygmy*:
Jeffrey Hudson, Henrietta Maria's dwarf; he appeared, with other dwarves, in a
masque concerned with Gargantua (who is perhaps the 'canvas giant' of l. 15) performed at
Somerset House on 24 Nov. 1626 (Orgel and Strong, *Inigo Jones*, i. 389.–92); and—
again, with other dwarves—in *Chlor.* in 1631.   15 *channel*: gutter, artificial water-
course.   16 *Dowgate torrent*: which flowed after heavy rain down the steep Dowgate
Hill into the Thames west of London Bridge. Jonson is satirizing Jones's seascapes (in
Blackness, *Fort. Is.*, and *Love's Tr.*, etc.); for Jones's wave-machines, see A. Nicoll,
Stuart Masques and the Renaissance Stage (London, etc., 1937), pp. 59–60.   17 *fleet*:
such as that in *Fort. Is.* 619 ff. (see Nicoll, *Stuart Masques*, p. 79), and perhaps in
Love's Tr. (see Orgel and Strong, *Inigo Jones*, i. 407–8).

Your works thus differing, troth, let so your style:
Content thee to be Pancridge earl the while;   20
An earl of show, for all thy work is show.
But when thou turn'st a real Inigo,
Or canst of truth the least entrenchment pitch,
We'll have thee styled the Marquis of New Ditch.

36

To a Friend
An Epigram of Him

Sir Inigo doth fear it, as I hear,
And labours to seem worthy of that fear,
That I should write upon him some sharp verse
Able to eat into his bones, and pierce
The marrow. Wretch, I quit thee of thy pain: 5
Thou'rt too ambitious, and dost fear in vain!
The Libyan lion hunts no butterflies,
He makes the camel and dull ass his prize.
If thou be so desirous to be read
Seek out some hungry painter, that for bread   10
With rotten chalk or coal upon a wall
Will well design thee, to be viewed of all
That sit upon the common draught or strand:
Thy forehead is too narrow for my brand.

20 *Pancridge earl*: one of the mock-titles used amongst the Finsbury archers, who had an annual procession. 23 *pitch*: plan, lay out. 24 *New Ditch*: not an (unknown) city ditch, as H & S and others have suspected, but New River, Sir Hugh Middelton's 40-mile-long artificial river which supplied water to Londoners; it was opened in 1613 (see *Epig.* 133. 193–4 n.). Jonson is referring to Jones's similar interest in artificial waterways. (For the name, cf. the variants Fleet Ditch/Fleet River.)

36. From Martial, *Epig.* XII. lxi*. For the background, see *U.V.* 34, n. 13 *draught*: privy. 13 *strand*: sewer, gutter. 14 *brand*: perhaps remembered by Dryden, *Mac Flecknoe*, 177–8.

37

To My Detractor

My verses were commended, thou dar'st say,
   And they were very good: yet thou think'st nay.
For thou objectest (as thou hast been told)
   The envied return of forty pound in gold.
Fool, do not rate my rhymes: I've found thy vice   5
   Is to make cheap the lord, the lines, the price.
But bawl thou on; I pity thee, poor cur,
   That thou hast lost thy noise, thy foam, thy stir,
To be known what thou art, a blatant beast,
   By barking against me. Thou look'st at least   10
I now would write on thee? No, wretch, thy name
   Shall not work out unto it such a fame.
Thou art not worth it. Who will care to know
   If such a tyke as thou e'er wert or no,
A mongrel cur? Thou shouldst stink forth and die   15
   Nameless and noisome as thy infamy.
No man will tarry by thee as he goes
   To ask thy name, if he have half his nose,
But fly thee, like the pest! Walk not the street
   Out in the dog-days, lest the killer meet   20
Thy noddle, with his club; and, dashing forth
   Thy dirty brains, men smell thy want of worth.

---

37. *Detractor*: I.E. (probably John Eliot), who had written these lines on *Und.* 77 (addressed to Lord Weston): 'Your verses were commended, as 'tis true / That they were very good—I mean to you: / For they returned you, Ben, I have been told, / The seld seen sum of forty pound in gold. / These verses then being rightly understood, / His Lordship, not Ben Jonson, made them good.' Jonson's poem dates from 1631 or 1632.   9 *blatant beast*: from Spenser, *The Faerie Queene*, V. xii. 37, 41, etc.   20 *dog days*: in July or August, when dogs were thought liable to run mad.

38

From The Northern Lass, *1632*
To My Old Faithful Servant,
And, By His Continued Virtue, My Loving Friend,
The Author of This Work, Mr. Richard Brome

I had you for a servant once, Dick Brome,
    And you performed a servant's faithful parts;
Now you are got into a nearer room
    Of fellowship, professing my old arts.
And you do do them well, with good applause,   5
    Which you have justly gained from the stage
By observation of those comic laws
    Which I, your master, first did teach the age.
You learned it well, and for it, served your time
    A 'prenticeship, which few do nowadays.   10
Now each court hobby-horse will wince in rhyme;
    Both learned and unlearned, all write plays.
It was not so of old: men took up trades
    That knew the crafts they had been bred in, right:
An honest Bilbo smith would make good blades,   15
    And the physician teach men spew or shite;
      The cobbler kept him to his awl; but now
      He'll be a pilot, scarce can guide a plough.

39

An Answer to Alexander Gill

Shall the prosperity of a pardon still
Secure thy railing rhymes, infamous Gill,
At libelling? Shall no Star Chamber peers,
Pillory, nor whip, nor want of ears

38. 1 *servant*: cf. *B.F.* Ind. 8. R. J. Kaufmann, *Richard Brome, Caroline Dramatist* (New York & London, 1961), pp. 20 ff., thinks the term implies that 'Brome's initial status was a menial one'. 10 *'prenticeship*: Fleay (i. 37) assumes that Jonson refers to a formal seven-year apprenticeship, 1623–9: this is a guess. 12–16 Horace, *Epist.* II. i. 114–17*. 15 *Bilbo smith*: in Bilboa, Spain, where fine blades were made. 18 Persius, *Sat.* v. 102–4.

39. Milton's friend Alexander Gill the younger was son of the headmaster of St. Paul's School, whom Jonson had attacked in 1623 in *Time Vind.* (171–88). In 1628 the Star Chamber found the son guilty of speaking disrespectfully of James, Charles, and the

(All which thou hast incurred deservedly) 5
Nor degradation from the ministry
To be the Denis of thy father's school,
Keep in thy barking wit, thou bawling fool?
Thinking to stir me, thou hast lost thy end:
I'll laugh at thee, poor wretched tyke; go send 10
Thy blatant muse abroad, and teach it rather
A tune to drown the ballads of thy father;
For thou hast nought in thee to cure his fame,
But tune and noise, the echo of thy shame.
A rogue by statute, censured to be whipped, 15
Cropped, branded, slit, neck-stocked: go, you are stripped!

40

From Meditation of Man's Mortality, *1634*
To Mrs. Alice Sutcliffe,
on Her Divine Meditations

When I had read
Your holy *Meditations*,
And in them viewed
The uncertainty of life,
The motives and true spurs 5
To all good nations;
The peace of conscience
And the godly's strife,

recently murdered Duke of Buckingham: he was degraded from the ministry and from
his degree in divinity, fined £2,000, and condemned to lose both his ears: after
interventions on his behalf, the penalties were remitted. Jonson himself had at one
time been suspected of writing libellous verses about Buckingham's murder; D. L.
Clark suggests that he 'had to detest Gil[l] very hard indeed to avoid being blamed
along with him': *John Milton at St Paul's School* (New York, 1948), p. 96. Jonson's
poem is also a counter-attack upon Gill's satirical lines on *M.L.* in 1631 (printed by H
& S, xi. 346–8). 7 *Denis*: tyrant: after Dionysius the younger, tyrant of Syracuse, who
kept a school after being deposed (H & S). Gill had been formally dismissed from his
post as usher at St. Paul's School in 1628, but evidently lingered on. 11 *blatant*: noisy,
clamorous; cf. *U.V.* 37. 9 and n. 12 *ballads of thy father*: satirical songs about him:
see Clark, op. cit., pp. 79–83, 91–3.

40. Alice Sutcliffe (*née* Woodhouse) was the wife of John Sutcliffe of Yorkshire,
Esquire of King James and later Groom of the Privy Chamber of King Charles.

The danger of delaying
      To repent,   10
And the deceit of pleasures
      By consent;
The comfort of weak Christians,
      With their warning
From fearful back-slides;   15
      And the debt we're in
To follow goodness
      By our own discerning
Our great reward,
      The eternal crown to win: 20
I said, Who'd supped so deep
      Of this sweet chalice
Must Celia be:
      The anagram of Alice.

41

From The Female Glory, *1635*
The Garland of the Blessed Virgin Marie

Here are five letters in this blessed name,
    Which, changed, a five-fold mystery design:
The M, the myrtle, A, the almonds claim,
    R, rose, I, ivy, E, sweet eglantine.

Jonson may not have known her personally. The first edition of her work appeared probably in 1633: the dedicatory verses of Jonson and others were prefixed to the second edition of 1634: see Ruth Hughey, *RES*, x (1934), 156–64. Jonson's poem alludes to the headings in the book's table of contents.

41. The author of *The Female Glory* was Anthony Stafford (1587–?1645), who wrote a number of books on religious and philosophical subjects. Jonson's authorship of these verses (signed B.I.) is not certain. It is usually assumed that Jonson wrote them during his Catholic period, i.e. between 1598 and 1610; but the evidence is not particularly convincing. (Stafford himself was not a Catholic, though he was attacked for his apparent Catholic leanings.) The 'garland' is the poem itself (cf. *Und.* 70. 77), which in turn may be, as Newdigate suggests, 'the interpretation of some emblematic picture of the Holy Child and His Mother, crowned with a garland'. Paul Cubeta suggests the emblem may be a pendant on the Bridgettine rosary: *JEGP*, lxii (1963), 98–101. 3–4 Cubeta (p. 99) takes the myrtle as the symbol of joyous love, the almond, the rose (and the lily) as symbols of hope, the ivy as the symbol of friendship in adversity, and the

These form thy garland. Whereof myrtle green,   5
    The gladdest ground to all the numbered five,
Is so implexed and laid in between
    As love here studied to keep grace alive.

The second string is the sweet almond bloom
    Ymounted high upon Selinus' crest;   10
As it alone, and only it, had room
    To knit thy crown, and glorify the rest.

The third is from the garden culled, the rose,
    The eye of flowers, worthy for his scent
To top the fairest lily now that grows   15
    With wonder on the thorny regiment.

The fourth is humble ivy, intersert,
    But lowly laid, as on the earth asleep,
Preserved in her antique bed of vert:
    No faith's more firm or flat than where 't doth creep.   20

But that which sums all is the eglantine,
    Which of the field is clept the sweetest briar,
Inflamed with ardor to that mystic shine
    In Moses' bush, unwasted in the fire.

Thus love and hope and burning charity   25
    (Divinest graces) are so intermixed
With odorous sweets and soft humility,
    As if they adored the head whereon they're fixed.

The Reverse: on the Back Side

These mysteries do point to three more great
    On the reverse of this your circling crown,   30
        All pouring their full shower of graces down,
The glorious Trinity in Union met.

---

eglantine of poetry.   7 *implexed*: entwined.   9–10 Cf. Spenser, *The Faerie Queene*, I.
vii. 32: 'Like to an almond tree ymounted high / On top of green Selinus all alone';
echoed by Marlowe, *Tamburlaine the Great*, Pt. II, IV. iii. 119–21. Selinus is in
Sicily.   14 *eye*: choicest; cf. *Und.* 65. 8.   17 *intersert*: inserted, interpolated.   20 *flat*:
plain, absolute.   24 *Moses' bush*: Exod. 3 : 2 ff.

Daughter and mother and the spouse of God,
   Alike of kin to that most blessed Trine
   Of persons, yet in Union one, divine;   35
How are thy gifts and graces blazed abroad!

Most holy and pure virgin, blessed maid,
   Sweet tree of life, King David's strength and tower,
   The house of gold, the gate of heaven's power,
The morning-star, whose light our fall hath stayed.   40

Great queen of queens, most mild, most meek, most wise,
   Most venerable, cause of all our joy;
   Whose cheerful look our sadness doth destroy,
And art the spotless mirror to man's eyes.

The seat of sapience, the most lovely mother,   45
   And most to be admired of thy sex,
   Who mad'st us happy all, in thy reflex,
By bringing forth God's only Son, no other.

Thou throne of glory, beauteous as the moon,
   The rosy morning, or the rising sun,   50
   Who, like a giant, hastes his course to run,
Till he has reached his two-fold point of noon.

How are thy gifts and graces blazed abroad,
   Through all the lines of this circumference
   To imprint in all purged hearts this virgin sense,   55
Of being daughter, mother, spouse of God!

42

From The Shepherds' Holiday, *1635*
To My Dear Son and Right-Learned Friend,
Master Joseph Rutter

   You look, my Joseph, I should something say
   Unto the world in praise of your first play;
   And truly, so I would, could I be heard.
   You know I never was of truth afeared

38–40 These attributes of the Virgin, as Newdigate points out, are to be found in the
Litany of Loretto, trs. into English in 1620 as *The Paradise of Delights.*   47 *in thy
reflex*: in thy reflection; i.e. vicariously.   52 *two-fold point of noon*: when the two hands
of the clock point upright.

And less ashamed; not when I told the crowd   5
How well I loved truth: I was scarce allowed
By those deep-grounded, understanding men
That sit to censure plays, yet know not when
Or why to like; they found it all was new,
And newer than could please them, because true.   10
Such men I met withal, and so have you.
Now for mine own part, and it is but due
(You have deserved it from me), I have read
And weighed your play: untwisted every thread,
And know the woof and warp thereof; can tell   15
Where it runs round and even; where so well,
So soft and smooth it handles, the whole piece,
As it were spun by nature off the fleece:
This is my censure. Now there is a new
Office of wit, a mint (and this is true)   20
Cried up of late; whereto there must be first
A master-worker called, the old standard burst
Of wit, and a new made: a warden, then,
And a comptroller, two most rigid men
For order, and for governing the pyx;   25
A say-master, hath studied all the tricks
Of fineness and alloy: follow his hint,
You've all the mysteries of wit's new mint,
      The valuations, mixtures, and the same
      Concluded from a carat to a dram.   30

---

42. (opposite) Rutter was one of the 'sons of Ben'; he later wrote an elegy on Jonson's death (H & S, xi. 459–60). Jonson's verdict on Rutter's play is examined by Freda L. Townsend, *MP*, xliv (1947), 238–47.   6 *How well I loved truth*: perhaps referring to *B.F.* Ind. 127–32; cf. *Sej.* 'To the Readers', 18–19.   7 *deep-grounded, understanding*: cf. the word-play of *B.F.* Ind. 49–50 ('the understanding gentlemen o' the ground') and 76 ('the grounded judgements and understandings').   14–15 For the metaphor, see *U.V.* 26. 49 n.   17 *the whole piece*: cf. *Epig.* 56. 14.   19 *censure*: judgement.   19 ff. Cf. the 'rhyming club' of *U.V.* 30. 8 and the 'play club' of *Songs*, 14. 26; its members cannot now be identified.   26 *say-master*: assay-master.   27 *fineness*: comparative freedom from alloy.   28 *wit's new mint*: cf. *Love's Labour's Lost*, I. i. 164.

43

From Annalia Dubrensia, *1635*
An Epigram to My Jovial Good Friend
Mr. Robert Dover,
on His Great Instauration of His Hunting and Dancing
at Cotswold

I cannot bring my muse to drop her vies
'Twixt Cotswold and the Olympic exercise;
But I can tell thee, Dover, how thy games
Renew the glories of our blessed James:
How they do keep alive his memory   5
With the glad country, and posterity;
How they advance true love, and neighbourhood,
And do both church and commonwealth the good,
In spite of hypocrites, who are the worst
Of subjects: let such envy, till they burst.   10

44

A Song of Welcome to King Charles

Fresh as the day, and new as are the hours,
Our first of fruits, that is the prime of flowers
Bred by your breath on this low bank of ours,

---

43. Captain Robert Dover (?1575–1641) was the reviver (c. 1612) of the Cotswold Olympic Games, which took place annually at Whitsun on Dover's Hill, near Chipping Campden and the Vale of Evesham in Gloucestershire. The Games were discontinued during the Commonwealth but resumed at the Restoration, and were an annual event until the middle of the nineteenth century. Over thirty writers contributed to *Annalia Dubrensia*, including Drayton, Randolph, Davenant, and T. Heywood: some of the verses had been written considerably earlier. See Christopher Whitfield's edition, *Robert Dover and the Cotswold Games* (London & New York, 1962). 1 *drop her vies*: i.e. refrain from comparing; a 'vie' is a challenge or sum ventured in card-playing. 9 *hypocrites*: Puritans. Dover's Games, performed with royal licence, 'were a protest against the rising puritanical prejudices' (*DNB*; cf. Whitfield, op. cit., pp. 15 ff.). James had defended holiday games against Puritan attack in his *Declaration of Sports* in 1617/18.

44. An untitled fragment from a lost entertainment: H & S suggest that it belongs to the early years of Charles's reign. 1 Cf. *For.* 2. 40.

Now in a garland by the graces knit,
Upon this obelisk advanced for it,   5
We offer as a circle the most fit
To crown the years which you begin, great king,
And you with them, as father of our spring.

45

A Song of the Moon

To the wonders of the Peak
I am come to add and speak,
Or, as some would say, to break
    My mind unto you:
And I swear by all the light   5
At my back, I am no sprite,
But a very merry wight
        Pressed in to see you.

I had somewhat else to say,
But have lost it by the way,   10
I shall think on't ere 't be day.
    The moon commends her
To the merry beards in hall,
Those turned up, and those that fall,
Morts and merkins that wag all,   15
        Tough, foul, or tender.

And as either news or mirth
Rise or fall upon the earth,
She desires of every birth
        Some taste to send her.   20

---

7 *begin*: cf. *Und*. 64. 5 n.

45. From a lost entertainment, which apparently took place near the Peak in Derbyshire. The title is Gifford's.   1 *wonders of the Peak*: enumerated in *Ent. Welb.* 92 ff.   3–4 *break* / *My mind*: then a well-established phrase (see *OED* 22); 'as some would say' probably hints not at its newness but its possible ambiguity (see *OED* 12. † b).   13–15 Cf. the saying ' 'Tis merry in hall when beards wag all', *T. of T.* V. ix. 12, *Christmas*, 10–11, etc.   15 *Morts*: girls of easy virtue; gypsy girls.   15 *merkins*: artificial pubic hair; female private parts.

Specially the news of Derby:
For if there or peace or war be,
To the Peak it is so hard by,
    She soon will hear it.

If there be a cuckold major   25
That the wife heads for a wager
As the standard shall engage her,
    The moon will bear it.
Though she change as oft as she,
And of circle be as free,   30
Or her quarters lighter be,
    Yet do not fear it.

Or if any strife betide
For the breeches with the bride,
'Tis but the next neighbour ride   35
    And she is pleased.
Or if 't be the gossips' hap
Each to pawn her husband's cap
At Pem Waker's good ale-tap,
    Her mind is eased.   40

Or by chance if in their grease,
Or their ale, they break the peace,
Forfeiting their drinking lease:
    She will not seize it.

---

26 *heads*: gives horns to, cuckolds.   34 *For the breeches*: i.e. who rules the household:
cf. Tilley, B645.   39 *Pem Waker's*: cf. the alewife Pem in *Ent. Welb.* 122 ff.

46

To Mr. Ben Jonson in His Journey
By Mr. Craven

When wit and learning are so hardly set
That from their needful means they must be barred,
Unless by going hard they maintenance get,
Well may Ben Jonson say the world goes hard.

This Was Mr. Ben Jonson's Answer Of The Sudden

Ill may Ben Jonson slander so his feet,   5
For when the profit with the pain doth meet,
Although the gate were hard, the gain is sweet.

47

A Grace by Ben Jonson
Extempore before King James

Our king and queen the Lord God bless,
The Palsgrave and the Lady Bess,
And God bless every living thing
That lives and breathes and loves the king.
God bless the Council of Estate,   5
And Buckingham the fortunate;
God bless them all, and keep them safe:
And God bless me, and God bless Ralph.

46. Referring to Jonson's journey on foot to Scotland, 1618–19. The author of the original lines may be Sir William Craven (?1548–1618), Lord Mayor of London, 1610–11.   6 *profit*: cf. H & S, i. 76.   7 *gate*: going, journey.

47. H & S date this poem shortly after 1617, when Buckingham obtained his earldom. The second version, perhaps delivered at Lady Bedford's table, is dated in the margin '1613': but Villiers did not come to court until the following year.   2 *Palsgrave ... Lady Bess*: Frederick V, Elector Palatine, married Princess Elizabeth in 1613.   5 *the Council of Estate*: i.e the Council of State.   6 *Buckingham the fortunate*: George Villiers (1592–1628), court favourite: Viscount Villiers, 1616; Earl of Buckingham, 1617; Marquis of Buckingham, 1618; Duke of Buckingham, 1623; murdered, 1628. 8 *Ralph*: 'The king was mighty inquisitive to know who this Ralph was: Ben told him 'twas the drawer at the Swan tavern by Charing Cross, who drew him good Canary. For this drollery his majesty gave him an hundred pounds': John Aubrey, *Brief Lives*, ed. O. L. Dick (London, 1949), p. 179.

Another Version
A Form of a Grace

The king, the queen, the prince, God bless,
The Palsgrave and the Lady Bess;
God bless the Council and the State,
And Buckingham the fortunate;
God bless every living thing   5
That the king loves, and loves the king.
God bless us all, Bedford keep safe:
God bless me, and God bless Ralph.

48

Ode

If men and times were now
    Of that true face
As when they both were great, and both knew how
    That fortune to embrace
By cherishing the spirits that gave their greatness grace: 5
    I then could raise my notes
    Loud to the wondering throng
And better blazon them than all their coats,
That were the happy subject of my song.

But clownish pride hath got   10
    So much the start
Of civil virtue, that he now is not
    Nor can be of desert
That hath not country impudence enough to laugh at art:
    Whilst like a blaze of straw   15
    He dies with an ill scent
To every sense, and scorn to those that saw
How soon with a self-tickling he was spent.

Another Version.   7 *Bedford*: 'Countess of Bedford' (MS. marginal note).   8 *Ralph*: 'The Countess's man who won the race' (MS. marginal note); H & S cf. 'light-foot Ralph' of *S. of N.* II. iii. 11.

48. Established as Jonson's on internal evidence by W. D. Briggs, *The Athenaeum*, 13 June 1914.   2 *face*: condition, appearance.   16–17 He dies with a bad smell that is perceptible to everyone, and to the scorn of those who saw (etc.).

Break then thy quills, blot out
    Thy long-watched verse,   20
And rather to the fire than to the rout
    Thy laboured tunes rehearse,
Whose air will sooner hell, than their dull senses, pierce:
    Thou that dost spend thy days
    To get thee a lean face   25
And come forth worthy ivy or the bays,
And, in this age, canst hope no other grace.

Yet since the bright and wise
    Minerva deigns
Upon so humbled earth to cast her eyes,   30
    We'll rip our richest veins
And once more strike the ear of time with those fresh strains
    As shall, besides delight
    And cunning of their ground,
Give cause to some of wonder, some despite,   35
But unto more, despair to imitate their sound.

Throw, holy virgin, then
    Thy crystal shield
About this isle, and charm the round, as when
    Thou mad'st in open field   40
The rebel giants stoop, and gorgon envy yield;
    Cause reverence, if not fear,
    Throughout their general breasts,
And, by their taking, let it once appear
Who worthy win, who not, to be wise Pallas' guests.   45

19–21 Cf. *Poet.* Apol. Dial. 209–11.   24–7 Ibid. 233–6.   22 *Thy*: this edition; conj.
emendation of 'Their'.   32–6 Cf. *Poet.* Apol. Dial. 228–32.   34 *ground*: a play on
words: earth, melody.   38 *crystal shield*: Minerva's mirror-shield, which she lent to
Perseus to enable him to kill the gorgon.   39 *charm the round*: i.e. those who fall
within the circle of Minerva's power; playing on the musical and dancing senses of the
words (cf. *Macbeth*, IV. i. 129–30: 'I'll charm the air to give a sound, / While you
perform your antic round').

49

An Epistle to a Friend

'Censure not sharply then, but me advise,
Before I write more verse, to be more wise.'

So ended your epistle; mine begins:
He that so censureth or adviseth, sins;
The empty carper scorn, not credit, wins.   5

I have, with strict advantage of free time,
O'er-read, examined, tried, and proved your rhyme
As clear and distant as yourself from crime;

And though your virtue, as becomes it, still
Deigns mine the power to find, yet want I will   10
Or malice to make faults, which now is skill.

Little know they that profess amity,
And seek to scant her comely liberty,
How much they lame her in her property;

And less they know, that being free to use   15
That friendship which no chance, but love, did choose,
Will unto licence that free leave abuse.

It is an act of tyranny, not love,
In course of friendship wholly to reprove,
And flattery, with friends' humours still to move.   20

From each of which I labour to be free;
Yet if with either's vice I tainted be,
Forgive it as my frailty, and not me.

For no man lives so out of passion's sway
But sometimes shall be tempted to obey   25
Her fury, though no friendship he betray.

49. 12–26 Repeated from *Und.* 37. 19–33: see notes there.

50

A Speech out of Lucan

Just and fit actions, Ptolemy (he saith),
Make many hurt themselves; a praised faith
Is her own scourge, when it sustains their states
Whom fortune hath depressed. Come near the fates
And the immortal gods: love only those   5
Whom thou seest happy; wretches flee as foes.
Look how the stars from earth or seas from flames
Are distant: so is profit from just aims.
The main command of sceptres soon doth perish
If it begin religious thoughts to cherish;   10
Whole armies fall, swayed by those nice respects.
It is a licence to do ill protects
Even states most hated, when no laws resist
The sword, but that it acteth what it list.
Yet 'ware: thou mayst do all things cruelly;   15
Not safe, but when thou dost them thoroughly;
He that will honest be may quit the court:
Virtue and sovereignty, they not consort.
That prince that shames a tyrant's name to bear,
Shall never dare do anything but fear.   20

50. A translation of the eunuch Photinus's speech to Ptolemy, urging the murder of Pompey: Lucan's *Pharsalia*, viii. 484–95. Cf. 'Shame, no statist', Herrick, p. 182, and Massinger's *The False One*, I. i. 302–12.   9–16 Cf. *Sej*. II. 180–7.   19–20 Cf. *Sej*. II. 178–9.

SONGS AND POEMS FROM THE PLAYS AND MASQUES

SONGS AND POEMS

I

From Cynthia's Revels, *I. ii. 65–75*
Echo's Song

Slow, slow, fresh fount, keep time with my salt tears;
Yet slower yet, O faintly gentle springs;
List to the heavy part the music bears:
    Woe weeps out her division when she sings.
       Droop, herbs and flowers,   5
       Fall, grief, in showers;
       Our beauties are not ours:
          Oh, could I still
(Like melting snow upon some craggy hill)
       Drop, drop, drop, drop   10
Since nature's pride is now a withered daffodil.

2

From Cynthia's Revels, *II. v (F & Q)*
Beggar's Song

    Come follow me, my wags, and say as I say:
    There's no riches but in rags: hey day, hey day!
    You that profess this art, come away, come away,
    And help to bear a part: hey day, hey day!

On the music of Jonson's songs, see W. McC. Evans, *Ben Jonson and Elizabethan Music* (New York, 1965), H & S, xi. 605–9, E. and L. Pissarro, *Songs by Ben Jonson* (London, 1906).

1. For a musical setting of this song (perhaps that used in the first performance) see Henry Youll's *Canzonets To Three Voices* (1608), no. viii. The song is discussed by W. McC. Evans, pp. 48–51. Jonson's song is remembered by Marianne Moore, *Complete Poems* (New York, 1967), p. 189. 4 *division*: 'a rapid melodic passage, originally conceived as the dividing of each of a succession of long notes into several short ones . . .'. (*OED*).

Bear-wards and blacking-men,   5
Corn-cutters and car-men,
Sellers of marking-stones,
Gatherers-up of marrow-bones,
Pedlars and puppet-players,
Sow-gelders and sooth-sayers,   10
Gypsies and gaolers,
Rat-catchers and railers,
Beadles and ballad-singers,
Fiddlers and fadingers,
Thomalins and tinkers,   15
Scavengers and skinkers:
There goes the hare away!
                    Hey day, hey day!

Bawds and blind doctors,
Paritors and spittle proctors,   20
Chemists and cuttlebungs,
Hookers and horn-thumbs,
With all cast commanders
Turned post-knights or panders;
Jugglers and jesters,   25
Borrowers of testers,
And all the troop of trash
That are allied to the lash:
Come and join with your jags,
Shake up your muscle-bags,   30
For beggary bears the sway;
Then sing: cast care away!
                    Hey day, hey day!

2. 5 *blacking-men*: who sold blacking.
6 *Corn-cutters*: chiropodists; cf. *B.F.* II. iv.   6 *car-men*: carriers, carters.   7 *marking-stones*: for marking cattle.   14 *fadingers*: those who danced the fading, an Irish dance.
15 *Thomalins*: 'beggarly itinerants' (H & S).   16 *skinkers*: tapsters.   17 *There goes the hare away!*: a proverbial saying: Tilley, H157.   20 *Paritors*: apparitors, summoning officers of the ecclesiastical courts.   21 *cuttlebungs*: cut-purses.   22 *Hookers*: thieves, pilferers, who used hooks.   22 *horn-thumbs*: cut-purses (who used horn thimbles).   23 *cast*: cashiered.   24 *post-knights*: men who lived by giving false evidence in the courts.   26 *testers*: sixpences.   29 *jags*: rags, tatters.   30 *muscle-bags*: thighs.

3
From Cynthia's Revels, *IV. iii. 242–53*
Hedon's Song

Oh, that joy so soon should waste!
　　Or so sweet a bliss
　　　　As a kiss
　Might not for ever last!
So sugared, so melting, so soft, so delicious:　　5
　　The dew that lies on roses
　　When the morn herself discloses
　　　Is not so precious.
Oh, rather then I would it smother,
Were I to taste such another;　　10
　　It should be my wishing
　　That I might die, kissing.

4
From Cynthia's Revels, *IV. iii. 305–16*
Amorphus's Song

Thou more than most sweet glove
　Unto my more sweet love,
　Suffer me to store with kisses
　This empty lodging, that now misses
　　The pure rosy hand that ware thee,　　5
　　Whiter than the kid that bare thee.
　　　Thou art soft, but that was softer;
　　　Cupid's self hath kissed it ofter
　　Than e'er he did his mother's doves,
　　Supposing her the queen of loves　　10
　　　That was thy mistress,
　　　　Best of gloves.

3. This song (like the next) is gently parodic: l. 5 glances at Ford and Dekker's *The Sun's Darling*, II. i. 139–40, and the final 'die' of l. 12 (as subsequent dialogue reveals) was elaborately drawn out in the singing.

4. The song is probably an ironical imitation of Dorus's song on Pamela's glove in Sidney's *The Countess of Pembroke's Arcadia*, Bk. iii; see his *Poems*, ed. W. A. Ringler (Oxford, 1962), p. 70.　10–11 Cf. Cupid's confusion in *Und.* 2. v. 10 ff.

5

From Cynthia's Revels, *V. vi. 1–18*
Hymn to Cynthia

Queen and huntress, chaste and fair,
Now the sun is laid to sleep,
Seated in thy silver chair,
State in wonted manner keep:
    Hesperus entreats thy light,   5
    Goddess excellently bright.

Earth, let not thy envious shade
Dare itself to interpose;
Cynthia's shining orb was made
Heaven to clear, when day did close: 10
    Bless us then with wished sight,
    Goddess excellently bright.

Lay thy bow of pearl apart,
And thy crystal-shining quiver;
Give unto the flying hart   15
Space to breathe, how short soever:
    Thou that mak'st a day of night,
    Goddess excellently bright.

6

From Poetaster, *II. ii. 163–72, 179–88*
Crispinus' and Hermogenes' Song

Crispinus   If I freely may discover
           What would please me in my lover,
              I would have her fair and witty,
              Savouring more of court than city:
              A little proud, but full of pity: 5

5. Cynthia (Diana), goddess of the moon, of chastity, and of hunting; the name, and hence the song itself, also refer to Queen Elizabeth.   5 *Hesperus*: the name given to Venus when it appears after sunset.   7 *envious*: cf. *Und.* 65. 9.   17 *day of night*: Earl Miner (*The Cavalier Mode. From Jonson to Cotton* (Princeton, 1971), pp. 197–8) compares Isa. 9 : 2.

6. For Henry Lawes's setting, see W. McC. Evans, op. cit. p. 9. From Martial, *Epig.* I. lvii*.

Light and humorous in her toying,
Oft building hopes, and soon destroying,
Long but sweet in the enjoying,
Neither too easy, nor too hard:
All extremes I would have barred.   10

Hermogenes She should be allowed her passions,
So they were but used as fashions;
Sometimes froward, and then frowning,
Sometimes sickish, and then swowning,
Every fit with change still crowning.   15
Purely jealous I would have her,
Then only constant when I crave her:
'Tis a virtue should not save her.
Thus, nor her delicates would cloy me,
Neither her peevishness annoy me.   20

7

From Poetaster, *III. i. 8–12*
Horace's Ode

Swell me a bowl with lusty wine,
Till I may see the plump Lyaeus swim
            Above the brim:
I drink as I would write,
In flowing measure, filled with flame and spright.   5

8

From Poetaster, *IV. v. 176–83, 188–99*
Feasting Song

Wake, our mirth begins to die;
Quicken it with tunes and wine:
Raise your notes—you're out: fie, fie!
This drowsiness is an ill sign.

19 *delicates*: delights.

7. Jonson quoted these lines to Drummond, *Conv. Dr.* 96.   2 *Lyaeus*: Bacchus.

We banish him the choir of gods   5
>> That droops again:
>> Then all are men,
> For here's not one but nods.

Hermogenes	Then in a free and lofty strain
	Our broken tunes we thus repair; 10
Crispinus	And we answer them again,
	Running division on the panting air:
Both	To celebrate this feast of sense,
	As free from scandal as offence.
Hermogenes	Here is beauty for the eye; 15
Crispinus	For the ear, sweet melody;
Hermogenes	Ambrosiac odours for the smell;
Crispinus	Delicious nectar for the taste;
Both	For the touch, a lady's waist,
	Which doth all the rest excel! 20

9

From Volpone, *I. ii. 66–81*
The Fools' Song

Fools, they are the only nation
Worth men's envy or admiration;
Free from care or sorrow-taking,
Selves and others merry making:
All they speak or do is sterling.   5
Your fool, he is your great man's darling,
And your lady's sport and pleasure;
Tongue and bauble are his treasure.
E'en his face begetteth laughter,
And he speaks truth free from slaughter;   10

8. 12 *division*: see *Songs*, 1. 4 n.
13 *feast of sense*: cf. Chapman's poem, *Ovid's Banquet of Sense*, 1595; ll. 15–20 reverse the Ficinian hierarchy of the senses, which begins with the most physical sense, touch, and proceeds to the most spiritual, sight. See J. F. Kermode, *BJRL*, xliv (1961), 68–99.

9. Sung by Volpone's dwarf, eunuch, and hermaphrodite: Nano, Castrone, and Androgyno; J. D. Rea in his edition of *Volpone* (New Haven, 1919) indicates a source in Erasmus's *Praise of Folly*, 436 c. 10 *slaughter*: retribution; i.e. with impunity. For the rhyme, cf. *Songs*, 16. 20–1.

He's the grace of every feast,
And, sometimes, the chiefest guest:
Hath his trencher and his stool,
When wit waits upon the fool.
    Oh, who would not be   15
      He, he, he?

10

From Volpone, *II. ii. 191–203*
The Mountebank's Song

You that would last long, list to my song:
Make no more coil, but buy of this oil.
Would you be ever fair and young,
Stout of teeth and strong of tongue?
Tart of palate, quick of ear,   5
Sharp of sight, of nostril clear?
Moist of hand and light of foot
(Or I will come nearer to 't)
Would you live free from all diseases?
Do the act your mistress pleases,   10
Yet fright all aches from your bones?
Here's a medicine for the nonce.

11

From The Silent Woman, *I. i. 91–102*
Clerimont's Song

Still to be neat, still to be dressed,
As you were going to a feast;
Still to be powdered, still perfumed:
Lady, it is to be presumed,
Though art's hid causes are not found,   5
All is not sweet, all is not sound.

---

12 *chiefest guest*: a fool often sat at the head of the table: cf. *B.F.* III. iv. 121–4.

10. Sung by Volpone in the guise of a mountebank.   5 *Tart*: keen.   11 *aches*: the word was disyllabic: 'aitches'.

11. Modelled on a poem in the *Anthologia Latina*, and imitated by Herrick, 'Delight in Disorder', Herrick, p. 28.   1–2 *dressed . . . feast*: contemporary pronunciation made the rhyme possible.

Give me a look, give me a face,
That makes simplicity a grace;
Robes loosely flowing, hair as free:
Such sweet neglect more taketh me   10
Than all the adulteries of art:
They strike mine eyes, but not my heart.

12

From Bartholomew Fair, *III. v. 69 ff.*
Nightingale's Song

My masters and friends and good people, draw near,
And look to your purses, for that I do say:
And though little money in them you do bear,
It cost more to get than to lose in a day.
    You oft have been told,   5
    Both the young and the old,
    And bidden beware of the cut-purse so bold:
Then if you take heed not, free me from the curse,
Who both give you warning for and the cut-purse.
Youth, youth, thou hadst better been starved by thy nurse   10
Than live to be hanged for cutting a purse.

It hath been upbraided to men of my trade
That oftentimes we are the cause of this crime:
Alack and for pity, why should it be said?
As if they regarded or places or time.   15
    Examples have been
    Of some that were seen
In Westminster Hall, yea, the pleaders between:
Then why should the judges be free from this curse,
More than my poor self, for cutting the purse?   20
Youth, youth, &c.

---

12. Sung to the tune of Packington's Pound: see William Chappell, *Old English Popular Music*, revised H. Ellis Wooldridge (London & New York, 1893), i. 259. During the course of the song, Bartholomew Cokes (a rapt listener) is robbed of his purse. 9 *for and*: and moreover.

At Worcester 'tis known well, and even i' the gaol,
A knight of good worship did there show his face
Against the foul sinners, in zeal for to rail:
And lost, *ipso facto*, his purse in the place.   25
    Nay, once from the seat
    Of judgement so great
A judge there did lose a fair pouch of velvet.
O Lord for thy mercy, how wicked or worse
Are those that so venture their necks for a purse!   30
Youth, youth, &c.

At plays and at sermons and at the sessions
'Tis daily their practice such booty to make:
Yea, under the gallows at executions
They stick not the stare-abouts' purses to take.   35
    Nay, one without grace
    At a far better place
At court, and in Christmas, before the king's face:
Alack then for pity, must I bear the curse
That only belongs to the cunning cut-purse?   40
Youth, youth, &c.

But O you vile nation of cut-purses all,
Relent and repent and amend and be sound:
And know that you ought not by honest men's fall
Advance your own fortunes, to die above ground;   45
    And though you go gay
    In silks as you may,
It is not the highway to heaven, as they say:
Repent then, repent you, for better, for worse,
And kiss not the gallows for cutting a purse: 50
Youth, youth, thou hadst better been starved by thy nurse
Than live to be hanged for cutting a purse.

26–8 Sir Thomas More, with the help of a professional cut-purse, had played this trick on a judge who had rebuked others for their simplicity in allowing themselves to be robbed. 38 *At court*: the offence was committed by John Selman, executed 7 Jan. 1612.

13

From The New Inn, *IV. iv. 4–13*
A Vision of Beauty

It was a beauty that I saw
So pure, so perfect, as the frame
Of all the universe was lame;
To that one figure, could I draw,
Or give least line of it a law!   5

A skein of silk without a knot!
A fair march made without a halt!
A curious form without a fault!
A printed book without a blot!
All beauty, and without a spot!   10

14

On The New Inn
Ode. To Himself

Come, leave the loathed stage,
    And the more loathsome age,
Where pride and impudence, in faction knit,
    Usurp the chair of wit:
Indicting and arraigning every day   5
    Something they call a play.
    Let their fastidious, vain
    Commission of the brain
Run on and rage, sweat, censure, and condemn:
They were not made for thee, less thou for them.   10

---

13. Discussed by Barbara Everett, *Crit. Q.* i (1959), 238–44.

14. Jonson's ode on the failure of *The New Inn* in 1629 provoked a number of immediate replies (see H & S, xi. 333–40) and subsequent imitations, e.g. Rochester's 'Leave this gawdy, gilded stage' (see J. Treglown, *RES*, xxiv (1973), 43). On Jonson's hostility to the stage, see Jonas Barish in *A Celebration of Ben Jonson*, ed. W. Blissett et al. (Toronto, 1973), pp. 27–53. 5 *Indicting and arraigning*: cf. *B.F.* Ind. 105. 7 *fastidious*: proud, scornful.

Say that thou pour'st them wheat,
  And they will acorns eat:
'Twere simple fury still thyself to waste
    On such as have no taste:
To offer them a surfeit of pure bread   15
    Whose appetites are dead.
  No, give them grains their fill,
  Husks, draff to drink, and swill;
If they love lees, and leave the lusty wine,
Envy them not, their palate's with the swine.   20

  No doubt some mouldy tale
  Like *Pericles*, and stale
As the shrieve's crusts, and nasty as his fish-
    Scraps out of every dish,
Thrown forth, and raked into the common tub,   25
    May keep up the play club:
  There sweepings do as well
  As the best-ordered meal.
For who the relish of these guests will fit
Needs set them but the alms-basket of wit.   30

  And much good do 't you then:
  Brave plush and velvet men
Can feed on orts; and safe in your stage-clothes
    Dare quit, upon your oaths,
The stagers and the stage-wrights too (your peers)   35
    Of larding your large ears
  With their foul comic socks,
  Wrought upon twenty blocks:
Which, if they're torn and turned and patched enough,
The gamesters share your guilt, and you their stuff.   40

---

22 Pericles: printed in 1609; additionally 'mouldy', no doubt, because of its use of older romance elements.   25 *common tub*: left-overs from city and court feasts were collected in a tub for the poor.   26 *play club*: cf. *U.V.* 30. 8, 42. 19 ff.   27 *There sweepings*: replacing an earlier reading, 'Brome's sweepings': i.e. Richard Brome's.   30 *alms-basket of wit*: cf. *Love's Labour's Lost*, V. i. 35–6, 'the alms-basket of words'.   33 *stage-clothes*: cf. Fitzdottrell's theatre-going finery in *D. is A.*; and *Und.* 15. 108–10.   35 *stagers*: stage-players.   36 *stage-wrights*: *OED*'s only example; probably a coinage, like 'playwright' in *Epig.* 49: see n. there.   40 *gamesters*: players.

Leave things so prostitute,
And take the Alcaic lute,
Or thine own Horace, or Anacreon's lyre;
Warm thee by Pindar's fire:
And though thy nerves be shrunk and blood be cold   45
Ere years have made thee old,
Strike that disdainful heat
Throughout, to their defeat:
As curious fools, and envious of thy strain,
May, blushing, swear no palsy's in thy brain.   50

But when they hear thee sing
The glories of thy king,
His zeal to God, and his just awe o'er men:
They may, blood-shaken, then
Feel such a flesh-quake to possess their powers,   55
As they shall cry: Like ours
In sound of peace or wars
No harp e'er hit the stars,
In tuning forth the acts of his sweet reign:
And raising Charles's chariot 'bove his wain.   60

15

From The Sad Shepherd, *I. v. 65–80*
Karolin's Song

Though I am young and cannot tell
Either what death or love is well,
Yet I have heard they both bear darts,
And both do aim at human hearts;

---

42 *Alcaic*: after Alcaeus of Lesbos (seventh–sixth century B.C.), imitated by Horace; Alcaic metre is named after him.   45 *nerves*: sinews; referring to Jonson's paralytic strokes.   58 *hit the stars*: Horace, *Odes*, I. i. 35–6.   60 *wain*: Charles's Wain, the Great Bear.

15. For Nicholas Lanier's setting, see Playford's *Select Musical Airs and Dialogues* (1652), ii. 24. On the association of the figures of Cupid and Death, see E. Panofsky, *Studies in Iconology* (New York, etc., 1972), pp. 124–5, and B. A. Harris's introduction to Shirley's *Cupid and Death* in *A Book of Masques* (Cambridge, 1967), pp. 373–7; and

And then again I have been told   5
  Love wounds with heat, as death with cold:
So that I fear they do but bring
  Extremes to touch, and mean one thing.

As in a ruin, we it call
  One thing to be blown up, or fall;   10
Or to our end like way may have
  By a flash of lightning, or a wave;
So love's inflamed shaft or brand
  May kill as soon as death's cold hand:
Except love's fires the virtue have   15
  To fright the frost out of the grave.

16

From The Entertainment at Althorp, *53–100*
The Satyr's Verses

Satyr   This is Mab the mistress-fairy,
           That doth nightly rob the dairy,
           And can hurt, or help, the churning,
           As she please, without discerning.

Elf   Pug, you will anon take warning?   5

Satyr   She that pinches country wenches
           If they rub not clean their benches,
           And with sharper nails remembers
           When they rake not up their embers;
           But if so they chance to feast her,   10
           In a shoe she drops a tester.

Elf   Shall we strip the skipping jester?

---

cf. 'To Cupid', Herrick, p. 333.   5–6 Cf. Donne, 'The Paradox', 7–8, *Elegies*, p. 38.
16. Much of Jonson's fairy-lore may be parallelled in Corbett's 'The Fairies'
Farewell', *Poems*, ed. Bennett and Trevor-Roper, pp. 49–52, and in Herrick's 'The
Fairies', Herrick, p. 201.   10–11 *feast her . . . tester*: on the rhyme, see *Songs*, ll. 1–
2 n. A tester is sixpence.

Satyr   This is she that empties cradles,
               Takes out children, puts in ladles,
               Trains forth midwives in their slumber   15
               With a sieve the holes to number;
               And then leads them from her boroughs
               Home through ponds and water-furrows.

Elf   Shall not all this mocking stir us?

Satyr   She can start our franklins' daughters   20
               In their sleep, with shrieks and laughters;
               And on sweet St Anne's night
               Feed them with a promised sight,
               Some of husbands, some of lovers,
               Which an empty dream discovers.   25

Elf   Satyr, vengeance near you hovers.

Satyr   And in hope that you would come here
               Yester-eve, the Lady Summer
               She invited to a banquet:
               But, in sooth, I can you thank yet   30
               That you could so well deceive her
               Of the pride which 'gan up-heave her,
               And, by this, would so have blown her,
               As no wood-god should have known her.

Here he skipped into the wood.   35

Elf   Mistress, this is only spite:
               For you would not yester-night
               Kiss him in the cock-shut light.

And came again.

· 20 *franklins'*: landowners not of noble birth.   22 *St. Anne's night*: i.e. St. Agnes eve,
20 Jan.: Keats's poem remembers the same tradition. H & S suggest that the form
'Anne's' here may have been by way of compliment to Queen Anne, in whose honour
the entertainment was given. ('Agnes' was then pronounced 'an-yas'.)   25 *empty*: i.e.
fasting; but the primary sense, 'worthless' is also present.   28 *Lady Summer*: 'For she
was expected there on Midsummer Day at night, but came not till the day following'
(Jonson's note).   38 *cock-shut light*: twilight.

Satyr	By Pan, and thou hast hit it right.

There they laid hold on him, and nipped him.

Fairy	Fairies, pinch him black and blue:
	Now you have him, make him rue.

Satyr	Oh, hold, Mab: I sue.

Elf	Nay, the devil shall have his due.

40

45

17

From The Masque of Queens, *75–94*
The Witches' Charm

The owl is abroad, the bat and the toad,
   And so is the cat-o'-mountain;
The ant and the mole sit both in a hole,
   And frog peeps out o' the fountain;
The dogs they do bay, and the timbrels play,   5
   The spindle is now a-turning;
The moon it is red, and the stars are fled,
   But all the sky is a-burning:
The ditch is made, and our nails the spade,
With pictures full, of wax and of wool;   10
Their livers I stick with needles quick:
That lacks but the blood, to make up the flood.
   Quickly, dame, then, bring your part in,
   Spur, spur upon little Martin,
   Merrily, merrily, make him sail,   15
   A worm in his mouth, and a thorn in's tail,

---

17. 2 *cat-o'-mountain*: leopard or wild cat.   6 *spindle*: 'All this is but a periphrasis of the night in their charm, and their applying themselves to it with their instruments, whereof the spindle in antiquity was the chief, and . . . of special act to the troubling of the moon . . .' (Jonson's note).   9 *ditch*: 'This rite also of making a ditch with their nails is frequent with our witches . . .' (Jonson's note).   14 *Martin*: 'Their little Martin is he that calls them to their conventicles, which is done in a human voice; but coming forth, they find him in the shape of a great buck-goat, upon whom they ride to their meetings . . .' (Jonson's note).

Fire above and fire below,
With a whip i' your hand to make him go.
Oh, now she's come!
Let all be dumb.   20

18

From Oberon, The Fairy Prince, *210–17*
Satyrs' Catch

Buzz, quoth the blue fly,
    Hum, quoth the bee:
Buzz and hum, they cry,
    And so do we.
In his ear, in his nose,   5
    Thus, do you see?
He ate the dormouse,
    Else it was he.

19

From Mercury Vindicated, *6–17*
Cyclope's Song

Soft, subtle fire, thou soul of art,
    Now do thy part
On weaker nature, that through age is lamed.
    Take but thy time, now she is old,
    And the sun her friend grown cold,   5
She will no more in strife with thee be named.

Look but how few confess her now
    In cheek or brow!
From every head, almost, how she is frighted!
    The very age abhors her so,   10
    That it learns to speak and go
As if by art alone it could be righted.

19. 4 *old*: cf. *Disc.* 124–8, *For.* 4. 14.

20

From The Vision of Delight, *237–47*
Epilogue

Aurora   I was not wearier where I lay
By frozen Tithon's side tonight,
Than I am willing now to stay
And be a part of your delight:
    But I am urged by the day,   5
Against my will, to bid you come away.

The Choir   They yield to time, and so must all:
As night to sport, day doth to action call,
    Which they the rather do obey
Because the morn with roses strews the way.   10

21

From Pleasure Reconciled to Virtue, *13–36*
Song for Comus

Room, room, make room for the bouncing belly!
First father of sauce, and deviser of jelly:
Prime master of arts and the giver of wit,
That found out the excellent engine, the spit,
The plough and the flail, the mill and the hopper,   5
The hutch and the bolter, the furnace and copper:
The oven, the bavin, the mawkin and peel,
The hearth and the range, the dog and the wheel.
He, he first invented both hogshead and tun,
The gimlet and vice, too, and taught 'em to run;   10

20. Stephen Dedalus murmurs these lines by the Liffey in Joyce's *A Portrait of the Artist as a Young Man.* 2 *Tithon*: Tithonus, Aurora's husband, to whom Aurora had granted eternal life, but not eternal youth and vigour.

21. On Comus ('the god of cheer, or the belly') and on this song, see Stephen Orgel, *The Jonsonian Masque* (Cambridge, Mass. 1965), pp. 152–9. 3 *Prime master of arts*: Persius, Prol. *Sat.* 8–11. 7 *bavin*: brushwood bundle used to light fires, esp. for bakers' ovens. 7 *mawkin*: mop for cleaning bakers' ovens. 8 *the dog and the wheel*: to operate the spit.

And since, with the funnel, an hippocras bag
He's made of himself, that now he cries swag.
Which shows, though the pleasure be but of four inches,
Yet he is a weezle, the gullet that pinches,
Of any delight; and not spares from the back   15
Whatever, to make of the belly a sack.
Hail, hail, plump paunch! O the founder of taste
For fresh meats or powdered or pickle or paste;
Devourer of broiled, baked, roasted or sod,
And emptier of cups, be they even or odd.   20
All which have now made thee so wide i' the waist
As scarce with no pudding thou art to be laced:
But eating and drinking until thou dost nod,
Thou break'st all thy girdles, and break'st forth a god.

22

From Pleasure Reconciled to Virtue, *253–72*
Daedalus's Song

Come on, come on, and where you go
    So interweave the curious knot
As even th'observer scarce may know
    Which lines are pleasure's, and which not.
First, figure out the doubtful way   5
    At which a while all youth should stay,
Where she and virtue did contend
    Which should have Hercules to friend.

11 *hippocras bag*: conical strainer for wine.   12 *cries swag*: proclaims himself a pot-belly.   14 *weezle*: weasand, windpipe.   19 *sod*: boiled.   22 *pudding . . . laced*: a double play on words: for a 'pudding' is also, in nautical terminology, a thick rope or tow binding; and 'laced' means both 'strapped up' and 'fortified' (of food or drink).   24 *break'st forth*: with a scatological *double entendre*.

22. 7–8 Hercules in early manhood is said to have met two women, Virtue and Vice (*alias* Happiness); Vice showed him an easy road to travel, Virtue a difficult one: Hercules chose the latter. See Xenophon, *Memorabilia*, II. i. 21–34.

Then, as all actions of mankind
   Are but a labyrinth or maze,   10
So let your dances be entwined,
   Yet not perplex men unto gaze;
But measured, and so numerous, too,
   As men may read each act you do.
And when they see the graces meet,   15
   Admire the wisdom of your feet.
For dancing is an exercise
   Not only shows the mover's wit
But maketh the beholder wise,
   As he hath power to rise to it.   20

23

From The Gypsies Metamorphosed, *121–44*
The Jackman's Song

From the famous Peak of Derby
   And the Devil's Arse there hard by
Where we yearly keep our musters,
   Thus the Egyptians throng in clusters.

Be not frighted with our fashion,   5
   Though we seem a tattered nation:
We account our rags our riches,
   So our tricks exceed our stitches.

Give us bacon, rinds of walnuts,
   Shells of cockles and of small nuts,   10
Ribbons, bells, and saffron linen,
   All the world is ours to win in.

---

10 *labyrinth*: Daedalus had constructed a labyrinth in Crete. 13 *numerous*: rhythmical.

23. *Jackman*: a learned beggar. 2 *Devil's Arse*: a cavern near Castleton; see *Und*. 43. 202.

Knacks we have that will delight you,
   Sleights of hand that will invite you
   To endure our tawny faces,   15
   And not cause you cut your laces.

All your fortunes we can tell ye,
   Be they for your back or belly,
   In the moods too, and the tenses,
   That may fit your fine five senses.   20

Draw but then your gloves, we pray you,
   And sit still, we will not fray you:
   For though we be here at Burley,
   We'd be loath to make a hurly.

24

From The Gypsies Metamorphosed, *262–71*
The Patrico's Song

The fairy beam upon you,
The stars to glister on you,
   A moon of light
   In the noon of night,
Till the firedrake hath o'er-gone you.   5

The wheel of fortune guide you,
The boy with the bow beside you
   Run aye in the way
   Till the bird of day
And the luckier lot betide you.   10

16 *cut your laces*: the standard way of reviving fainting ladies.   23 *Burley*: Burley-on-the-Hill, the Marquis of Buckingham's estate.
24. *Patrico*: a beggars' priest.   5 *firedrake*: meteor, will-o'-the-wisp, or fiery dragon.

25

From The Gypsies Metamorphosed, *1329–90*
Blessing the Sovereign and His Senses

From a gypsy in the morning,
   Or a pair of squint eyes turning,
From the goblin and the spectre,
   Or a drunkard, though with nectar;
From a woman true to no man,   5
   And is ugly, beside common,
A smock rampant, and that itches
   To be putting on the breeches:
Wheresoe'er they have their being,
Bless the sovereign and his seeing.   10

From a fool, and serious toys,
   From a lawyer three-parts noise,
From impertinence, like a drum
   Beat at dinner in his room;
From a tongue without a file,   15
   Heaps of phrases, and no style,
From a fiddle out of tune,
   As the cuckoo is in June;
From the candlesticks of Lothbury,
   And the loud, pure, wives of Banbury,   20
Or a long pretended fit
   Meant for mirth, but is not it,
Only time and ears out-wearing:
Bless the sovereign and his hearing.

From a strolling tinker's sheet,   25
   And a pair of carrier's feet,
From a lady that doth breathe
   Worse above than underneath;

---

25. 6 *common*: promiscuous; cf. Dol Common in *Alch.*   7 *smock rampant*: shrew; the same phrase is used of Dol Common, *Alch.* V. iv. 126.   15 *without a file*: i.e. rough.   18 *cuckoo*: cf. Tilley, A309: 'In April the cuckoo can sing her song by rote; in June, out of tune, she cannot sing a note'.   19 *Lothbury*: street in London where candlesticks were made: a noisy trade.   20 *Banbury*: a Puritan centre.

From the diet, and the knowledge,
　　Of the students in Bears' College;　　　　　30
From tobacco, with the type
　　Of the devil's clyster-pipe;
Or a stink all stinks excelling,
　　A fishmonger's dwelling:
Bless the sovereign and his smelling.　　　　　35

From an oyster and fried fish,
　　A sow's baby in a dish,
Any portion of a swine;
　　From bad venison, and worse wine,
Ling, what cook soe'er it boil,　　　　　40
　　Though with mustard sauce and oil;
Or what else would keep man fasting:
Bless the sovereign and his tasting.

Both from birdlime and from pitch,
　　From a doxy and her itch,　　　　　45
From the bristles of a hog,
　　Or the ringworm in a dog;
From the courtship of a briar,
　　From St Anthony's old fire;
From a needle or a thorn　　　　　50
　　I' the bed at even or morn,
Or from any gout's least grutching,
Bless the sovereign and his touching.

Bless him, too, from all offences
　　In his sports, as in his senses:　　　　　55
From a boy to cross his way,
　　From a fall, or a foul day.

Bless him, O bless him, heaven, and lend him long
　　To be the sacred burthen of all song:
The acts and years of all our kings to outgo,　　　　　60
　　And, while he's mortal, we not think him so.

---

30 *Bears' College*: Paris Garden, on the Bankside, where bear-baiting was held: the 'students' are the bears. Cf. *Epig.* 133. 117.　32 *clyster-pipe*: see *Epig.* 50. 2 n. James's *Counterblast To Tobacco* was published in 1604. Tobacco, pig, and ling were his special aversions: hence the benedictions in this and the following stanza.　49 *St. Anthony's old fire*: erysipelas.　52 *grutching*: complaint.　57 *fall*: James had twice fallen while riding.

26

From Neptune's Triumph, *472–503*
Song to the Ladies

Proteus   Come, noble nymphs, and do not hide
The joys for which you so provide.

Saron   If not to mingle with the men
What do you here? Go home again.

Portunus   Your dressings do confess   5
By what we see, so curious parts
Of Pallas' and Arachne's arts,
    That you could mean no less.

Proteus   Why do you wear the silkworm's toils,
Or glory in the shellfish spoils,   10
Or strive to show the grains of ore
That you have gathered on the shore
    Whereof to make a stock
To graft the greener emerald on,
Or any better-watered stone?   15

Saron   Or ruby of the rock?

Proteus   Why do you smell of ambergris,
Of which was formed Neptune's niece,
The queen of love, unless you can,
Like sea-born Venus, love a man?   20

Saron   Try, put yourselves unto 't.

Chorus   Your looks, your smiles and thoughts that meet,
Ambrosian hands and silver feet,
    Do promise you will do 't.

26. 7 *Pallas' and Arachne's arts*: of weaving.

LEGES CONVIVALES

Leges Convivales
Quod felix, faustumque in Apolline sit.

1. Nemo asymbolus, nisi umbra, huc venito.
2. Idiota insulsus, tristis, turpis, abesto.
3. Eruditi, urbani, hilares, honesti, adsciscuntor.
4. Nec lectae foeminae repudiantor.
5. In apparatu, quod convivis corruget nares, nil esto.
6. Epulae delectu potiùs, quam sumptu, parantor.
7. Opsonator et coquus, convivarum gulae periti sunto.
8. De discubitu non contenditor.
9. Ministri à dapibus oculati et muti; à poculis auriti, et celeres, sunto.
10. Vina puris fontibus ministrantor; aut vapulet hospes.
11. Moderatis poculis provocare sodales, fas esto.
12. At fabulis magis, quàm vino, velitatio fiat.
13. Convivae nec muti, nec loquaces sunto.
14. De seriis, aut sacris, poti et saturi ne disserunto.
15. Fidicen, nisi accersitus, non venito.
16. Admisso; risu, tripudiis, choreis, cantu, salibus, omni gratiarum festivitate, sacra celebrantor.
17. Joci sine felle sunto.
18. Insipida poëmata nulla recitantor.
19. Versus scribere, nullus cogitor.
20. Argumentationis totus strepitus abesto.
21. Amatoriis querelis, ac suspiriis, liber angulus esto.
22. Lapitharum more scyphis pugnare, vitrea collidere, fenestras excutere, supellectilem dilacerare, nefas esto.
23. Qui foràs vel dicta, vel facta, eliminat, eliminator.
24. Neminem reum pocula faciunto.
Focus perennis esto.

The *Leges Convivales* are a set of rules for tavern behaviour drawn up by Jonson for the benefit of the intimate circle of friends over which he presided in the Apollo Room of the Devil and St. Dunstan Tavern, near Temple Bar. They were engraved in gold letters on a marble tablet over the mantelpiece. See Percy Simpson's full account in *MLR*, xxiv (1939), 367–73 (largely repeated in H & S, xi. 294–300). John Buxton (*MLR*, xlviii (1953), 52–4) has shown that John Chamberlain's statement in 1624 that the Apollo Room was 'lately built' is unreliable; it is therefore probable that Drayton's ode of 1619, 'The Sacrifice To Apollo', *Works*, ed. J. W. Hebel (Oxford, 1961), ii. 357–8, derives from the *Leges Convivales*, rather than the other way about. The *Leges* owe something to Horace, *Epist.* I. v, and Martial, *Epig.* X. xlviii.

Rule 1 *umbra*: 'the Roman term for a guest not invited by the host, but brought by a guest whom he had invited' (Simpson, p. 369, comparing Horace, *Epist.* I. v. 28). 5 Cf. Horace, ibid., 23. 17, 24 Cf. Martial, *Epig.* X. xlviii. 21, 24; Jonson, *Epig.* 101. 39–41. 18 Cf. *Epig.* 101. 24. 22 Cf. Horace, *Odes*, I. xviii. 7–9, I. xxvii. 1–4. The

Alexander Brome's Translation of the Leges Convivales
Ben Jonson's Sociable Rules for the Apollo

Let none but guests or clubbers hither come;
Let dunces, fools, sad, sordid men keep home;
Let learned, civil, merry men be invited,
And modest, too; nor the choice ladies slighted.
Let nothing in the treat offend the guests;   5
More for delight than cost prepare the feasts;
The cook and purveyor must our palates know;
And none contend who shall sit high or low.
Our waiters must quick-sighted be, and dumb;
And let the drawers quickly hear and come.   10
Let not our wine be mixed, but brisk and neat,
Or else the drinkers may the vintners beat.
And let our only emulation be
Not drinking much, but talking wittily.
Let it be voted lawful to stir up   15
Each other with a moderate chirping cup;
Let none of us be mute, or talk too much;
On serious things or sacred let's not touch
With sated heads and bellies. Neither may
Fiddlers, unasked, obtrude themselves to play.   20
With laughing, leaping, dancing, jests, and songs
And whate'er else to grateful mirth belongs
Let's celebrate our feasts. And let us see
That all our jests without reflection be;
Insipid poems let no man rehearse,   25
Nor any be compelled to write a verse.
All noise of vain disputes must be forborne,
And let no lover in a corner mourn.
To fight and brawl like Hectors let none dare,
Glasses or windows break, or hangings tear.   30
Whoe'er shall publish what's here done or said
From our society must be banished.
Let none by drinking do or suffer harm,
And while we stay let us be always warm.

Lapithae were a tribe of mountaineers in Thessaly; they disturbed the wedding of their
king by fighting with the Centaurs (cf. *S.W.* IV. v. 45–7).   23 Cf. Horace, *Epist.* I. v.
24–5. *Focus perennis esto*: Martial, *Epig.* X. xlvii. 4 (trs. Jonson, *Und.* 90).

[Brome's translation]
Alexander Brome (1620–66), poet and attorney.   16 *chirping*: cheering.

Over the Door at the Entrance into the Apollo

Welcome, all who lead or follow
To the oracle of Apollo:
Here he speaks out of his pottle,
Or the tripos, his tower bottle;
All his answers are divine: 5
Truth itself doth flow in wine.
Hang up all the poor hop-drinkers!
Cries old Sim, the king of skinkers;
He the half of life abuses
That sits watering with the muses. 10
Those dull girls no good can mean us;
Wine, it is the milk of Venus
And the poets' horse accounted:
Ply it, and you all are mounted.
'Tis the true Phoebeian liquor, 15
Cheers the brains, makes wit the quicker,
Pays all debts, cures all diseases,
And at once three senses pleases.
Welcome, all who lead or follow
To the oracle of Apollo! 20

[Over the Door]
The verses were painted on a panel still preserved in the dining-room of Williams and
Glyn's Bank Ltd., 1 Fleet St.; a bust of Apollo also surmounted the door. See K. A.
Esdaile, *E & S*, xxix (1943), 93–100. 2 *oracle of Apollo*: H & S cf. Rabelais's 'Oracle of
the bottle', *Pantagruel*, V. 34, and *S. of N.* IV. ii. 8. 3 *pottle*: two-quart tankard.
4 *tripos*: drinking-bowl. 4 *tower*: i.e. tower-shaped. 6 *Truth itself: in vino veritas*.
8 *Sim*: Simon Wadloe, keeper of the inn; cf. the drinking song, 'Old Sir Simon the
King'. 8 *skinkers*: tapsters. 13 *poets' horse*: Pegasus. 15 *Phoebeian*: of Phoebus
Apollo.

DUBIA

DUBIA

I

An Elegy

To make the doubt clear that no woman's true,
    Was it my fate to prove it full in you?
Thought I but one had breathed the purer air,
    And must she needs be false, because she's fair?
Is it your beauty's mark, or of your youth,   5
    Or your perfection, not to study truth?
Or think you heaven is deaf, or hath no eyes,
    Or those it has wink at your perjuries?
Are vows so cheap with women, or the matter
    Whereof they are made, that they are writ in water,   10
And blown away with wind? Or doth their breath,
    Both hot and cold at once, threat life and death?
Who could have thought so many accents sweet
    Tuned to our words, so many sighs should meet,
Blown from our hearts: so many oaths and tears   15
    Sprinkled among, all sweeter by our fears,
And the divine impression of stolen kisses,
    That sealed the rest, could now prove empty blisses?
Did you draw bonds to forfeit? Sign, to break?
    Or must we read you quite from what you speak,   20
And find the truth out the wrong way? Or must
    He first desire you false, would wish you just?
Oh, I profane! Though most of women be
    The common monster, love shall except thee,
My dearest love, however jealousy   25

Dubia: The poems in this section fall into two categories: (i) Poems which were printed in the 1640 Folio, or in H & S's *Ungathered Verse*, and which now appear to be probably not by Jonson. (ii) Other poems for which a plausible but inconclusive case for Jonsonian authorship may be made out. Several of the poems printed in H & S's Appendix XVI, *Poems Ascribed to Jonson* (viii. 424–52), fall into a third category of implausibly or incorrectly attributed miscellaneous pieces, and are not reprinted here.

1. Printed as an elegy, 'The Expostulation', in Donne's *Poems* of 1633, and subsequently in Jonson's 1640 Folio (*Und.* 39 in H & S). On the question of authorship, see *Und.* 38, n.   4 *false, because she's fair*: cf. Tilley, F3.   10–11 Cf. Catullus, *Carm.* lxx. 3–4; Tilley, W698.   19 *break*: go bankrupt.   19–22 Cf. 'Woman's Constancy', Donne, *Elegies*, pp. 42–3.

With circumstance might urge the contrary.
Sooner I'll think the sun would cease to cheer
  The teeming earth, and that forget to bear;
Sooner that rivers would run back, or Thames
  With ribs of ice in June would bind his streams;   30
Or nature, by whose strength the world endures,
  Would change her course, before you alter yours.
But, oh, that treacherous breast to whom weak you
  Did trust our counsels! And we both may rue,
Having his falsehood found too late. 'Twas he   35
  That made me cast you guilty, and you me;
Whilst he, black wretch, betrayed each simple word
  We spake unto the cunning of a third!
Cursed may he be, that so our love hath slain,
  And wander wretched on the earth as Cain: 40
Wretched as he, and not deserve least pity;
  In plaguing him, let misery be witty.
Let all eyes shun him, and he shun each eye,
  Till he be noisesome as his infamy;
May he without remorse deny God thrice,   45
  And not be trusted more, on his soul's price;
And after all self-torment, when he dies,
  May wolves tear out his heart, vultures his eyes,
Swine eat his bowels; and his falser tongue,
  That uttered all, be to some raven flung;   50
And let his carrion corse be a longer feast
  To the king's dogs than any other beast.
Now I have cursed, let us our love revive;
  In me the flame was never more alive.
I could begin again to court and praise,   55
  And in that pleasure lengthen the short days
Of my life's lease, like painters that do take
  Delight, not in made works, but whilst they make.
I could renew those times when first I saw
  Love in your eyes that gave my tongue the law   60
To like what you liked; and at masques or plays
  Commend the self-same actors, the same ways;

27–32 Cf. Ovid, *Tristia*, I. viii. 1–10. 36 *cast*: condemn. 39 ff. Cf. Catullus, *Carm.* cviii; Donne, 'The Curse', and 'The Bracelet', 91–110 (*Elegies*, pp. 40–1, 4). 57–8 Cf. *Und.* 5. 13–16, and n. 59–64 Perhaps recalling Ovid, *Amores*, III. ii. 1–7.

Ask how you did? And often, with intent
    Of being officious, grow impertinent:
All which were such soft pastimes, as in these   65
    Love was as softly catched as a disease.
But being got, it is a treasure sweet,
    Which to defend is harder than to get;
And ought not be prophaned on either part,
    For though 'tis got by chance, 'tis kept by art.   70

2

Fair friend, 'tis true your beauties move
    My heart to a respect,
Too little to be paid with love,
    Too great for your neglect.

I neither love nor yet am free,   5
    For though the flame I find
Be not intense in the degree
    'Tis of the purest kind.

It little wants of love, but pain:
    Your beauty takes my sense;   10
And lest you should that price disdain,
    My thoughts, too, feel the influence.

'Tis not a passion's first access
    Ready to multiply,
But like love's calmest state it is   15
    Possessed with victory.

64 *officious*: obliging.   68–70 Cf. Ovid, *Ars Amatoria*, ii. 13–14.

2. Printed in the 1640 Folio (*Und.* 80 in H & S), but attributed to Sidney Godolphin
in one MS. W. D. Briggs wavers over the attribution (*Anglia*, xxxix (1915), 41–4),
which is nevertheless accepted by William Dighton in his edition of Godolphin's
Poems (Oxford, 1931), pp. 60–1, and by H & S. F. R. Leavis—perhaps unaware of its
disputed authorship—finds the poem 'characteristic Jonson', *Revaluation* (London,
1953), pp. 22–4; the verdict is echoed by F. W. Bradbrook ('typical of the Jonsonian
lyric'), *From Donne To Marvell*, ed. Boris Ford (Harmondsworth, 1956), p. 138. The
stylistic evidence in fact favours Godolphin: cf. 'Or love me less, or love me more', and
'No more unto my thoughts appear', *Poems*, pp. 8–9, 22–4.

It is like love to truth reduced,
   All the false values gone,
Which were created and induced
   By fond imagination.   20

'Tis either fancy or 'tis fate
   To love you more than I:
I love you at your beauty's rate,
   Less were an injury.

Like unstamped gold, I weigh each grace,   25
   So that you may collect
Th'intrinsic value of your face
   Safely from my respect.

And this respect would merit love,
   Were not so fair a sight   30
Payment enough: for who dare move
   Reward for his delight?

3

On the King's Birthday

Rouse up thyself, my gentle muse,
   Though now our green conceits be grey,
And yet once more do not refuse
   To take thy Phrygian harp, and play
In honour of this cheerful day: 5
      Long may they both contend to prove
      That best of crowns is such a love.

---

3. Printed in the 1640 Folio (*Und.* 81 in H & S), but attributed to Sir Henry Wotton in two MSS. and in *Reliquiae Wottonianae* (London, 1651), p. 521, where it is said to have been written on the king's return from his coronation in Scotland (in 1633). H & S consider it to be by Wotton, and quote from a letter of Wotton's of 1628 which parallels the expression in l. 2 of the poem. But Briggs (*Anglia*, xxxix (1915), pp. 213–15) draws attention to other Jonsonian parallels, noted at ll. 18 and 21.   7 *best of crowns*: cf. *Und.* 76. 2, 79. 7.

Make first a song of joy and love
   Which chastely flames in royal eyes,
Then tune it to the spheres above,   10
   When the benignest stars do rise
   And sweet conjunctions grace the skies.
      Long may, &c.

To this let all good hearts resound,
   Whilst diadems invest his head;   15
Long may he live, whose life doth bound
   More than his laws, and better lead
By high example than by dread.
      Long may, &c.

Long may he round about him see   20
   His roses and his lilies blown:
Long may his only dear and he
   Joy in ideas of their own,
   And kingdom's hopes, so timely sown.
      Long may they both contend to prove   25
      That best of crowns is such a love.

4

From The Touchstone of Truth, *1624*

Truth is the trial of itself,
   And needs no other touch;
And purer than the purest gold,
   Refine it ne'er so much.
It is the life and light of love,   5

---

18 *example . . . dread*: see *Epig.* 35. 2 n.   21 *His roses and his lilies*: see *Und.* 65. 3 n.

4. These verses, signed B.I., appeared in the second edition of *The Touchstone of Truth*—a Puritan compilation—in 1624. H & S, with considerable misgiving, print them as *U.V.* 27. Jonson's authorship seems improbable both on grounds of style (esp. the concluding lines) and of Puritan context; 'warrant of the word' (l. 9) was a Puritan catch-phrase which Jonson had ridiculed in *B.F.* IV. i. 109.

The sun that ever shineth,
And spirit of that special grace,
    That faith and love defineth.
It is the warrant of the word,
    That yields a scent so sweet,   10
As gives a power to faith, to tread
    All falsehood under feet.
It is the sword that doth divide
    The marrow from the bone,
And in effect of heavenly love   15
    Doth show the holy one.
This, blessed Warre, thy blessed book
    Unto the world doth prove:
A worthy work, and worthy well
    Of the most worthy love.   20

5

Another Epigram On the Birth of the Prince

Another phoenix! though the first is dead:
A second's flown from his immortal bed,
To make this our Arabia to be
The nest of an eternal progeny.
Choice nature framed the former but to find   5
What error might be mended in mankind;
Like some industrious workmen, which affect
Their first endeavours only to correct:
So this the building, that the model was,
The type of all that now is come to pass: 10
That but the shadow, this the substance is.
All that was but a prophecy of this:

17 *Warre*: the Dedication of the work is signed James Warre.

5. On the birth of the future Charles II, 29 May 1630; cf. *Und.* 65 and n. The poem
appears in the 1640 Quarto and Duodecimo, but not in the Folio or Newcastle MS.
G. B. Johnston (*Poems*, p. 344) is inclined to accept the poem as Jonson's; Briggs
(*Anglia*, xxxviii (1914), 119) and H & S suspend judgement. 1 *the first is dead*:
Charles's and Henrietta Maria's first son had died on 13 May 1629; see *Und.* 63.

And when it did this after-birth fore-run
'Twas but the morning-star unto this sun;
The dawning of this day, when Sol did think   15
We having such a light, that he might wink
And we ne'er miss his lustre: nay, so soon
As Charles was born, he and the pale-faced moon
With envy then did copulate, to try
If such a birth might be produced i' the sky.   20
What heavenly favour made a star appear
To bid wise kings to do their homage here
And prove him truly Christian? Long remain
On earth, sweet prince, that when great Charles shall reign
In heaven above, our little Charles may be   25
As great on earth, because as good, as he.

6

A Petition of the Infant Prince Charles
The Prince's Verses for One of His Rockers

Read, royal father, mighty king,
What my little hands do bring:
I, whose happy birth imparts
Joy to all good subjects' hearts,
Though an infant, do not break   5
Nature's laws whilst thus I speak

---

14 *sun*: cf. the pun of *Und*. 65. 12.   16 *wink*: an eclipse of the sun occurred two days after Charles's birth; cf. *Und*. 65. 9–12.   17–20 G. B. Johnston detects an allusion to the language of alchemy, in which gold (called 'sol') and silver ('luna') were spoken of as father and mother of the elixir or philosopher's stone.   19 *envy*: cf. *Und*. 65. 9. 24–5 *On earth ... In heaven*: cf. *Und*. 70. 96.   24–5 *great Charles ... little Charles*: cf. *Und*. 67. 54.   26 *great ... good*: a favourite Jonsonian conjunction: see *Epig*. Ded. 15–16 n.

6. Ascribed to Jonson in one MS. Briggs (*Anglia*, xxxix (1915), 244) is inclined to accept the attribution, H & S reject it, chiefly because it 'is unlike any work of Jonson's at that date [after the birth of Charles, 29 May 1630], when his powers showed symptoms of decline'. This test does not seem decisive. The identity of the 'rocker' and the nature of her offence are not known.   5 *Though an infant*: playing on the etymology of the word, 'unable to speak'.

By this interpreter for one
Whose face doth blush, whose heart doth groan
For her acknowledged offence,
And only hath my innocence   10
To gain her mercy. Though thus bold
Yet some proportion it may hold
That to the father she may run
Through mediation of the son.
If therefore now, O royal sir,   15
My first request may purchase her
Restoring unto grace, to me
(Though prince) it will an honour be
That in my cradle 'twill be said
I Master of Requests was made.   20

7

Ode

Scorn, or some humbler fate
Light thick, and long endure
On the ridiculous state
Of our pied courtlings and secure
Race of self-loving lords,   5
That wallow in the flood
Of their great birth and blood;
Whiles their whole life affords
No other graces
But pride, lust, oaths, and faces: 10
And yet would have me deem
Of them at that high rate
As they themselves esteem.
Perish such surquedry,
O'erwhelmed with dust: 15
'Tis only virtue must
Blazon nobility.

7. The tone and style of assertion seem characteristically Jonsonian; Briggs thinks
Jonson's authorship 'quite probable' (*Anglia*, xxxix (1915), 250).   4 *courtlings*: a
favourite word of Jonson's: cf. *Epig.* 52 & 72; *OED* credits him with the first usage, in
C.R. V. iv. 33. For 'pied', cf. *U.V.* 2. 4.   4 *secure*: smug.   10 *faces*: cf. *Epig.* 11. 4, 'It
made me a great face', etc.; *Und.* 70. 23.   14 *surquedry*: pride.   16–17 Cf. *Und.* 84.
viii. 20–1 and n.

8

Horace, Odes, II. iii

Remember when blind fortune knits her brow,
Thy mind be not dejected over-low;
Nor let thy thoughts too insolently swell,
Though all thy hopes do prosper ne'er so well.
For drink thy tears, with sorrow still oppressed,   5
Or taste pure vine, secure and ever blessed,
In those remote and pleasant shady fields
Where stately pine and poplar shadow yields,
Or circling streams that warble, passing by;
All will not help, sweet friend: for thou must die.   10
    The house thou hast thou once must leave behind thee,
And those sweet babes thou often kissest kindly:
And when thou'st gotten all the wealth thou can,
Thy pains is taken for another man.
    Alas! What poor advantage doth it bring   15
To boast thyself descended of a king!
When those that have no house to hide their heads
Find in their grave as warm and easy beds.

8. Printed in John Ashmore's *Certain Selected Odes of Horace, Englished*, 1621 (i.e. during Jonson's lifetime), with a note by Ashmore that the translation 'came unto my hands under the name of Mr. Ben Jonson . . .'. H & S consider the freedom of the translation uncharacteristic of Jonson; G. B. Johnston (*Poems*, p. 341) argues that 'it is no more free than several of his adaptations of classical poems'.

SOURCE MATERIAL

CATULLUS, *Carm.* v. 7–13
Give me a thousand kisses, then a hundred, then another thousand, then a second hundred, then yet another thousand, then a hundred. Then, when we have made up many thousands, we will confuse our counting, that we may not know the reckoning, nor any malicious person blight them with evil eye, when he knows that our kisses are so many.

For. 6. 6–11

Ibid. vii
You ask how many kissings of you, Lesbia, are enough for me and more than enough. As great as is the number of the Libyan sand that lies on silphium-bearing Cyrene, between the oracle of sultry Jove and the sacred tomb of old Battus; or as many as are the stars, when night is silent, that see the stolen loves of men,—to kiss you with so many kisses, Lesbia, is enough and more than enough for your mad Catullus; kisses, which neither curious eyes shall count up nor an evil tongue bewitch.

For. 6. 12–22

CICERO, *De Oratore*, II. ix. 36
...History ... bears witness to the passing of the ages, sheds light upon reality, gives life to recollection and guidance to human existence, and brings tidings of ancient days ...

Und. 24. 17–18

CLAUDIAN, *De Consulatu Stilichonis*, i. 49–50
The silent suffrage of the people had already offered thee all the honours the court was soon to owe.

Epig. 67. 7–10

Ibid. ii. 160–2
Nay, pride itself is far removed from thee, pride, a vice so familiar in success, ungracious attendant on the virtues.

Epig. 76. 10

FLORUS, *De Qualitate Vitae*, ix
New consuls and proconsuls are made every year, only kings and poets are not born every year.

Epig. 4. 3

HORACE, *Epist.* I. xvi. 52–3
The good hate vice because they love virtue; you will commit no crime because you dread punishment.

For. 11. 87–90

Ibid. II. i. 114–17
A man who knows nothing of a ship fears to handle one, no one dares to give

southernwood to the sick unless he has learnt its use; doctors undertake a doctor's work, carpenters handle carpenters' tools, but, skilled or unskilled, we scribble poetry, all alike.

U.V. 38. 12–16

Sat. II. vii. 15–18
Volanerius, the jester, when the gout he had earned crippled his finger-joints, kept a man, hired at a daily wage, to pick up the dice for him and put them in a box.

Und. 15. 139–40

Ibid. II. vii. 68–71
Suppose you have escaped: then, I take it, you will be afraid and cautious after your lesson. No, you will seek occasion so as again to be in terror, again to face ruin, O you slave many times over! But what beast, having once burst its bonds and escaped, perversely returns to them again?

For. 4. 25 ff.

Ibid. II. vii. 86–8
[The wise man] in himself is a whole, smoothed and rounded, so that nothing from outside can rest on the polished surface, and against whom fortune in her onset is ever maimed.

Epig. 98. 3

JUVENAL, *Sat.* x. 141–2
For who would embrace virtue herself if you stripped her of her rewards?

Epig. 66. 18

Ibid. x. 349–59
. . . in place of what is pleasing, [the gods] will give us what is best. Man is dearer to them than he is to himself. . . . Still, that you may have something to pray for, . . . you should pray for a sound mind in a sound body; ask for a stout heart that has no fear of death, and deems length of days the least of nature's gifts . . .

For. 3. 95–106

Ibid. xi. 176–8
In men of moderate position, gaming and adultery are shameful; but when those others do these same things, they are called gay fellows and fine gentlemen.

Und. 15. 57–8

LUCIAN, (attrib.) *In Praise of Demosthenes*, 13
. . . the two impulses of love that come upon men, the one that of a love like the sea, frenzied, savage, and raging like stormy waves in the soul, a veritable

sea of Earthly Aphrodite surging with the fevered passions of youth, the other the pull of a heavenly cord of gold that does not bring with fiery shafts afflicting wounds hard to cure, but impels men to the pure and unsullied form of absolute beauty, inspiring with a chaste madness . . .

For. 11. 37 ff.

'Laws for Banquets', *Saturnalia*, 17
All shall drink the same wine, and neither stomach trouble nor headache shall give the rich man an excuse for being the only one to drink the better quality. All shall have their meat on equal terms.

For. 2. 61 ff.

MARTIAL, *Epig.* I. lvii
Do you ask, Flaccus, what sort of girl I like or dislike? I dislike one too yielding, and one too coy. That middle type between the two I approve: I like not that which racks me, nor like I that which cloys.

Songs, 6

Ibid. I. lxiii
You ask me to recite to you my epigrams. I decline. You don't wish to hear them, Celer, but to recite them.

Epig. 81

Ibid. I. xcviii
Diodorus goes to law, and suffers, Flaccus, from gout in the feet. But he offers his advocate no fee: this is gout in the hand.

Epig. 31. 1

Ibid. III. lviii. 33–44
Nor does the country visitor come empty-handed: that one brings pale honey in its comb, and a pyramid of cheese from Sassina's woodland; that one offers sleepy dormice; this one the bleating offspring of a shaggy mother; another capons debarred from love. And the strapping daughters of honest farmers offer in a wicker basket their mothers' gifts. When work is done, a cheerful neighbour is asked to dine; no niggard table reserves a feast for the morrow; all take the meal, and the full-fed attendant need not envy the well-drunken guest.

For. 2. 49–71

Ibid. III. lx. 1–2, 9
Since I am asked to dinner, no longer, as before, a purchased guest, why is it not the same dinner served to me as to you? . . . why do I dine without you, although, Ponticus, I am dining with you?

For. 2. 61 ff.

Ibid. IV. xxvii
Oft are you wont to praise my poems, Augustus. See, a jealous fellow denies

it: are you wont to praise them the less for that? Have you not besides given me, honoured not in words alone, gifts that none other could give? See, the jealous fellow again gnaws his filthy nails! Give me, Caesar, all the more, that he may writhe!

Und. 76

Ibid. V. xxxiv. 9–10
And let not hard clods cover her tender bones, nor be thou heavy upon her, O earth; she was not so to thee.

Epig. 22. 12

Ibid. V. xxxvii. 4–6
... before whom thou wouldst not choose Eastern pearls, nor the tusk new polished of India's beast, and snows untrodden, and the untouched lily ...

Und. 2. iv. 21–4

Ibid. VII. xcvii
If you knew well, little book, Caesius Sabinus, the pride of hilly Umbria, fellow-townsman of my Aulus Pudens, you will give him these, though he be engaged. Though a thousand duties press on and distract him, yet he will be at leisure for my poems. For he loves me, and next to Turnus' famous satires, reads me. Oh, what a reputation is being stored up for you! Oh, what glory! How many an admirer! With you banquets, with you the forum will echo, houses, by-ways, colonnades, bookshops! You are being sent to one, by all will you be read.

Und. 78. 19 ff.

Ibid. VIII. xxxv
Seeing that you are like one another, and a pair in your habits, vilest of wives, vilest of husbands, I wonder you don't agree.

Epig. 42

Ibid. VIII. lxx. 3–4
Though he might have drained sacred Permessis in full draughts, he chose to slake his thirst with diffidence ...

Epig. 79. 3–4

Ibid. IX. lxi. 11–16
Ofttimes under this tree sported fauns flown with wine, and a late-blown pipe startled the still house; and, while o'er lonely fields she fled by night from Pan, oft under these leaves the rustic dryad nestled hid. And fragrant has the dwelling been when Lyaeus held revel, and more luxuriant grown the tree's shade from spilth of wine ...

For. 2. 10 ff.

Ibid. X. xxx. 21–4
... the fishpond feeds turbot and home-reared bass; to its master's call swims the dainty murry; the usher summons a favourite grey mullet, and, bidden to appear, aged surmullets put forth their heads.

For. 2. 25 ff.

Ibid. X. xlviii. 21–4
To crown these there shall be jests without gall, and a freedom not to be dreaded the next morning, and no word you would wish unsaid; let my guest converse of the Green and the Blue [rival factions of charioteers]; my cups do not make any man a defendant.

Epig. 101. 37–42

Ibid. XII. xxxiv. 8–11
If you wish to shun some bitternesses and to beware of sorrows that gnaw the heart, to no man make yourself too much a comrade: your joy will be less and less will be your grief.

Epig. 45. 12

Ibid. XII. lxi
You are afraid, Ligurra, I should write verses on you, and some short and lively poem, and you long to be thought a man that justifies such fear. But vain is your fear, and your longing is vain. Against bulls Libyan lions rage, they are not hostile to butterflies. Look out, I advise you, if you are anxious to be read of, for some dark cellar's sottish poet, one who with coarse charcoal or crumbling chalk scrawls poems which people read in the jakes. Your brow is not one to be marked by my brand.

U.V. 36

OVID, *Amores*, III. iv. 3–4
If she is pure when freed from every fear, then first she is pure; she who sins not because she may not—she sins.

For. 11. 87–90

PHILOSTRATUS, *Epistles*
(33) So set the cups down and leave them alone, especially for fear of their fragility; and drink to me only with your eyes; 'twas such a draught that Zeus too drank—and took to himself a lovely boy to bear his cup. And, if it please you, do not squander the wine, but pour in water only, and, bringing it to your lips, fill the cup with kisses and so pass it to the thirsty. Surely nobody is so ignorant of love as to yearn for the gift of Dionysus any longer after the vines of Aphrodite. (32) I first and foremost, when I see you, feel thirst, and against my will stand still, and hold the cup back; and I do not bring it to my lips, but I know that I am drinking of you. (60) And if ever you sip from the cup all that is left becomes warmer with your breath and sweeter than nectar. At all events it slips by a clear passage down to the throat, as if it

were mingled not with wine but with kisses.   (2) I have sent you a garland of roses, not to honour you (though I would fain do that as well), but to do a favour to the roses themselves, so that they may not wither.   (46) If you wish to do a favour for a lover, send back what is left of [my roses], since they now breathe a fragrance, not of roses only, but also of you.

For. 9

PLINY, *Epist*. I. xvi
But you will view him with still higher pleasure in the character of an historian, where his narrative style is by turns concise, clear, smooth, or actually glowing and sublime . . .

Epig. 95. 29–30

Ibid. I. xxii
How consummate is his knowledge both in the political and civil laws of his country! How thoroughly conversant is he in history, precedents, antiquity! There is no article, in short, you would wish to be informed of, in which he cannot enlighten you.

Epig. 14. 7–9

Ibid. I. xxii
[Titius Aristo] places no part of his happiness in ostentation, but refers the whole of it to conscience; and seeks the reward of a virtuous action, not in the applauses of the world, but in the action itself.

Epig. 63. 5–6

Ibid. V. xiv
I congratulate myself, therefore, no less than him, and as much upon public as private grounds, that virtue is now no longer, as formerly, the road to danger, but to office.

Epig. 64. 11–12

Ibid. V. xvi
With what forbearance, patience, nay courage, did she endure her last illness! She complied with all the directions of her physicians; she encouraged her sister and her father; and when all her strength of body was exhausted, supported herself by the single vigour of her mind.

Und. 83. 47 ff.

Panegyricus, I. 6
. . . and my vote of thanks be as far removed from a semblance of flattery as it is from constraint.

Epig. 43. 10–12

Ibid. xlvi. 1
Could any terror have had the power to effect what has been accomplished
through our regard for you?

Epig. 91. 18

PLUTARCH, *How to Tell a Flatterer*, xxiv
Whenever, then, the flatterer, who is but a light and deceptive plated-ware, is
examined and closely compared with genuine and solid-wrought friendship,
he does not stand the test, but he is exposed, and so he does the same thing
as the man who had painted a wretched picture of some cocks. For the
painter bade his servant scare all real cocks as far away as possible from the
canvas; and so the flatterer scares all real friends away, and does not allow
them to come near.

Und. 69. 9–13

Lysander, viii. 4
He who overreaches his enemy by means of an oath, confesses that he fears
that enemy, but despises God.

Epig. 16. 10

SALLUST, *Catiline*, xx. 4
... for agreement in likes and dislikes—this, and this only, is what con-
stitutes true friendship.

Epig. 42. 16–18

Jugurtha, lxxxv. 15
For my part, I believe that all men have one and the same nature, but that
the bravest is the best born ...

Epig. 116. 9–10

SENECA, *Ad Marciam de Consolatione*, xix. 1
... we have sent them on their way—nay, we have sent them ahead and shall
soon follow.

Epig. 33. 2–3

Ad Polybium de Consolatione, ix. 9
The way thither is the same for us all. Why do we bemoan his fate? He has
not left us, but has gone before.

Epig. 33. 2–3

Ibid. xviii. 2
... among human achievements, this is the only work that no storm can
harm, nor length of time destroy. All others, those that are formed by piling
up stones and masses of marble, or rearing on high huge mounds of earth, do
not secure a long remembrance, for they themselves will also perish; but the
fame of genius is immortal.

For. 12. 43 ff.

De Brevitate Vitae, iii. 5
How late is it to begin to live just when we must cease to live!

Epig. 70. 1–2

Ibid. ix. 1
... yet postponement is the greatest waste of life ... the greatest hindrance to living is expectancy, which depends upon the morrow and wastes today.

Epig. 70. 5

De Clementia, I. xiv. 1–3
What, then, is his duty? It is that of a good parent who is wont to reprove his children sometimes gently, sometimes with threats, who at times admonishes them even by stripes. Does any father in his senses disinherit a son for his first offence? Only when great and repeated wrong-doing has overcome his patience, only when what he fears outweighs what he reprimands, does he resort to the final pen [= disinheritance]; but first he makes many an effort to reclaim a character that is still unformed, though inclined now to the more evil side; when the case is hopeless, he tries extreme measures. No one resorts to the exaction of punishment until he has exhausted all the means of correction. ... Slow would a father be to sever his own flesh and blood; aye, after severing he would yearn to restore them, and while severing he would groan aloud, hesitating often and long; for he comes near to condemning gladly who condemns swiftly, and to punishing unjustly who punishes unduly.

Und. 38. 40–50

Ibid. I. xvii. 2
It is a poor physician that despairs of his ability to cure ... the aim of the prince should be not merely to restore the health, but also to leave no shameful scar.

Und. 38. 49–52

De Tranquillitate Animi, vii. 5
... for both classes were necessary in order that Cato might be understood—he needed to have good men that he might win their approval, and bad men that he might prove his strength.

Epig. 102. 3–4

Epist. xciii. 3
What benefit does this older man derive from the eighty years he has spent in idleness? A person like him has not lived; he has merely tarried awhile in life. Nor has he died late in life; he has simply been a long time dying. He has lived eighty years, has he? That depends upon the date from which you reckon his death.

Und. 70. 25–32

Epist. xciii. 4

Your other friend, however, departed in the bloom of his manhood. But he fulfilled all the duties of a good citizen, a good friend, a good son; in no respect had he fallen short. His age may have been incomplete, but his life was complete (*licet aetas eius inperfecta sit, vita perfecta est*). The other man has lived eighty years, has he? Nay, he has existed eighty years, unless perchance you mean by 'he has lived' what we mean when we say that a tree 'lives'.

Und. 70. 43–52, 58–9, 65

Epist. xciii. 5

We should therefore praise, and number in the company of the blest, that man who has invested well the portion of time, however little, that has been allotted to him; for such a one has seen the true light. He has not been one of the common herd. He has not only lived, but flourished. Sometimes he enjoyed fair skies; sometimes, as often happens, it was only through the clouds that there flashed to him the radiance of the mighty star [the sun]. Why do you ask, 'How long did he live?' He still lives! At one bound he has passed over into posterity and has consigned himself to the guardianship of memory.

Und. 70. 78–84

Ibid. cxv. 10

. . . we fulfil duties if it pays, or neglect them if it pays, and we follow an honourable course as long as it encourages our expectations, ready to veer across to the opposite course if crooked conduct shall promise more.

Epig. 102. 9–10

VALERIUS MAXIMUS, VIII. xv. 2

. . . and became great through his own personal merit rather than by favour of fortune.

Epig. 63. 2–3

VIRGIL, *Georgics*, ii. 458–71

O happy husbandmen! too happy, should they come to know their blessings! for whom, far from the clash of arms, most righteous earth, unbidden, pours forth from her soil an easy sustenance. What though no stately mansion with proud portals disgorges at dawn from all its halls a tide of visitors, though they never gaze at doors inlaid with lovely tortoise-shell. . . . Yet theirs is repose without care, and a life that knows no fraud, but is rich in treasures manifold. Yea, the ease of broad domains, caverns, and living lakes, and cool vales, the lowing of the kine, and soft slumbers beneath the trees—all are theirs. They have woodland glades and the haunts of game . . .

For. 3. 13 ff.

Ibid. ii. 501–2
He plucks the fruits which his boughs, which his ready fields, of their own
free will, have borne. . .

For. 2. 25 ff.

Ibid. iii. 66–7
Life's fairest days are ever the first to flee, for hapless mortals . . .

Epig. 70. 6

Index of Titles
and First Lines

(Titles are indicated in italic, first lines in roman.)

After many scorns like these	131
Alchemists, To	9
Alleyn, To Edward	46
All men are worms: but this no man. In silk	13
All to Whom I Write, To	10
Amorphus's Song	347
And art thou born, brave babe? Blest be thy birth	228
And must I sing? What subject shall I choose	101
And why to me this, thou lame lord of fire	193
An elegy? No, muse; it asks a strain	103
Another Lady's Exception, Present at the Hearing	139
Another phoenix! though the first is dead	380
Ask not to know this man. If fame should speak	165
At court I met it, in clothes brave enough	11
Aubigny, Epistle to Katherine, Lady	111
—, *To Esmé, Lord*	73
Away, and leave me, thou thing most abhorred	35
A woman's friendship! God whom I trust in	164
Bacon, see *Lord Bacon*	
Bank feels no lameness of his knotty gout	20
Bank the Usurer, On	20
from Bartholomew Fair: *Nightingale's Song*	352
Bawds and Usurers, On	31
from Beaumont, Sir John, Bosworth Field: *On the Honoured Poems of His Honoured Friend*	318
Beaumont, To Francis	30
Bedford, On Lucy, Countess of (This morning, timely rapt)	40
—, *To Lucy, Countess of* (Lucy, you brightness of our sphere)	50
—, — (Madam, I told you late)	43
Beggar's Song	345
Begging Another, on Colour of Mending the Former	136
Be safe, nor fear thyself so good a fame	41
Blessing the Sovereign and His Senses	365
Boast not these titles of your ancestors	267
Brain-Hardy, To	14
Brave infant of Saguntum, clear	233
from Breton, Nicholas, Melancholic Humours: *In Authorem*	284
from Brome, Richard, The Northern Lass: *To My Old Faithful Servant . . . The Author*	328
from Brooke, Christopher, The Ghost of Richard the Third: *To His Friend the Author*	303
from Browne, William, Britannia's Pastorals, The Second Book: *To My Truly-Beloved Friend*	304

Bulstrode, Epitaph on Cecilia   295
Burgess, To Master John (Father John Burgess)   221
—, — (Would God, my Burgess)   220
Burghley, An Epigram, On William, Lord   178
Burlase, A Poem Sent Me by Sir William   216
But 'cause thou hear'st the mighty king of Spain   325
Buzz, quoth the blue fly   360
By those bright eyes, at whose immortal fires   163

Camden, most reverend head, to whom I owe   13
Camden, To William   13
Can beauty that did prompt me first to write   162
Captain Hazard the Cheater, On   45
Captain Hungry, To   60
Cary, Sir Lucius, and Sir H. Morison. To the Immortal Memory . . . of That
   Noble Pair   233
—, *To His Lady, then Mistress*   73
—, *To Sir Henry*   36
Cashiered Captain Surly, On   43
Cavendish, Charles to His Posterity   305
Celebration of Charis in Ten Lyric Pieces, A   129
Celia, Song To (Come, my Celia)   97
—, — (Drink to me only)   101
        To the Same (Kiss me, sweet)   98
Censorious Courtling, To   29
Censure not sharply then, but me advise   340
from Chapman, George, The Georgics of Hesiod: *To My Worthy and Honoured*
   Friend   305
Charis, guess, and do not miss   135
Charis one day in discourse   137
Cheverel cries out my verses libels are   30
Cheverel, On (Cheverel cries out)   30
Cheverel the Lawyer, On (No cause nor client fat)   23
Chuff, Bank's the Usurer's Kinsman, On   26
Chuff, lately rich in name, in chattels, goods   26
Chute, An Epitaph on Elizabeth   182
Claiming a Second Kiss by Desert   135
Clerimont's Song   351
Cob, thou nor soldier, thief, nor fencer art   37
Coke, An Epigram on Sir Edward   206
Come follow me, my wags, and say as I say   345
Come, leave the loathed stage   354
Come, let us here enjoy the shade   182
Come, my Celia, let us prove   97
Come, noble nymphs, and do not hide   367
Come on, come on, and where you go   362
Come, with our voices let us war   140
Corbett, An Epitaph on Master Vincent   146
from Coryate, Thomas, Crambe: *Certain Verses Written Upon Coryate's*
   Crudities   298
—, — Crudities: *Certain Opening and Drawing Distichs*   295

from Coryate, Crudities: *To the Right Noble Tom* 298
Courtling, I rather thou shouldst utterly 29
Courtling, To 38
Court-Parrot, On 38
Court Pucelle, An Epigram on the 212
Court-Worm, On 13
Covell, Epistle to my Lady 220
Crispinus' and Hermogenes' Song 348
Cyclope's Song 360
from Cynthia's Revels:
   Amorphus's Song 347
   Beggar's Song 345
   Echo's Song 345
   Hedon's Song 347
   Hymn to Cynthia 348

Daedalus's Song 362
Death, Of 22
Dedication of Her Crädle, The 260
Dedication of the King's New Cellar, The 210
Desmond, An Ode to James, Earl of 170
Digby, Elegy on My Muse ... the Lady Venetia 268
—, *Epigram to My Muse, the Lady* 251
—, *the Lady Venetia; Eupheme; or the Fair Fame Left to Posterity of*
—, *To Kenelm, John, George* 267
Do but consider this small dust 144
Doctor Empiric, To 12
Does the court pucelle then so censure me 212
Doing a filthy pleasure is, and short 279
Donne, To John (Donne, the delight of Phoebus) 17
—, — (Who shall doubt, Donne) 52
Donne, the delight of Phoebus and each muse 17
Don Surly, On 19
Don Surly, to aspire the glorious name 19
Dorset, An Epistle to Sir Edward Sackville, Now Earl of 148
from Dover, Robert, Annalia Dubrensia: *An Epigram* 334
Do what you come for, Captain, with your news 60
from Drayton, Michael, The Battle of Agincourt: *The Vision of Ben Jonson* 313
Dream, The 146
Drink to me only with thine eyes 101

Echo's Song 345
Edmondes, To Clement (Not Caesar's deeds) 63
    To the Same (Who, Edmondes, reads thy book) 64
Egerton, To Thomas, Lord Chancellor 39
    (see also *Ellesmere, Thomas Lord*)
Elegies:
   An Elegy (By those bright eyes) 163
   — (Can beauty that did prompt) 162
   — (Let me be what I am) 190
   — (Since you must go, and I must bid farewell) 190

Elegies—(cont.):
  An Elegy (That love's a bitter sweet I ne'er conceive)   188
  — (Though beauty be the mark of praise)   166
  — ('Tis true, I'm broke)   185
  — (To make the doubt clear)   375
  Elegy on My Muse ... the Lady Venetia Digby   268
  —, on the Lady Jane Paulet, Marchioness of Winton   256
Ellesmere, An Epigram to Thomas, Lord (So, justest lord)   179
    Another to Him (The judge his favour)   179
    (see also Egerton, Thomas, Lord Chancellor)
from England's Parnassus   285
English Monsieur, On   45
from The Entertainment at Althorpe: The Satyr's Verses   357
Envious and foul disease, could there not be   181
Epigrams:
  An Epigram (That you have seen the pride)   225
  — Consolatory to King Charles and Queen Mary   226
  — on Sir Edward Coke   206
  — on the Court Pucelle   212
  — on the Prince's Birth (And art thou born, brave babe)   228
    Another (Another phoenix)   380
  — on William, Lord Burghley   178
  — to a Friend   326
  — to a Friend, and Son   232
  — to King Charles, for a Hundred Pounds He Sent Me   226
  — to My Muse, the Lady Digby   251
  — to My Bookseller   222
  — to Our Great and Good King Charles   227
  — to the Counsellor that Pleaded and Carried the Cause   180
  — to the Honoured Elizabeth, Countess of Rutland   214
  — to the Household   231
  — to the King, on His Birthday   239
  — to the Queen, then Lying In   229
  — to the Right Honourable, the Lord Treasurer   250
  — to the Smallpox   181
  — to Thomas, Lord Ellesmere (So, justest lord)   179
    Another to Him (The judge his favour timely then)   179
  — to William, Earl of Newcastle (They talk of fencing)   223
  —, — (When first, my lord)   218
Epilogue to The Vision of Delight   361
Epistles:
  Epistle Answering to One that Asked to be Sealed of the Tribe of Ben   207
  Epistle Mendicant, An: To the Lord High Treasurer   238
  Epistle to a Friend, An (Censure not sharply then)   340
  —, — (Sir, I am thankful)   183
  —, — (They are not, sir, worst owers)   162
  —, — to Persuade Him to the Wars   155
  Epistle to Elizabeth, Countess of Rutland   107
  — to Katherine, Lady Aubigny   111
  — to Mr. Arthur Squib (I am to dine)   219
  —, — (What I am not)   205

Epistle to Master John Selden   152
— *to My Lady Covell*   220
— *to Sir Edward Sackville, now Earl of Dorset*   148
Epitaphs:
    Epitaph on Cecilia Bulstrode   295
    — *on Elizabeth Chute*   182
    — *on Elizabeth, L. H.*   71
    — *on Henry, Lord La Warr*   224
    — *on Katherine, Lady Ogle*   316
    — *on Master Philip Gray*   161
    — *on Master Vincent Corbett*   146
    — *on Salomon Pavy*   69
Epithalamion; or, a Song, Celebrating the Nuptials of . . . Mr. Jerome Weston   242
Epode   104
Ere Cherries ripe! and Strawberries! be gone   48
Eupheme; or, the Fair Fame   260
Execration upon Vulcan, An   193
Expostulation with Inigo Jones, An   319

Fair fame, who art ordained to crown   260
Fair friend, 'tis true your beauties move   377
False world, good night. Since thou hast brought   95
Famous Voyage, On the   77
Farewell, thou child of my right hand, and joy   26
Father John Burgess   221
Feasting Song   349
Ferrabosco, To Alphonso (To urge, my loved Alphonso)   75
    To the Same (When we do give, Alphonso)   76
from Filmer, Edward, French Court Airs: To My Worthy Friend   318
Fine Grand, To   39
Fine Lady Would-Be, To   33
Fine Madam Would-Be, wherefore should you fear   33
Fit of Rhyme against Rhyme, A   176
from Fletcher, John, The Faithful Shepherdess: To the Worthy Author 293
Follow a shadow, it still flies you   98
Fool or Knave, To   33
Fool's Song   350
Fools, they are the only nation   350
For all night-sins with others' wives, unknown   23
Forbear to tempt me, Prowl, I will not show   43
For his mind I do not care   139
For Love's sake, kiss me once again   136
Fresh as the day, and new as are the hours   334
Friend, To a   43
Friend, To a: An Epigram of Him   326
From a gypsy in the morning   365
From death and dark oblivion (near the same)   170
From the famous Peak of Derby   363

Ghost of Martial, To the   22
Giles and Joan, On   25

Gill, An Answer to Alexander 328
Good and great God, can I not think of thee 117
Goodyere, I'm glad and grateful to report 44
Goodyere, To Sir Henry (Goodyere, I'm glad) 44
To the Same (When I would know thee, Goodyere) 44
Grace by Ben Jonson, Extempore before King James, A 337
Another Version 338
Gray, An Epitaph on Master Philip 161
Great Charles, among the holy gifts of grace 226
Groin, come of age, his 'state sold out of hand 68
Groin, On 68
Groom Idiot, On 32
Guilty, because I bade you late be wise 23
Guilty, be wise; and though thou know'st the crimes 20
Gut eats all day, and lechers all the night 68
Gut, On 68
from The Gypsies Metamorphosed:
   Blessing the Sovereign 365
   The Jackman's Song 363
   The Patrico's Song 364
Gypsy, new bawd, is turned physician 24
Gypsy, On 24

Hail, happy genius of this ancient pile 215
Hail Mary, full of grace! it once was said 229
Hang up those dull and envious fools 142
Happy is he, that from all business clear 274
Hardy, thy brain is valiant, 'tis confessed 14
Hear me, O God 127
Heaven, To 117
Hedon's Song 347
Helen, did Homer never see 174
Herbert, To Sir Edward 60
Here are five letters in this blessed name 330
Here lies, to each her parents' ruth 16
Here, like Arion, our Coryate doth draw 296
Her Man Described by Her Own Dictamen 138
Her Triumph 132
He that fears death, or mourns it in the just 22
He that should search all glories of the gown 206
High-spirited friend 173
His bought arms Mong not liked; for his first day 28
His Discourse with Cupid 133
His Excuse for Loving 129
from Holland, Hugh, Pancharis: Odε ἀλλτγορικὴ 287
Horace, A Speech according to 201
Horace, Epode ii: The Praises of a Country Life 274
——, *Odes, II. iii* 383
——, *Ode ix, 3rd Book: To Lydia* 278
——, *Ode the First, the Fourth Book: To Venus* 277
Horace's Ode 349

Hornet, thou hast thy wife dressed for the stall   41
Hornet, To   41
Hour-Glass, The   144
How, best of kings, dost thou a sceptre bear   8
How blest art thou canst love the country, Wroth   91
How happy were the subject, if he knew   227
How He Saw Her   130
How I do love thee, Beaumont, and thy muse   30
How like a column, Radcliffe, left alone   49
How well, fair crown of your fair sex, might he   57
Humble Petition of Poor Ben, The   249
from The Husband: *To the Worthy Author*   304
Hymn on the Nativity of My Saviour, A   128
Hymn to Cynthia   348
Hymn to God the Father, A   127

I am to dine, friend, where I must be weighed   219
I beheld her, on a day   130
I cannot bring my muse to drop her vies   334
I cannot think there's that antipathy   40
I could begin with that grave form, *Here lies*   311
Idiot, last night I prayed thee but forbear   32
I do but name thee, Pembroke, and I find   57
If all you boast of your great art be true   9
If, as their ends, their fruits were so the same   31
If I freely may discover   348
If I would wish, for truth and not for show   71
If men and times were now   338
If men get name for some one virtue, then   60
If, my religion safe, I durst embrace   51
If, passenger, thou canst but read   224
If Rome so great, and in her wisest age   46
If, Sackville, all that have the power to do   148
If thou wouldst know the virtues of mankind   178
If to admire were to commend, my praise   77
If to my mind, great lord, I had a state   250
I grieve not, Courtling, thou are started up   38
I had you for a servant once, Dick Brome   328
I have my piety too, which could   146
I know to whom I write. Here, I am sure   152
Ill may Ben Jonson slander so his feet   337
I'll not offend thee with a vain tear more   21
I love and he loves me again   143
I must believe some miracles still be   66
Inigo, Marquis Would-Be, To   325
I now think Love is rather deaf than blind   145
In picture, they which truly understand   292
In place of scutcheons that should deck thy hearse   18
In the Person of Womankind   141
Inviting a Friend to Supper   55
I sing the birth was born tonight   128

I sing the just and uncontrolled descent 262
Is there a hope that man would thankful be 73
Is this the sir, who, some waste wife to win 27
It fits not only him that makes a book 304
I, that have been a lover, and could show it 175
It hath been questioned, Michael, if I be 313
It was a beauty that I saw 354
I was not wearier where I lay 361
It will be looked for, book, when some but see 7

Jealousy, Against 145
Jephson, thou man of men, to whose loved name 68
Jephson, To Sir William 68
Jones, An Expostulation with Inigo 319
[Jones], To Inigo, Marquis Would-Be 325
Jonson, To Mr. Ben, in His Journey 337
Just and fit actions, Ptolemy (he saith) 341

Karolin's Song 356
King, To My Lord the: On the Christening His Second Son 255
—, To the: Epigram Anniversary on His Birthday 239
King Charles, An Epigram to 226
—, A New Year's Gift Sung to 253
—, A Song of Welcome to 334
—, To Our Great and Good 227
King Charles and Queen Mary: an Epigram Consolatory, To 226
King James, A Grace By Ben Jonson Extempore Before 337
—, To (How, best of kings) 8
—, — (That we thy loss might know) 29
—, — (Who would not be thy subject) 22
King's Birthday, On the 378
Kiss me, sweet: the wary lover 98

La Warr, An Epitaph on Henry, Lord 224
Learned Critic, To the 14
Leave, Cod, tobacco-like, burnt gums to take 28
Leges Convivales 370
—, Alexander Brome's Translation of 371
Let it not your wonder move 129
Let me be what I am: as Virgil cold 190
Let none but guests or clubbers hither come 371
L. H., Epitaph on Elizabeth 71
Liber, of all thy friends, thou sweetest care 280
Lieutenant Shift, On 11
Life and Death, Of 42
Lippe, the Teacher, On 40
Little Shrub Growing By, A 165
Long-gathering Old-End, I did fear thee wise 30
Look up, thou seed of envy, and still bring 240
Lord Bacon's Birthday 215
Lord, how is Gamester changed! His hair close cut 16

from Love's Martyr:
 Ode ἐθουσιαστικὴ 286
 The Phoenix Analysed 286
Lo, what my country should have done—have raised 32
Lucy, you brightness of our sphere, who are 50

from Mabbe, James (tr.), The Rogue: *On the Author, Work, and Translator* 306
Madam, had all antiquity been lost 59
Madam, I told you late how I repented 43
Madam, / Whilst that for which all virtue now is sold 107
Marble, weep, for thou dost cover 24
Martial, Epigram xlvii, Book X 280
—, *Epigram lxxvii, Book VIII* 280
Martial, thou gav'st far nobler epigrams 22
from The Masque of Queens: *The Witches' Charm* 359
Master Surveyor, you that first began 319
May none whose scattered names honour my book 10
May others fear, fly, and traduce thy name 14
from May, Thomas (tr.), Lucan's Pharsalia: *To My Chosen Friend, the Learned
 Translator* 312
Men, if you love us, play no more 141
Men that are safe and sure in all they do 207
from Mercury Vindicated: *Cyclope's Song* 360
Mill, My Lady's Woman, On 46
Mime, To 74
Mind, The 264
Mind of the Frontispiece to a Book, The 170
Mongrel Esquire, On 28
Monteagle, To William, Lord 32
Montgomery, To Susan, Countess of 58
Morison, Sir H., see *Cary, Sir Lucius* 233
Mountebank's Song 351
Musical Strife, The 140
My Book, To 7
My Bookseller, To (Thou that mak'st gain) 8
—, *Epigram to* (Thou, friend, wilt hear) 222
My Detractor, To 327
My First Daughter, On 16
My First Son, On 26
My Lord Ignorant, To 11
My Lord, / Poor wretched states, pressed by extremities 238
My masters and friends and good people, draw near 352
My Mere English Censurer, To 15
My Muse, To 35
My Picture Left in Scotland 145
My verses were commended, thou dar'st say 327

Nemo asymbolus, nisi umbra, huc venito 370
from Neptune's Triumph: *Song to the Ladies* 367
Neville, To Sir Henry 62
Newcastle, An Epigram to William, Earl of (They talk of fencing) 223

Newcastle, An Epigram to William, Earl of (When first, my lord)   218
New Cry, The   48
New Hot-House, On the   10
from The New Inn: *A Vision of Beauty*   354
on —, Ode. To Himself   354
New Motion, On the   53
New Years expect new gifts: sister, your harp   253
New Year's Gift Sung to King Charles, A   253
Nightingale's Song   352
Noblest Charis, you that are   133
No cause nor client fat will Cheverel leese   23
No more let Greece her bolder fables tell   77
Not Caesar's deeds, nor all his honours won   63
Not glad, like those that have new hopes or suits   34
Not he that flies the court for want of clothes   69
Not to know vice at all, and keep true state   104
Now, after all, let no man   286
Now that the hearth is crowned with smiling fire   115
Nymph's Passion, A   143

from Oberon, The Fairy Prince: *Satyrs' Catch*   360
Odes:
   An Ode (Helen, did Homer never see)   174
   — (High-spirited friend)   173
   — (If men and times)   338
   — (Scorn, or some humbler fate)   382
   —, *or Song, by All the Muses*   229
   — *To Himself* (Come, leave the loathed stage)   354
   —, — (Where dost thou careless lie)   167
   — *To James, Earl of Desmond*   170
   — *To Sir William Sidney*   115
   — *To the Right Honourable Jerome, Lord Weston*   167
Of your trouble, Ben, to ease me   138
Ogle, Epitaph on Katherine, Lady   316
Oh, do not wanton with those eyes   141
O holy, blessed, glorious Trinity   125
Oh, that joy so soon should waste   347
Old Colt, On   23
Old-End Gatherer, To   30
One that Desired Me Not to Name Him, To   41
Or scorn, or pity on me take   146
Our king and queen the Lord God bless   337
Overbury, To Sir Thomas   65
Over the Door at the Entrance into the Apollo   372

Painter, you're come, but may be gone   264
from Palmer, Thomas, The Sprite of Trees and Herbs   283
Parliament, To the   17
Patrico's Song, The   364
Pavy, Epitaph on Salomon   69
Pembroke, To William, Earl of   57

Penshurst, To   87
Person Guilty, To (Guilty, because I bade you)   23
— (Guilty, be wise)   20
Pertinax Cob, To   37
Petition of the Infant Prince Charles, A   381
Petronius Arbiter, A Fragment of   279
Picture of the Body, The   263
Playwright, by chance, hearing some toys I'd writ   55
Playwright, convict of public wrongs to men   37
Playwright me reads, and still my verses damns   28
Playwright, On (Playwright, convict of public wrongs)   37
— (Playwright, by chance, hearing some toys)   55
—, To (Playwright me reads)   28
from Pleasure Reconciled to Virtue:
    Daedalus's Song   362
    Song for Comus   361
Poems of Devotion   125
Poet-Ape, On   31
from Poetaster:
    Crispinus' and Hermogenes' Song   348
    Feasting Song   349
    Horace's Ode   349
Poor Poet-Ape, that would be thought our chief   31
Pray thee take care, that tak'st my book in hand   7
Prince's Birth, An Epigram on the   228
Proludium   103
Prowl the Plagiary, To   43

Queen and huntress, chaste and fair   348
Queen, Epigram to the   229

Radcliffe, On Margaret   24
Radcliffe, To Sir John   49
Reader, stay   161
Reader, To the   7
Read, royal father, mighty king   381
Reformed Gamester, On   16
Remember when blind fortune knits her brow   383
Retired, with purpose your fair worth to praise   73
Rhyme, the rack of finest wits   176
Ridway robbed Duncote of three hundred pound   10
Robbery, On a   10
Roe (and my joy to name) thou'rt now to go   74
Roe, On Sir John (In place of scutcheons)   18
—, — (What two brave perils)   21
    To the Same (I'll not offend thee)   21
Roe, To Sir Thomas (Thou hast begun well, Roe)   54
    To the Same (That thou hast kept thy love)   54
Roe, To William (Roe (and my joy to name))   74
—, — (When Nature bids us leave)   38
Room, room, make room for the bouncing belly   361

Rouse up thyself, my gentle muse 378
Rudyerd, as lesser dames to great ones use 70
Rudyerd, To Benjamin (Rudyerd, as lesser dames) 70
   To the Same (If I would wish) 71
   — (Writing thyself or judging others' writ) 71
Rutland, Epigram to the Honoured Elizabeth, Countess of 214
—, *Epistle to Elizabeth, Countess of* 107
—, *To Elizabeth, Countess of* 42
from Rutter, Joseph, The Shepherds' Holiday: *To My Dear Son and Right-
  Learned Friend* 332

from The Sad Shepherd: *Karolin's Song* 356
Salisbury, To Robert, Earl of (What need hast thou of me) 25
—, — (Who can consider thy right courses run) 34
   To the Same (Not glad, like those) 34
Satirical Shrub, A 164
Satyrs' Catch 360
Satyr's Verses 357
Savile, To Sir Henry 51
Scorn, or some humbler fate 382
See the chariot at hand here of Love 132
See you yond motion? Not the old fa-ding 53
Selden, An Epistle to Master John 152
from Shakespeare, William, Comedies, Histories, and Tragedies: *To the Memory
  of . . . the Author* 307
—, — *To the Reader* 307
Shall the prosperity of a pardon still 328
Sheldon, To Sir Ralph 69
Shift, here in town not meanest among squires 11
Shrewsbury, To the Memory of Lady Jane . . . Countess of 311
Sickness, To 99
Sidney, Ode to Sir William 115
—, *To Mistress Philip* 66
from The Silent Woman: *Clerimont's Song* 351
Since, Bacchus, thou art father 210
Since men have left to do praiseworthy things 36
Since you must go, and I must bid farewell 190
Sinner's Sacrifice, The 125
Sir Annual Tilter, To 20
Sir Cod, To (Leave, Cod, tobacco-like) 28
— *the Perfumed, On* (That Cod can get no widow) 15
—, *To the Same* (The expense in odours) 15
Sir, I am thankful, first to heaven for you 183
Sir Inigo doth fear it, as I hear 326
Sir Luckless, troth, for luck's sake pass by one 27
Sir Luckless Woo-All, To (Is this the sir) 27
   To the Same (Sir Luckless, troth) 27
Sir Voluptuous Beast, On (While Beast instructs) 17
   On the Same Beast (Than his chaste wife) 18
Sitting, and ready to be drawn 263
Slow, slow, fresh fount, keep time with my salt tears 345

Smallpox, An Epigram to the   182
Soft, subtle fire, thou soul of art   360
So, justest lord, may all your judgements be   179
Some act of Love's bound to rehearse   87
Some men, of books or friends not speaking right   304
Somerset, To the Most Noble ... Robert, Earl of   302
Something that Walks Somewhere, On   11
Son, and my friend, I had not called you so   232
Songs:
  Song (Come, let us here enjoy the shade)   182
  — (Oh, do not wanton)   141
  — *Apologetic, A: In the Person of Womankind*   141
     Another: in Defence of Their Inconstancy   142
  Song for Comus   361
  — *of Her Descent*   262
  — *of the Moon*   335
  — *of Welcome to King Charles*   334
  — *That Women Are but Men's Shadows*   98
  — *To Celia* (Come, my Celia)   97
  —, — (Drink to me only)   101
  To the Same (Kiss me, sweet)   98
  — *to the Ladies*   367
Sonnet to the Noble Lady, the Lady Mary Worth, A   175
Sons, seek not me among these polished stones   305
So Phoebus makes me worthy of his bays   65
Speech according to Horace, A   201
Speech out of Lucan, A   341
Speech Presented unto King James at a Tilting, A   301
Spies, On   32
Spies, you are lights in state, but of base stuff   32
Splendour, O more than mortal   286
Squib, Epistle to Mr. Arthur (I am to dine)   219
—, — (What I am not)   205
from Stafford, Anthony, The Female Glory: *The Garland of the Blessed Virgin
    Marie*   330
Stay, view this stone; and if thou beest not such   295
from Stephens, John, Cynthia's Revenge: *To ... the Author*   301
Still to be neat, still to be dressed   351
Strength of my country, whilst I bring to view   62
Such pleasure as the teeming earth   241
Suffolk, To Thomas, Earl of   36
Surly's old whore in her new silks doth swim   43
from Sutcliffe, Alice, Man's Mortality: *To Mrs. Alice Sutcliffe*   329
Swell me a bowl with lusty wine   349
Sylvester, To Mr. Joshua   77

Than his chaste wife though Beast now know no more   18
That Cod can get no widow, yet a knight   15
That I, hereafter, do not think the Bar   180
That love's a bitter sweet I ne'er conceive   188
That neither fame nor love might wanting be   36

That not a pair of friends each other see 74
That poets are far rarer births than kings 42
That thou art loved of God, this work is done 255
That thou hast kept thy love, increased thy will 54
That we thy loss might know, and thou our love 29
That Women are but Men's Shadows, Song 98
That you have seen the pride, beheld the sport 225
The expense in odours is a most vain sin 15
The fairy beam upon you 364
The judge his favour timely then extends 179
The king, the queen, the prince, God bless 338
The owl is abroad, the bat and the toad 359
The ports of death are sins; of life, good deeds 42
There's reason good that you good laws should make 17
The things that make the happier life are these 280
The wisdom, madam, of your private life 214
The wise and many-headed bench that sits 293
They are not, sir, worst owers, that do pay 162
They are not those, are present with their face 302
They talk of fencing and the use of arms 223
This book will live: it hath a genius; this 318
This figure that thou here seest put 307
This is King Charles's day. Speak it, thou Tower 239
This is Mab the mistress-fairy 357
This morning, timely rapt with holy fire 40
Those that in blood such violent pleasure have 285
Thou art not, Penshurst, built to envious show 87
Thou call'st me poet, as a term of shame 11
Thou, friend, wilt hear all censures; unto thee 222
Though beauty be the mark of praise 166
Though, happy muse, thou know my Digby well 251
Though I am young and cannot tell 356
Though thou hast passed thy summer standing, stay 242
Thou hast begun well, Roe, which stand well to 54
Thou more than most sweet glove 347
Thou that mak'st gain thy end, and wisely well 8
Thou that wouldst find the habit of true passion 284
Thy praise or dispraise is to me alike 33
Tilter, the most may admire thee, though not I 20
'Tis a record in heaven. You, that were 316
'Tis grown almost a danger to speak true 111
'Tis true, I'm broke! Vows, oaths, and all I had 185
Today old Janus opens the New Year 253
To draw no envy, Shakespeare, on thy name 307
To make the doubt clear that no woman's true 375
Tonight, grave sir, both my poor house and I 55
To paint thy worth, if rightly I did know it 216
To pluck down mine, Poll sets up new wits still 38
To put out the word 'whore' thou dost me woo 43
To thee my way in epigrams seems new 15
To the wonders of the Peak 335

Touched with the sin of false play in his punk   45
from The Touchstone of Truth   379
To urge, my loved Alphonso, that bold fame   75
Town's Honest Man, On the   66
Treasurer, An Epigram to the Right Honourable, the Lord   250
True Soldiers, To   62
Truth is the trial of itself   379
Try and trust Roger was the word, but now   298
'Twere time that I died too, now she is dead   268
Two noble knights, whom true desire and zeal   301

Union, On the   9
Up public joy, remember   229
Urging Her of a Promise   137
Uvedale, thou piece of the first times, a man   72
Uvedale, To Sir William   72

Venus, again thou mov'st a war   277
Vere, To Sir Horace   47
Vision of Beauty, A   354
from The Vision of Delight: *Epilogue*   361
from Volpone:
   The Fools' Song   350
   The Mountebank's Song   351

Wake, friend, from forth thy lethargy; the drum   155
Wake, our mirth begins to die   349
from Warre, James, The Touchstone of Truth   379
Weak Gamester in Poetry, To a   64
Weep with me all you that read   69
Welcome, all who lead or follow   372
Were they that named you prophets? Did they see   58
Weston, Mr. Jerome, Epithalamion Celebrating the Nuptials of   242
—, *On the Right Honourable and Virtuous Lord*   240
—, *To the Right Honourable Jerome, Lord: An Ode Gratulatory*   241
What beauty would have lovely styled   182
What can the cause be, when the king hath given   231
What charming peals are these   318
What gentle ghost, besprent with April dew   256
What He Suffered   131
What I am not, and what I fain would be   205
What is't, fine Grand, makes thee my friendship fly   39
What need hast thou of me, or of my muse   25
What two brave perils of the private sword   21
When first, my lord, I saw you back your horse   218
When I had read   329
When I would know thee, Goodyere, my thought looks   44
When late, grave Palmer, these thy grafts and flowers   283
When men a dangerous disease did 'scape   12
When Mill first came to court, the unprofiting fool   46
When Nature bids us leave to live, 'tis late   38
When, Rome, I read thee in thy mighty pair   312

When these, and such, their voices have employed   303
When was there contract better driven by fate   9
When we do give, Alphonso, to the light   76
When wit and learning are so hardly set   337
Where art thou, genius? I should use   170
Where dost thou careless lie   167
Where lately harboured many a famous whore   10
Which of thy names I take, not only bears   47
While Beast instructs his fair and innocent wife   17
Whilst, Lydia, I was loved of thee   278
Whilst thy weighed judgements, Egerton, I hear   39
Who can consider thy right courses run   34
Who dares deny that all first fruits are due   226
Who, Edmondes, reads thy book and doth not see   64
Whoever he be, would write a story at   299
Who now calls on thee, Neville, is a muse   62
Who saith our times nor have, nor can   287
Who says that Giles and Joan at discord be   25
Whose work could this be, Chapman, to refine   305
Who shall doubt, Donne, whe'er I a poet be   52
Who takes thy volume to his virtuous hand   301
Who tracks this author's or translator's pen   306
Who would not be thy subject, James, to obey   22
Why, disease, dost thou molest   99
Why I Write Not of Love   87
Why? though I seem of a prodigious waist   217
Why yet, my noble hearts, they cannot say   201
Winton, An Elegy on the Lady Jane Pawlet, Marchioness of 256
Witches' Charm, The   359
With thy small stock, why art thou venturing still   64
World, To the   95
Would God, my Burgess, I could think   220
Wouldst thou hear what man can say   71
Would you believe, when you this monsieur see   45
Wretched and foolish jealousy   145
from Wright, Thomas, The Passions of the Mind in General:
    To the Author   292
Writing thyself or judging others' writ   71
Wroth, To Mary, Lady (How well, fair crown)   57
—, — (Madam, had all antiquity been lost)   59
—, *A Sonnet to the Noble Lady, the Lady Mary*   175
—, *To Sir Robert*   91

You look, my Joseph, I should something say   332
You that would last long, list to my song   351
You wonder who this is, and why I name   66
You won not verses, madam, you won me   220